HISTORY OF PHYSIOLOGY

By
Karl E. Rothschuh, M.D.
Professor
of the History of Medicine
University of Münster, West-Germany

Translated and edited,
with a new English bibliography,
by

Guenter B. Risse, M.D., Ph.D.
Associate Professor and Chairman,
Department of the History of Medicine
The University of Wisconsin, Madison

Robert E. Krieger Publishing Company
Huntington, New York
1973

Dedicated to my wife and my collaborators.

First English edition, 1973
Based upon an original German edition titled
GESCHICHTE DER PHYSIOLOGIE, Springer Verlag 1953

Printed and Published by
Robert E. Krieger Publishing Co., Inc.
Box 542, Huntington, N.Y. 11743

Library of Congress Card #74-158126
SBN#088275-069-0

Printed in the United States of America.

Contents

List of Illustrations vii
Introduction xi
Preface to English Edition xiii
Preface to German Edition xv
Translator's Introduction xvii

CHAPTER I
Physiology in Antiquity

1. The Beginnings of Physiological Thought with the Greek Philosophers of Nature and the Hippocratic Physicians 1
2. The Physiology of Aristotle and the School of Alexandria 8
3. Galen of Pergamon and Roman Physiology 14

CHAPTER II
Physiology During the Middle Ages

1. Early Medieval Times — Islamic Physiology — Salerno 27
2. Scholasticism — Renaissance — Humanism and the End of the Middle Ages 33

CHAPTER III
Foundation and Development of Physiology in the Sixteenth and Seventeenth Century

1. The Renaissance of Anatomy in the Sixteenth Century 47
2. The Application of Chemical Principles in Physiology by the Iatrochemists 55
3. The Discovery of the Circulation of the Blood 62
4. The Application of Mechanical Principles for the Solution of Physiological Problems: The Seventeenth-Century Iatrophysicists 73
5. Further Physiological Developments in the Seventeenth Century, Especially in Great Britain 85
6. The Beginnings of Microscopical Observations and their Importance for the Solution of Physiological Problems 97

CHAPTER IV
The Physiology of the Enlightenment

1. Herman Boerhaave, Friedrich Hoffmann, Georg E. Stahl 112
2. Albrecht von Haller and the State of Physiology in the Middle of the Eighteenth Century 123
3. Physiology at the end of the Eighteenth Century 132

CHAPTER V

Nineteenth Century Physiology: The Beginnings

1. General Lines of Development in Nineteenth Century Physiology 150
2. The So-called "Romantic" Interlude in Physiology 155
3. Empirical Physiology in France and Germany 1800-1850 (Bichat — Magendie — Purkinje — Wagner — Weber) 166
4. The Development of British Physiology 1800-1848 180
5. American Physiology from its beginnings until 1860 187

CHAPTER VI

Johannes Müller, Carl Ludwig and their Circle of Students

1. Johannes Müller 195
2. Carl Ludwig 204
3. Hermann von Helmholtz and his Disciples 212
4. Emil Du Bois-Reymond and his Disciples 220
5. Ernst Brücke and his Disciples 234
6. Other Disciples of Johannes Müller 242
7. The School of Carl Ludwig 248

CHAPTER VII

Nineteenth and Twentieth Century Physiology: Western Europe-America-Russia

1. Claude Bernard and French Physiology Between 1848-1914 264
2. The Chemical Current in Physiology, Especially in Germany 275
3. Other German Schools of Physiology: Carl Voit, Otto Frank, Friedrich L. Goltz and Ewald Hering 292
4. British Physiology since 1848 305
5. American Physiology: 1870-1930 318
6. The Development of Russian Physiology in the Nineteenth and Twentieth Centuries 326
7. The Development of Scandinavian, Dutch and Belgian Physiology in the Nineteenth Century with an Appendix on Japan 334
8. Physiology in the Twentieth Century 348
9. Review — Prospects — Epilogue 357

List of Illustrations

CHAPTER 1 & 2

Figure 1: The Eastern Mediterranean in Antiquity 1
Figure 2: Elements, qualities and humors of the body 6
Figure 3: A brief sketch of Galen's physiological ideas 19
Figure 4: Main pathways of transmission of ancient medical knowledge to the Middle Ages 28
Figure 5: Albertus Magnus 35
Figure 6: Drawings of Leonardo da Vinci regarding the mechanical function of the semilunar cusps in the valves of the heart 41

CHAPTER 3

Figure 7: Depiction of a skeleton from the 14th century 50
Figure 8: Depiction of a skeleton from Andreas Vesalius' De Humani Corporis Fabrica 51
Figure 9: Physiology of J. Fernel 54
Figure 10: Theophrastus von Hohenheim 57
Figure 11: Johannes Baptista van Helmont 59
Figure 12: Frans de le Boë 60
Figure 13: Regnier de Graaf's method of creating an experimental pancreatic fistula 62
Figure 14: Michael Servetus 65
Figure 15: William Harvey 68
Figure 16: Experiments carried out by Jan de Waal in 1640 to ascertain the correctness of Harvey's circulation scheme 71
Figure 17: Bartholinus' sketch of the blood circulation, 1685 72
Figure 18: Santorio's balance-chair used for metabolic studies 77
Figure 19: Descartes' diagram indicating the hypothetical relationship between sensory organs and muscular movements 78
Figure 20: Giovanni Alfonso Borelli 79
Figure 21: Borelli: Geometrical and mathematical approach to the mechanical contraction of the muscles 80
Figure 22: Johannes Kepler 84
Figure 23: Thomas Willis 87
Figure 24: Richard Lower 89
Figure 25: Experiments of John Mayow demonstrating that the animal uses a certain portion of air during respiration 91
Figure 26: Robert Boyle, — Low pressure chamber 92
Figure 27: Niels Stensen 93
Figure 28: Stensen — Geometrical theory of muscular contraction 94
Figure 29: Stephen Hales 96
Figure 30: Passavant — The work of the heart (from his book De Vi Cordis) 96
Figure 31: Marcello Malpighi 100
Figure 32: Malpighi — Capillaries of the lungs observed in 1660 in the lungs of a frog 101
Figure 33: Leeuwenhoek: Crystals from various solutions observed under the microscope 103

CHAPTER 4

Figure 34: Lecture room of Boerhaave in Leyden 116
Figure 35: Herman Boerhaave 117
Figure 36: Title page of Boerhaave's Institutiones Medicae 118
Figure 37: Georg E. Stahl 120
Figure 38: Albrecht von Haller 124
Figure 39: René A. F. de Réaumur 132
Figure 40: Lazzaro Spallanzani 133
Figure 41: John Hunter 134

Figure 42: Antoine L. Lavoisier 135
Figure 43: The calorimeter of Lavoisier and Laplace for the measurement of combustion heat. 137
Figure 44: Alexander Stuart and his experiments on frogs concerning spinal reflexes. 139
Figure 45: Georg Prochaska 139
Figure 46: Galvani's experiments about the origin and transmission of "animal electricity" 142
Figure 47: Voltaic pile 144

CHAPTER 5

Figure 48: Scheme of the fundamental ideas of the romantic philosophy of nature. 159
Figure 49: Johann B. Wilbrand 161
Figure 50: Carl G. Carus 162
Figure 51: Cosmic polarity of forces acting on man, the microcosm. Interpretation of Joseph Görres 163
Figure 52: Marie Jean Pierre Flourens 167
Figure 53: François Magendie 168
Figure 54: Karl Friedrich Burdach 170
Figure 55: Johannes Evangelista Purkinje 171
Figure 56: Microscopical observations of ciliary motion according to Purkinje and Valentin 173
Figure 57: Gabriel G. Valentin 174
Figure 58: The use of Czermak's laryngoscope 174
Figure 59: Rudolph Wagner 176
Figure 60: Ernst Heinrich Weber 178
Figure 61: Alfred W. Volkmann 179
Figure 62: Arnold Adolph Berthold 179
Figure 63: Charles Bell 182
Figure 64: Marshall Hall 185
Figure 65: William Beaumont 188
Figure 66: Austin Flint 189

CHAPTER 6

Figure 67: Johannes Müller 195
Figure 68: Carl Ludwig 206
Figure 69: Ludwig's kymograph of 1846 206
Figure 70: Ludwig's "Stromuhr" for the measurement of blood flow 209
Figure 71: Ludwig's most prominent foreign collaborators 210
Figure 72: Hermann Helmholtz 212
Figure 73: Bernstein's "Membrane Theory" 1912. Uninjured fibre with double electrical layer 218
Figure 74: Johannes von Kries 218
Figure 75: Emil Du Bois-Reymond 221
Figure 76: Du Bois-Reymond's measuring devices: 222
1. Simple switch
2. Mercury switch
3. Non-polarizable electrodes
Figure 77: Eduard F. W. Pflüger 225
Figure 78: Nathan Zuntz and his portable breathing apparatus 227
Figure 79: Ludimar Hermann together with Martin Gildemeister (middle) and Otto Weiss (left) in the Physiological Institute. 228
Figure 80: Rudolph Heidenhain 231
Figure 81: Ernst Brücke 234
Figure 82: Alexander Rollet 237
Figure 83: Willy Kühne 239
Figure 84: Jacob Johann von Uexküll 241
Figure 85: Georg Meissner 243
Figure 86: Jacob Henle 244

Figure 87: Adolf Fick 248
Figure 88: Jacques Loeb 250
Figure 89: Hugo Kronecker 251
Figure 90: Ludwig's Physiological Institute at Leipzig around 1880 253
Figure 91: Edmund Drechsel 254
Figure 92: Max von Frey 255
Figure 93: Luigi Luciani 257
Figure 94: Angelo Mosso 258

CHAPTER 7

Figure 95: Claude Bernard 265
Figure 96: Title page of Bernard's Introduction to the Study of Experimental Medicine. 270
Figure 97: Charles E. Brown-Séquard 273
Figure 98: Etienne J. Marey 273
Figure 99: Marey's double sphygmograph 274
Figure 100: Charles Richet 275
Figure 101: Justus von Liebig 278
Figure 102: Liebig's laboratory at Giessen 279
Figure 103: Felix Hoppe-Seyler 284
Figure 104: Friedrich Miescher 285
Figure 105: Eugen Baumann 286
Figure 106: Franz Hofmeister 288
Figure 107: Emil Abderhalden 291
Figure 108: Carl Voit 292
Figure 109: Pettenkofer's metabolic apparatus 293
Figure 110: Max Rubner 294
Figure 111: Otto Frank 296
Figure 112: Friedrich L. Goltz 297
Figure 113: Ewald Hering 299
Figure 114: Max Verworn and his circle of collaborators at the Physiological Institute in Bonn. 302
Figure 115: Michael Foster 306
Figure 116: John Burdon-Sanderson 307
Figure 117: Victor A. H. Horsley 308
Figure 119: Frederick Gowland Hopkins 312
Figure 120: Joseph Barcroft 313
Figure 121: Bayliss and Starling with Sir Henry Dale 314
Figure 122: Charles Scott Sherrington 316
Figure 123: Henry Pickering Bowditch 319
Figure 124: Henry Newell-Martin 320
Figure 125: William Henry Howell 321
Figure 126: Warren Plimpton Lombard 322
Figure 127: Walter Bradford Cannon 323
Figure 128: Samuel James Meltzer 324
Figure 129: Filip W. Ovsianikov 327
Figure 130: Vladimir M. Bechterev 328
Figure 131: Ivan M. Setchenov 329
Figure 132: Victor W. Pachutin 329
Figure 133: Nicolai E. Wedensky 330
Figure 134: Aleksej A. Uchtomsky 331
Figure 135: Ilya I. Metchnikoff 331
Figure 136: Ivan P. Pavlov 332
Figure 137: Christian Bohr 335

Figure 138: Alarik Frithjof Holmgren 336
Figure 139: Robert Armand Tigerstedt 338
Figure 140: Jöns Erik Johansson 339
Figure 141: Franciscus C. Donders 341
Figure 142: Theodor Wilhelm Engelmann 342
Figure 143: Willem Einthoven 343
Figure 144: Rudolf Magnus 344
Figure 145: Theodor Schwann 345
Figure 146: Leon Frédéricq 346
Figure 147: The growth of periodicals dealing with physiological subjects 350
Figure 148: Otto H. Warburg 353
Figure 149: Bernardo Alberto Houssay 354
Figure 150: Alan Lloyd Hodgkin 356
Figure 151: Walter R. Hess 356

Introduction to K.E. Rothschuh, *History of Physiology*

Physiology, as a subject of inquiry has a long and remarkably continuous history beginning with studies and speculations of the Greeks in the fifth century B.C. Dr. Rothschuh's survey of the history of physiology was first published in German in 1953. He has now greatly revised and extended the German text and in translating it into English Dr. Guenter B. Risse has added certain supplementary information and references for English readers. The resultant book is probably the most complete survey of the history of physiology that has yet been written.

In his approach Dr. Rothschuh has tended to write a history of physiologists rather than of the development of physiological ideas. Yet he is also concerned with ideas and suggests how they were expressed, grew and changed in the work of successive individual physiologists and groups of physiologists. Dr. Rothschuh is particularly concerned with the relation of teacher to pupil in physiology and in the nineteenth century, particularly in his treatment of nineteenth century German physiologists, his approach becomes especially valuable for English-speaking readers.

From about 1830 Germany began to assume a leading position in medical research and teaching, and in no area of medical research were the Germans more prominent and influential than in physiology. The Germans were the first to exploit for the study of living tissues, the potentialities of the new and powerful achromatic compound microscopes available after 1830. They developed apparatus to measure and record physiological phenomena and were particularly ingenious in their development of electrical apparatus for physiological use. They made physiology more completely a quantitative science.

Dr. Rothschuh's account of the growth and spread of physiological research in Germany is extremely thorough. He describes the influence of Johannes Müller, professor of anatomy and physiology at Berlin from 1831 to 1858, on his many pupils and follows the careers of those pupils until they became professors and had pupils, and so on to the third and fourth generation of physiologists who traced their academic descent from Müller. For nineteenth century Germany Dr. Rothschuh's history therefore tends to be a sociological study of a population of scientists, their relationships to each other and to the institutions in which they studied and taught. He also shows how the German physiologists influenced the development of physiology in Great Britain and the United States through the migration of British and American students to Germany for periods of study, and traces the rise of physiological research in the English speaking countries after 1870.

Dr. Rothschuh is interested too in the beliefs and assumptions which have guided and sustained physiologists in the course of their work. The work of Müller's students and their successors was an effort to determine the nature of the physical and chemical processes going on within the living body in the belief that the totality of physical and chemical processes would be a description of life. Although this line of investigation has proven extremely successful, it has led to the discovery of physical and chemical systems so complex and multifarious that Dr. Rothschuh is inclined to believe that some additional modes of study will be needed in future physiological research.

Thus Dr. Rothschuh's book is not only a survey of the efforts of a multitude of physiologists in various countries throughout the course of history but it is a thoughtful summary of the meaning of their joint efforts.

Leonard G. Wilson
Department of the History of Medicine
University of Minnesota
March 1972.

Preface To The English Edition

The possibility of publishing an English translation of my *Geschichte der Physiologie* in the United States after almost twenty years, has given me the welcome opportunity to effect some changes of limited scope in the text. I have made some corrections and inserted the results of important new contributions made by modern physiological research. Since the new edition is primarily directed to a circle of English-speaking readers, a certain number of references to German, French and Italian sources had to be omitted for lack of space. Above all, the section of the book dealing with the history of American and British physiology has been considerably expanded, both in actual text and illustrations. Furthermore, in the epilogue I have dealt much more with the development of recent physiology than previously. While this *History of Physiology* proceeds essentially in a chronological fashion, I have considered the history of physiological ideas in another publication: *Physiologie, Wandel ihrer Konzepte, Probleme und Methoden vom 16.-20. Jahrhundert* (Karl Alber Verlag, Freiburg 1968).

K.E. Rothschuh
Münster, Westphalia
Spring 1972

Preface To The First German Edition

"History is the intellectual form in which a given culture renders an account of its past".[1] In this sense the history of physiology here submitted describes the fundamental trends and various detours, the achievements and errors which have shaped the historical development of this science.

Considering the limited space available, only broad outlines of the historical development could be sketched, while many details had to be excluded. Therefore, the possibility of portraying in detail the evolution of individual and special problems — as F. Lieben did in his history of physiological chemistry[2] — had to be abandoned.

On the other hand, I could forego this more explicit treatment of the subject, because of my recent tabular history.[3] In that work I depicted step by step the growing perception of the causes and significance of individual physiological phenomena. Beginning with modern times, the tabulation lists authors, dates, subjects and places of publication.

In the present work, my task has been to point out, in the course of the historical development of physiology, the connections with the contemporary general cultural and intellectual trends. Moreover, physiology has been closely integrated with the history of science and medicine. Thus, the major emphasis of this presentation is directed towards the *evolution of physiological thought* in relation to the leading ideas of each historical period. Only in this way did it seem feasible to provide a clear perspective and recognize the driving forces of the scientific development. The latter should convey the belief that the "today" of a given situation is merely a transition between "yesterday" and "tomorrow."

Hence, I could not let the older historical periods largely recede into the background in favor of the investigation and interpretation of the nineteenth-century events. This has been the case of the only other history of physiology written in German by H. Boruttau.[4] Each change in the intellectual conception of nature and man is related to a distinct phase of the historical evolution, and this fact is also true for physiology. Only a thorough knowledge of the preceding step will make the following one truly understandable.

The external circumstances prevailing in Germany after the second World War considerably hindered the composition of those chapters in which new developments occurring in non-German and non-European countries were depicted. Unfortunately, in those cases, the pertinent literature was only partially available, and for the resulting gaps I would like to ask the indulgence of the reader.

Compared with these countries, the German physiology of the past hundred and fifty years has been presented in greater detail. The justification for such an approach was based on the accomplishments of German physiology as well as the fact that the history of English physiology has received special consideration by Michael Foster[5] and K.J. Franklin[6]. In the United States the works of C.J. Reed and John F. Fulton[7] dealt similarly with the achievements made in that country.

Often, the last hundred years seem an almost impenetrable jungle to the historian. There has been a tremendous expansion of the scientific methods and a change in the goals of scientific pursuit. As a result, the number of prominent researchers and their contributions has virtually skyrocketed. I have attempted to penetrate through this thick underbrush, using the broad connections between the great schools of physiology like the thread of Ariadne to escape from the Labyrinth. In conclusion, I hope that the recent history of physiology has thus become more articulated and structured. As a result of these efforts, the international relationships in the history of physiological thought tends to emerge clearly above the narrower limits imposed by the various nationalities.

References

[1]J. Huizinga, *Wege der Kulturgeschichte,* Studien, München. 1930.

[2]F. Lieben, *Geschichte der physiol. Chemie.,* Wien. 1935.

[3]K.E. Rothschuh, *Entwicklungsgeschichte physiol. Probleme in Tabellenform,* München-Berlin. 1952.

[4]H. Boruttau, "Geschichte der Physiologie u. ihrer Anwendung auf die Medizin bis zum Ende des 19. Jhs." *Hdb. d. Geschichte d. Med.* 2, 237, Jena 1903.

[5]M. Foster, *Lectures on the History of Physiology during Sixteenth, Seventeenth and Eighteenth Centuries,* Cambridge, 1901.

[6]K.J. Franklin, *A Short History of Physiology,* London, 1933.

[7]J.F. Fulton, *Selected Readings in the History of Physiology,* Springfield, Ill., 1930.
J.F. Fulton and L.G. Wilson, *Selected Readings in the History of Physiology.* 2. ed., Springfield, Ill., 1966.

Introduction

The first question facing those who attempt to write the history of some topic of human thought is how to outline and limit the subject. Regarding PHYSIOLOGY, the task could be simplified by considering solely what contemporary physiology deems its field of research and study. We could also view the history of physiology merely as the historical development of present physiological knowledge, an approach which has been used occasionally. In such circumstances, the subject matter available for historical reflection could be easily delineated, but physiology would emerge as a very young science, in contrast with the long history of human thought. In fact, scientific physiology is so recent, that we could only examine the last few centuries in order to determine its gradual growth and development.

However, not only physiological learning but also the contemporary concept of this science and its goals are the results of a long evolution. Therefore, the history of physiology not only encompasses the historical development of physiological knowledge per se, but it is also concerned with the evolution of *physiological thought* in general, and the various attitudes and approaches employed for solving the manifold problems connected with human vitality. It is precisely on the conceptual level that great changes have occurred from the time when vital phenomena first became objects of human reflection until the present, where we have exact ideas regarding them.

The questions about the general nature of the living processes and the special functions of certain organs are indeed as old as human thought. These problems occupied philosophers as much as physicians, since the answers to these riddles were of similar consequence for the philosophical view of man as well as for the practical knowledge and conduct at the bedside. Therefore, the subject of physiology which is vitality, has been variously interpreted and dealt with in different periods of intellectual and cultural history. Our present concept of life is only one among many views held at one time or another in the course of history.

Hence, when physiology is seen in a historical context, its goals and principles appear to be closely connected with philosophical and theoretical efforts. Because of this fact, the evolution of physiology as the most important component of medical theory can only be seen in relation to the changing intellectual climate of the successive historical periods. An example of this conceptual variability can be observed in the shifting viewpoint over a period of many centuries regarding physiological functions.

For the Hippocratic physicians, physiology was essentially the knowledge of man's place in the general cosmological order, while Galen saw it as the doctrine concerned with the "usefulness of the bodily parts."

Much later, Albrecht von Haller's physiology became an "anatomia animata," merely depicting processes which took place in the anatomical structures of the organism. Finally, with the growing recognition of physical and chemical laws, physiology is now understood as the sum of physico-chemical actions occurring in the body.

Thus, the function of physiology changed along with the philosophical and cosmological doctrines of each age, providing an understanding of the roles played by Aristotle, Democritus, Descartes, Bacon, Leibnitz and Schelling in the development of physiological thought. On the other hand, each new acquisition of knowledge in other sciences has been used to reinterpret the hidden causes (kriptomena) of the observable vital phenomena (phainomena), from Greek antiquity to our own time. The advances made in physics, mathematics, alchemy, etc., were employed along with comparative anatomy, histology, cybernetics and biochemistry. The contributions of these sciences to general physiology have been recently described in an excellent work in English.[1]

Simultaneously, with the increase of knowledge, there has always been a change in interpretation. However, this philosophical and theoretical aspect has been gradually receding since the middle of the nineteenth century, while the factual data accumulates at an increased rate thanks largely to modern methods of research. The history of physiology clearly exposes the continuous efforts of the human intellect to gather new facts concerning the processes and meaning of organic functions. It is truly a fascinating endeavor to follow these attempts through time and achieve a more comprehensive view of the living actions in the human and animal organism.

With these remarks I have intended to demonstrate the reasons for the constant alterations which occurred regarding the *subject* of physiology. Another problem deals with the source and changes of meaning which the term "physiology" experienced in the course of time. The word originated in antiquity, when the Greek philosopher *Aristotle* (384-322 B.C.) used it in the broad sense of "knowledge about nature" without limiting the term to the organic realm. To be sure, the Greek "physis" also meant the total and lawful nature, although it was somewhat restricted in medicine by the Hippocratic physicians as the "healing power of nature."

Galen used "physiology" only once, when he described the various aspects of medicine: "The parts comprising medicine are physiology, etiology, pathology, hygiene, semeiology and therapeutics. . . Physiology is thus that portion (of medicine) where we speak about the nature of man" (*Introductio seu medicus,* cap. VII). Somewhat later, a book entitled *Physiologos*[2] enjoyed great popularity during the Middle Ages, although it was solely a collection of stories, especially animal tales, transmitted from antiquity.

Subsequently, the word physiology re-emerged at the beginning of the modern era, already restricted to the actions of living beings. Such is the use it received in the work *De Naturali Parte Medicinae* (1542) of the French physician and mathematician Jean F. Fernel (1497-1558). However, only in Fernel's *Universa Medicina* (1554) was the term "Physiologia" explicitly used.[3,4]

Somewhat later the Swiss physician and anatomist, *Theodore Zwinger* (1533-1588) of Basel published a book under the title *Physiologia Medica,* wherein the gradual conceptual limitation of the term became apparent. Physiology was to be restricted to the vital actions as subjects of medical inquiry.

The same kind of restriction was apparent in one of the propositions of the German astronomer *Johannes Kepler* (1571-1630) in his great optical work *Dioptrice* (1610). After enunciating the sixty-fourth proposition dealing with myopia and presbyopia, Kepler added: "This statement belongs to physiology and is almost medical."

In later times the word physiology appeared in the announcements of lectures being offered at various universities. K. Bürger discovered the term for the first time in the list of subjects which were presented at the German university of Giessen. A professor of medicine, Hieronymous Röetel, printed the following announcement for the winter semester 1664/1665: "D. Deo Opt. Max. fav. absolutis, quae in Physiologia tractanda restant, Pathologiae operam dabit publice."

At about the same time, the word was used by the German chemist *Johannes Bohn* (1640-1718) in the title of his works *Exercitationes Physiologicae* (Leipzig, 1668-1677) and *Circulus Anatomico-physiologicus* (Leipzig, 1686). According to Foster, the Italian mathematician and physicist *Giovanni A. Borelli* (1608-1679) referred to physiology as being merely a part of general physics in the introduction to his famous work *De Motu Animalium* (Rome, 1680/81). Several decades later, another professor of the University of Giessen, the German botanist *Johann F. Dillenius* (1684-1747), announced his lectures for the winter semester 1701/1702 in the following way: "...oculorum illorum Naturae miraculorum, Physiologiam publicis praelectionibus absolverit."

The definite limitation of physiology as the science concerned with the vital and organic functions, which is very similar to our own present concept of this field of study, occurred in the eighteenth century under the influence of the Dutch physician and chemist *Herman Boerhaave* (1668-1738). To be sure, the concern with the question of organic activities was already a regular subject of medical education quite distinct from anatomy. However, a host of new facts had been uncovered since the beginnings of the seventeenth century, as the circulation of the blood, and the distribution of the lymphatic fluid. Moreover, recent knowledge concerning the role of the blood in respiration and the mechanics of muscular contraction had all immediate medical significance without clearly belonging to anatomy. The new chemical knowledge about the bodily humors, their distinction as alkali, acids, salts and metals, promised further progress for medicine. Finally, Stahl's theory of the phlogiston, Kepler's dioptrics and many other discoveries opened up new vistas of the bodily functions.

Herman Boerhaave attempted to acquaint his students at Leyden with many of these new and promising subjects in a series of special lectures. In 1708 he published the content of these presentations under the title *Institutiones Medicae, in usus annuae exercitationes domesticos* at Leyden. The first part of this work dealt systematically with the "Physiologia" and received wide diffusion and recognition as a textbook. Shortly thereafter, several chairs called "Institutiones Medicae" were established at various universities in England, while the German physician Friedrich Hoffmann (1660-1742) wrote an equivalent work entitled *Fundamenta Medicinae.*[5]

The great Swiss physiologist *Albrecht von Haller* (1708-1777) was born in the same year of Boerhaave's successful publication. He became a student of the great Dutch physician and after the death of his teacher published three further editions of the *Institutiones Medicae* in 1746, 1751 and 1755. Earlier, in 1714, Boerhaave's work had been translated into English, while a French edition appeared in 1740. A German translation was written by Johann P. Eberhard, and published under the title *Herman Boerhaaves Phisiologia* (Halle, 1754). Here for the first time the word physiology was used as a title for a textbook dealing with organic functions.

The numerous students of Boerhaave often lectured on physiology according to the example of their master. These presentations not only carried the same titles, but they also displayed a similar great interest in the functional aspects of life. Among those who taught physiology were Haller at Göttingen, the Scottish physicians *Andrew St. Clair* (1699-1760) *Robert Whytt* (1714-1766) and *William Cullen* (1710-1790) at Edinburgh, the French naturalist *Julien O. de LaMetrie* (1709-1751) in France and many more.

In 1751, Albrecht von Haller published at Göttingen his famous work *Primae Lineae Physiologiae* which, together with the later book *Elementa Physiologiae* (8 vols., Lausanne and Bern, 1757-1766), clearly delineated the subject to be called physiology and elevated it to the rank of an independent science. However, Haller's physiology, "anatomia animata" (animated anatomy) still contained a considerable amount of purely anatomical information which no longer is a part of modern physiology.

Hence, not until the nineteenth century did physiology detach itself generally from anatomy. Only in the hands of many famous investigators, especially the students of the great German physiologists *Johannes Müller* (1801-1858) and *Carl Ludwig* (1816-1895), did physiology finally become the physics and chemistry of the bodily functions.

Hence, even the changes which the term physiology experienced in history were the result of an intellectual evolution. The same was also true concerning the subject of physiology as knowledge about living organisms and their individual organs.

References

[1]Thomas S. Hall, *Ideas of Life and Matter,* 2 vols., Chicago, Ill. 1969.

[2]Der Physiologos, aus dem griechischen Original Übertragen von Emil Peters und mit einem Nachwort versehen v. Friedr. Würzbach, München. 1921.

[3]Charles S. Sherrington, *The Endeavor of Jean Fernel with a List of His Writings,* Cambridge, 1946.

[4]K.E. Rothschuh, "Der Begriff der Physiologie und sein Bedeutungswandel in der Geschichte der Wissenschaft," *Arch. Internat. d'Histoire des Sciences* 10 (1957), 217-225.

[5]Friedrich Hoffmann, *Fundamenta Medicinae,* trans. by Lester S. King, London, 1971.

CHAPTER I

Physiology In Antiquity

1.
THE BEGINNINGS OF PHYSIOLOGICAL THOUGHT WITH THE GREEK PHILOSOPHERS OF NATURE AND THE HIPPOCRATIC PHYSICIANS

If we look back and attempt to retrace our steps to the origins of the science of physiology, we reach the shores of the Aegean Sea, which laps the edges of the Greek peninsula and Asia Minor. (Fig. No. 1) Here, in the Greek settlements arose for the first time an emphatic desire to explain the appearances of nature as effects of *natural causes.* This was a truly essential characteristic of Western thought which eventually spread, from its eastern Mediterranean source, to large portions of the world.

To be sure, it is erroneous to assume that there were no significant scientific and medical developments prior to the Greeks. Hundreds and even thousands of years before the appearance of the Hellenic culture, other peoples had already collected many useful observations, particularly in Mesopotamia, Egypt and India. Nevertheless, the Greeks were especially endowed with the gift of bringing coherence to the available notions, then systematizing this knowledge and finally conceiving of the resulting natural order as the necessary expression of general laws.

Fig. 1. *The Eastern Mediterranean in Antiquity.*

This Hellenic ability was reflected in the crucial development from "mythos" to "logos" at the very dawn of Western culture.[1] Hence, Greece became the source of a rational and scientific medicine, as well as the origin of numerous other disciplines such as mathematics, astronomy, physics, etc. Undeniably, the Hellenic civilization in the Aegean Sea should be considered the birthplace of Western culture. Greek thought decisively influenced Rome, Byzantium and Islam, thereafter determining in great measure the development of medieval and modern scientific and medical conceptions in Europe. In turn, the typically Western approach to nature has been gradually recognized in almost all continents, thus making Greece the true cradle of modern medicine and physiology. Therefore, its accomplishments should be described at the onset of our historical considerations.[2]

However, we do not want to overlook entirely the achievements of those flourishing civilizations in Mesopotamia, Egypt and India, which exerted strong influences on Greek culture and thought as a result of commercial and intellectual exchanges between the various countries.[3] The Babylonian elements, astrological dominance of the stars over human affairs, and the power associated with mathematics as well as the idea of cosmic lawfulness, deeply influenced the Greek philosophers of nature. Moreover, the conception and importance of humors, prevalent in Mesopotamian medicine, and the Egyptian theory of air, were known in the Greek colonies.[4,5]

The crucial shift from a mythopoeic thought to a critical and rational approach was made for the first time by the oldest Greek philosophers of nature. Although the change was gradual and tentative, the conception of nature based on the effects of divine and magical forces, began to lose ground. Numerous attempts were made by these philosophers to delineate the boundaries between aspects susceptible to investigation, and the unexplainable and mysterious events which eluded human understanding.[6]

Among the Ionic philosophers of nature, *Thales of Miletos* (ca. 640-548 B.C.) attempted to explain the multiplicity of natural appearances as the result of one fundamental principle: water. According to another philosopher, *Anaximenes* (ca. 585-525 B.C.), also of Miletos, the basic element was air or vital breath. A third theory was propounded by the philosopher *Anaximander* (ca. 610-545 B.C.), also a citizen of Miletos, who saw in the "Apeiron" (the "Boundless") the primal world-principle from which everything else was derived. His abandonment of mythical explanations for natural events was evidenced in his explanation of an earthquake. Instead of blaming the event on the god Poseidon, Anaximander postulated the existence of some shifting tensions in the hollow recesses of the earth.

For the philosopher *Heraclitus of Ephesos* (ca. 550-475 B.C.) fire was the essential element pervading life and the entire universe. Like the flame, the world was a continuous creation and consumption, an eternal flux in which change was the only permanent feature. Almost at the same time another philosopher, *Pythagoras of Samos* (ca. 580-489 B.C.), taught at Kroton, Italy. He founded a school there which cultivated philosophy, mathematics and music. According to Pythagoras, the number was truly the origin and essence of all things. He discovered among other things that the pitch of a sound was proportional to the length of the musical chord. The world was a harmonious entity, entirely ordered and dominated by numbers, with the same laws applying to both the micro- and the macro-cosmos.

Among the students attributed to Pythagoras was a physician: *Alcmaeon of Croton* (fl. 500 B.C.).[7] He was the first to apply the theory of opposites to questions of health and disease. According to Alcmaeon, the proper balance of opposite qualities, such as the wet, dry, cold, warm, acid and sweet, determined the health of an individual, while any disturbance of this equilibrium caused disease. Alcmaeon was apparently the first one to undertake experiments on living animals. He considered the brain to be the central fireplace of the human microcosm, and an organ intimately associated with the organs of sense. He probably first recognized the ocular nerves[8] and viewed the hearing organ as a system of empty chambers without describing the tympanic membrane or the inner ear.[9]

Furthermore, it is assumed that Alcmaeon made the initial and fateful distinction between vessels containing blood and those which appeared empty. He has been called the earliest Greek physiologist since he pioneered animal experimentation, carrying out vivisection experiments in order to solve biological and medical questions.

The theories of the philosopher *Empedocles of Agrigentum* (ca. 495-435 B.C.) of Sicily were very important for the subsequent development of medical and physiological thought. Empedocles assumed that the world was formed through a constant combination and separation of some fundamental elements rather than originating from one single and essential principle. He conceived the existence of four basic substances: fire, water, air and earth, which were harmoniously blended through the action of the forces of love while opposite forces of hate or strife effected their separation or repulsion. Thus one can foresee in Empedocles efforts to illustrate the hitherto unexplained forces of nature by blending them with some of the better known mental qualities.

In Empedocles' primitive physiology everything occurred according to strict mechanical principles. The soul was located in the blood, and the heart was the seat of the thinking process. Health resulted from the proper balance of the four elements, disease in turn was due to an imbalance of these elements. Most of these ideas were later incorporated into the Hippocratic writings, where they were further elaborated and developed, giving rise to the theory of the four humors and their four qualities which explained health and disease for about two thousand years. It appears that some basic concepts of this doctrine resembled similar earlier ideas prevalent in ancient Egyptian philosophy.[10]

The philosopher *Anaxagoras of Clazomenae* (ca. 500-428 B.C.) also contributed to anatomical knowledge. A friend of Pericles, who lived and taught for thirty years in Athens, Anaxagoras is said to have dissected the brain and observed the lateral ventricles of this organ. Shortly before the death of Pericles in 429 B.C., another physician, *Diogenes of Apollonia* (fl. 435 B.C.) arrived in Athens. We are indebted to him for a fragment of his writings concerning the human arterial system and another work entitled *Of the Nature of Man.* He revived the ideas of Anaximenes by postulating that the air or "pneuma" was the basic cosmic element. Diogenes further declared that this pneuma was the origin of life and endowed with a capacity to think. In his scheme, the air reached the lungs from the surrounding atmosphere through respiration, mingling there with the blood and thus reaching the entire body, including the brain. Disease occurred when the pneuma present in the organism suffered changes both in quality and in quantity, while its complete disappearance from the arteries meant death.

The gradual fusion of speculative and philosophical views of nature and man under strict medical concepts resulted in the development of the first scientific ideas regarding the two realms.[11] In this sense another philosopher, *Democritus of Abdera* (ca. 460-362 B.C.) also made important contributions. He was the founder of atomism, and composed a series of works on epistemology, mathematics, physics and medicine. "Nothing does occur by chance," he wrote, "everything happens necessarily according to a reason or 'logos.'"[12]

The Greek philosophers of nature interpreted the cosmos in rational but speculative terms, without close correspondence to the world of experience. However, this approach constituted the first basis for future science while the purely empirical procedures and practices formed the second foundation.[13] The fusion of these two traditions, empiricism and the rationalism of the philosophers of nature, signaled the beginnings of Western science.[14,15]

Under those circumstances, medicine was the first discipline to come into its own. Heirs to the Greek philosophy of nature, medicine and physiology, between 450-350 B.C., merged with the practical healing tradition in order to establish a rational-empirical type of discipline known as *Hippocratic medicine.*[16] During the rule of Pericles and the civil war between Athens and Sparta, a physician named *Hippocrates* (ca. 460-377 B.C.) lived on the Island of Cos, where he was born. We generally consider him as the founder of Western scientific medicine. He was a contemporary of the Athenian philosopher *Socrates* (469?-399 B.C.), the historian *Thucydides* (ca. 460-400 B.C.), the dramatists *Sophocles* (ca. 495-406 B.C.) and *Euripides* (ca. 480-406 B.C.) as well as Democritus and Diogenes of Apollonia.

Hippocrates was the descendant of an Asclepiad family. We know that he traveled extensively and practiced in several areas of the Aegean. It is assumed that he spent some time in Athens completing his studies there, though, in any event, he was well acquainted with the teachings and writings of the philosophers of nature. Many ideas and concepts previously expressed by Alcmaeon, Diogenes and others, found their literary expression in the works of Hippocrates and his disciples.

These representatives of a healing art based on the teachings of Hippocrates, were called "Hippocratic" physicians. Their writings, together with those probably composed by their master, comprise the so-called "Corpus Hippocraticum," an anonymous collection of about fifty-three works written mainly before 350 B.C.[17] Among those attributed to Hippocrates himself are *On Airs, Waters and Places,* and *On the Sacred Disease.*[18]

The bulk of the *Corpus Hippocraticum*[19] was probably composed by other contemporary physicians and students of the great healer of Cos. Regarding the content of the collection, we shall solely concentrate on the physiological explanations presented in these writings, leaving aside the valuable ethical and practical aspects. Since the various treatises were written by different authors, there are a certain amount of contradictory views about the organic functions. Nevertheless, there are in the *Corpus* a number of recurring physiological explanations.

Like Democritus, the Hippocratic authors conceived of man as a microcosm, existing in close correspondence to the macrocosm which surrounded him. The human body, like its cosmic counterpart, consisted of fire, water, air and earth. Following the ideas of Alcmaeon, these elements were endowed with a number of opposing qualities, now, however, reduced to only four: the hot, cold, dry and moist. These antagonic properties could be harmoniously blended together in a proper organic mixture, thus providing health. Similar ideas had already been expressed by Pythagoras and Alcmaeon. In the Hippocratic treatise *On Diet,* the living organisms consisted only of fire and water.

Polybos, the son-in-law of Hippocrates, combined the four elements and qualities with the bodily humors in a monograph entitled *On the Nature of Man.* The blood was considered to be warm and moist, the phlegm was moist and cold, the black bile appeared to be cold and dry, and the yellow bile was warm and dry. The following diagram (Fig. No. 2) depicts the various relationships between the elements, qualities and humors. Their proper balance or "eucrasis" meant health, while disproportions or "dyscrasis" resulted in disease.[20]

With this theory of humoral combinations began the Western concept of vital phenomena. The healthy mixture was maintained by the body's own heat, which was regarded as the result of an internal fire occurring in the bloodless left ventricle of the heart. To make the combustion possible, the body needed a supply of air or "pneuma," and nourishment in the form of food and drink. Moreover, within the organism dominated the so-called "physis"[21] or all-embracing "nature" which directed the body according to its own unchanging laws.[22] In a narrower sense, the "physis" was viewed also as a healing force, unconciously governing the organic actions with the purpose of restoring a healthy balance. However, this goal-oriented performance of nature was considered to be of limited scope, and needed the assistance of the physician in order to effect the proper cures.

From the preceding summary it becomes obvious that the Hippocratic physiology was essentially dynamic in character and not appreciably connected to the anatomical structures. To some extent this was understandable, since the knowledge of the human body was still rudimentary, except for the skeletal and vascular systems, and based primarily on analogies taken from animal anatomy. Occasionally, some simple experiments had already been described[23,24] but their potential as sources of anatomical and physiological knowledge had as yet not been clearly realized.

All organs possessing cavities, like the uterus, heart, bladder and kidneys, were considered to exert a certain attraction upon the humors. Moreover, those sponge-like structures such as the lungs, spleen and mammary glands were thought to absorb fluids directly from the surrounding tissues. The

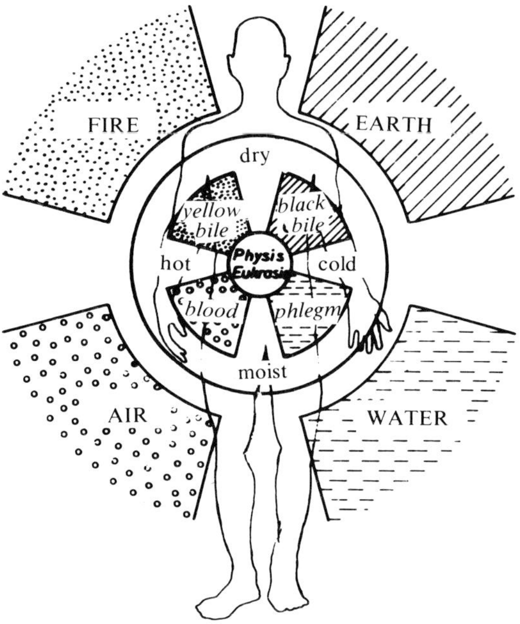

Fig. 2. *Elements, qualities and humors of the body:* The four humors and qualities were included in Hippocratic medicine. The complete schematization with the inclusion of the four elements occurred in late antiquity.

blood located in the vessels originated from the nourishment taken in by the body. All foodstuffs and drinks contained watery, phlegmatic, bilious and blood elements. The formation and growth of organs were conceived as a process of blood coagulation. In addition to all the necessary elements derived from food, the various bodily structures also received pneuma through the lungs and heat from the left cardiac ventricle.

In spite of the conflicting views expressed by various authors[25,26,27,28] we can presently assume that the Hippocratic physicians did not recognize the circulation of the blood.[29] Rather, these authors considered the blood movement to be merely a reciprocal ebb and flow motion. The heart was composed of two chambers with two auricles and numerous vessels which originated in the heart.[30] The presence of the semilunar valves was recognized and their action understood. The left cavities of the heart were thought to be empty of blood, the left ventricle being the seat of the body's temperature. During inspiration, the left ventricle received pneuma through the pulmonary veins from the lungs, in order to nourish the internal fire and at the same time cool the available blood. The inherent vital heat was believed to be responsible for the cardiac movements.

In the Hippocratic treatise *On Regimen* a sort of metabolic doctrine was developed. The author of the work distinguished between an anabolic function based on the action of nutritive forces, and an opposite phase of decomposition or combustion brought about by the body's temperature.

The brain was seen as the seat and the agent of the soul. Built like a gland, the cerebrum received blood from the liver and spleen, and pneuma from the heart through the carotid vessels. Injuries to the brain were thought to provoke a paralysis on the opposite side of the body. Moreover, lesions of the spinal cord — an extension of the brain — likewise produced paralysis, sensory disturbances, and problems with fecal and urinary excretion. There was no clear distinction between nerves and muscular tendons.

The olfactory function was explained as a type of absorption during which the moist brain descended into the nose and attracted the dry air with the smell through the nares. The Hippocratic physicians were the first to describe the tympanic membrane, which was considered to be another cerebral cover, while the real auditory organ remained unrecognized. The entire hearing process was conceived in close analogy to the views expressed by Empedocles and Alcmaeon. Following Diogenes of Apollonia, the followers of Hippocrates considered the visual process to be a phenomenon in which the object was reflected in the eye. This reflection occurred on a transparent skin located anteriorly in the eyeball itself, while the sensation actually occurred in the brain, thanks to a passage between the two structures.

Thus, we would like to conclude our considerations of the so-called "Hippocratic physiology." To be sure, it was primarily a rational theory of the observed vital occurrences, and it dealt exclusively with natural forces to explain the phenomena. The actual excess of hypothetical and speculative elements was primarily the result of poor anatomical knowledge. Furthermore, an analytical and experimental mode of thought and method was largely still alien to the Greeks. However, the largely speculative Hippocratic framework of physiological conceptions became the point of departure for further and greater progress in physiology in ensuing centuries.

2.
THE PHYSIOLOGY OF ARISTOTLE AND THE SCHOOL OF ALEXANDRIA

ARISTOTLE

After many years of unequaled political, economical and cultural flowering, especially under the great statesman *Pericles* (443-429 B.C.), the Attic confederacy disintegrated in the wake of the fraternal war between Athens and Sparta. Henceforth, Macedonians and Persians intervened in the political affairs occurring on the shores of the Aegean Sea, while the Greeks themselves concentrated on intellectual, social and legal matters. In the year 399 B.C. the Athenian philosopher Socrates was brought to trial on a charge of "impiety" and for "corrupting the youth," and forced to drink a cup of hemlock in fulfillment of the death sentence.

After this event, his student, the philosopher *Plato* (427-347 B.C.), abandoned a promising political career and founded an "Academy" in order to discover through scientific and philosophical efforts a new form of society supported by high ethical standards. According to Plato, the world of experience was merely a reflection or elaboration of the human mind, a copy of the real world of pure ideas. Thus, for Plato, the investigation of the world of appearances or visible nature, was only a peripheral concern. His late and mature work, the *Timaeus,*[31] contained a theory regarding the creation of the cosmos. Plato conceived of the universe as an enormous spherical and living creature, endowed with the highest perfection by the divine creator or Demiurgos, who had also fashioned the sacred human soul.

Plato's universe was viewed as a gradual self-unfolding of the supreme Reason into sensible appearances. The natural order was merely a reflection of eternity, and had to be judged for its goals, not its causes. Thus, Plato considered biological and physiological phenomena teleologically, that is, from the standpoint of their purposes within the general cosmic order.

Regarding specific physiological knowledge, the Platonic concepts did not go beyond the Hippocratic body of knowledge.[32] The human soul was divided into a sensual component located in the liver, an emotional part in the chest, and the eternal, rational soul which dwelled in the brain. The Platonic division became important for later physiological theories, especially the Galenic system, which also divided the pneuma into three portions: pneuma "psychikon," "zotikon" and "physikon."

Plato's Academy came into direct contact with medicine through the Sicilian physician Philistion of Lokroi, who resided in Athens between 367 and 363 B.C. It was probably Philistion who introduced in the Academy the doctrine of Empedocles concerning the four elements, and the idea of a supreme force which pervaded the entire organism, the pneuma.

Much later, long after the death of Plato, another famous physician came in contact with the Platonic Academy and the "Lyceum," then headed by the philosophers Aristotle and Theophrastus: *Diocles of Caristos* (ca. 340-260 B.C.)[33] Through him many of the teachings belonging to the Sicilian school became known to Aristotle, and were eventually incorporated into the philosopher's biological writings.[34] Pneuma, vital heat and the doctrine of the four elements were the most prominent aspects of the Sicilian system of physiological explanations. Moreover, the heart was considered to be the central organ of the entire body, concerned with the formation of blood. In all, Diocles totally subscribed to the teleological viewpoint of the Aristotelean school. A possible anatomical treatise which has been attributed to this physician has unfortunately been lost.

Aristotle (384-322 B.C.), great investigator, collector, observer and systematist, embodied in the most perfect sense the type of well-rounded savant which emerged on the Greek cultural scene since the times of Anaxagoras and Democritus. Until the death of Plato, Aristotle had been a member of the Academy and, for seven years beginning in 343 B.C., tutor and companion to Alexander the Great. In 325 B.C. the philosopher founded the Lyceum or Academy of the Peripatetic philosophers in Athens. The Aristotelean natural philosophy and his scientific views in many fields of knowledge, exerted a great influence on the concepts of vitality, not only in classical antiquity, but also for the Middle Ages. Thus, we will attempt to summarize briefly Aristotle's physiological ideas.[35]

Aristotle's most important biological works were *De Anima, Parva Naturalia, De Partibus Animalium, De Generatione Animalium* and *Historia Animalium.* In these writings the philosopher systematically incorporated all the known anatomical, physiological and zoological facts into a system of natural philosophy.[36] The best description of the Aristotelean physiological views is contained in C. Sievert.[37] For Aristotle, all being and becoming was based on form and matter. The latter was the passive aspect on which the active form was superimposed. Moreover, the form constituted the actuality, essence and force of every known being, while matter was only immobile and possessed unlimited potential.[38,39]

For Aristotle, the relationship between matter and form determined the specific appearances of all things. When matter came into contact with form, it received an impulse, possibilities and a certain goal. All organisms were conceived as consisting of matter and spirit. Matter was the material structure, while the form was the spiritual component or soul. The purpose of such a being was reflected in the particular force pervading and characterizing the entire organism: the "entelechy." It carried within itself the final goal of the organized body, and varied according to the different types of living creatures. Therefore, Aristotle conceived of the entelechy as a supramechanical, spiritual principle, which made life possible.

In addition, the plant kingdom was dominated by a vegetative and nourishing spirit, while the animals were also endowed with a sensitive soul. Man was the highest entity on earth and possessed a rational soul in addition to the sensitive and vegetative functions.

According to Aristotle, the heart was the central organ of the organism and the seat of the soul, instead of the brain, as propounded by Empedocles. Moreover, the heart was the body's blood-producing organ, the seat of the vital heat and the origin of the vascular system. In the Aristotelean conception, the pulse, cardiac contractions and respiratory movements were the result of the heat which accompanied the formation of the blood.

Actually, the heart action was viewed in part as the result of a sudden contraction of the warm, cardiac blood while being exposed to the cool, inspired air arriving in the left cardiac ventricle from the lungs. In turn, the blood expanded anew in the heart through the action of the innate heat, creating in the process the arterial pulse wave which coincided with cardiac diastole and thus facilitated the bloodflow through the arterial system down to the smallest branches. In the latter, Aristotle speculated that the blood was ultimately converted into flesh and tissues, while the fat, bone marrow, milk, sweat, sperm and urine were various secretions derived from the blood itself.

In the Aristotelean scheme, the lungs were thought to merely help in cooling the heart through their connections with the cardiac ventricles. The inspiration was precipitated by the expanding heat transmitted from the heart. As a result of this phenomenon, the chest cage expanded, allowing the entrance of fresh air into the lungs which not only was destined to cool the heart but also cause the expiratory movement of the thorax.

In the stomach, the vital heat liquified and separated the ingested food into true nourishment and disposable waste. While the latter was eliminated through the bowels, the nutritive juices were absorbed and conducted to the heart for the production of blood, a function in which the liver and spleen cooperated to some extent.

Aristotle was not aware of the existence of a systematized *nervous system,* including the brain and spinal cord. Nerves were not distinguished from muscular tendons, and no nervous conduction was envisioned. The spinal cord was equated with the bone marrow of the various bones, while the brain in Aristotle's opinion lacked arteries and blood, and was a cold, insensitive organ without connections to the organs of sense. In fact, the cerebrum was merely a place where the ascending heat of the organism could be tempered sufficiently to allow the production of mucous secretions which then escaped from the nose.[40] This cerebral thermoregulatory function even influenced the heart, considered as the organ of sensation and thought. According to the tenets of Aristotelean physiology, cardiac cooling resulted in periodic sleep.

The organs of sense perception were located in the head and benefited from the presence of more subtle and pure blood. Vision did not occur as a result of a flow of light from the object, but rather was the consequence of an impulse or motion which acted directly on the passive eye. Hearing on the other hand occurred when the atmospheric vibrations were communicated to the air present in the ear itself. Aristotle described the auditory canal, tympanic membrane, Eustachian tube, as well as the middle and inner ear. In his view there was no communication between the ear and the brain.

Aristotle's ideas regarding muscular function and bodily motions was rather sketchy. He felt that the heart was actually the cause of all movements through the alternate contraction and expansion of the cardiac fibres. Muscular contraction as such was still unknown — the muscles were merely seen as organs for the sensation of touch.

In summary, the physiological conceptions of Aristotle showed no perceptible progress over the previously existing ideas which had been expounded by Hippocrates' followers. Although the multitude of functional appearances achieved an unprecedented systematization in the hands of Aristotle, it is also true that a series of erroneous interpretations emerged. Prominent among these rather harmful and consequential ideas were the incorrect Aristotelean physiology of the heart and the brain. Likewise, Aristotle's doctrine of an "entelechy" represented a personification of the more vague Hippocratic "physis," and deeply influenced Galen's teleologically oriented physiology, as well as Western biology up to the vitalism of the German biologist Hans Driesch (1867-1941).[41]

The Physiology of the Alexandrian School

In the year 334 B.C. the great Macedonian conqueror Alexander the Great (356-323 B.C.) crossed the Hellespont with his army in order to attack the Persians. Shortly thereafter, the entire coastline of Asia Minor fell into his hands, and even Egypt subsequently surrendered to the advancing Greek forces. In order to better integrate the Egyptian territory into his growing empire, the Macedonian leader founded a city by the name of Alexandria in 332 B.C. on the Mediterranean coast. Located at the western end of the Nile Delta, this new settlement soon became a flourishing metropolis and one of the most important centers of the ancient world, boasting a population of about 900,000 at the height of its power.

Under the rule of the Ptolemies, Alexandria became the patron of Hellenistic art and a citadel of ancient learning. A large library with a rich collection of manuscripts numbering several hundred thousand offered optimal opportunities for scientific learning.[42] In addition, the so-called Museum — a combination of university, research academy and living quarters for scholars — brought a great number of distinguished men of learning from the Hellenic world under one roof. Among them were the mathematician *Euclid* (ca. 330-260 B.C.) and the astronomer *Aristarchus of Samos* (ca. 310-230 B.C.) who taught a heliocentric planetary system fifteen hundred years before the advent of the German astronomer *Nicolas Copernicus* (1473-1543). In addition, Alexandria attracted the famous botanist *Theophrastus* (ca. 372-288 B.C.), the physicists *Strato of Lampsacos* (fl. ca. 288 B.C.) and *Archimedes* (ca. 287-212 B.C.) and finally the founders of the Alexandrian medical tradition: *Herophilus* (ca. 335-280 B.C.) and *Erasistratus* (ca. 310-250 B.C.).[43]

Born in the city of Chalcedon, in Bithynia, Herophilus began his medical studies at Cos, under the physician *Praxagoras* (fl. 350-300 B.C.). The latter had concluded, as a result of observations in dead animals, that the arteries contained only air or pneuma, and were devoid of blood. Praxagoras also established a doctrine of specific types of pulses.[44] About 300 B.C., Herophilus was called to Alexandria to become personal physician to Egypt's ruler, Alexander's former general, Ptolemy I (ca. 367-283 B.C.). In addition to his practical duties, Herophilus carried out extensive anatomical studies. Except for a few isolated exceptions, he was the first and only physician in antiquity to carry out numerous dissections on human cadavers. However, the statement that Herophilus also performed vivisection experiments on some criminals who had been sentenced to death, is probably false.[45] According to the Carthaginian theologian *Tertullian* (ca. 160-230 A.D.), Herophilus dissected no less that 600 corpses during his career as an anatomist.

Unfortunately, almost all of Herophilus' writings have been lost, but from quotations and references which appeared in later authors, a fair picture of his views and knowledge has become available.[46] Most important was Herophilus' understanding of the nervous system. He described not only the ventricles of the brain, the meningeal vessels and the choroidal plexus, but distinguished the cerebrum from the cerebellum. In addition, Herophilus was the first to clearly separate nerves from tendons, and discern the functional relationship between the nerves and the brain.

Regarding the circulatory system, Herophilus distinguished between arteries and veins on the basis of their distinct anatomical structure. He recognized the cardiac auricles as independent cavities of the heart instead of mere ventricular appendages. Reflecting the views of his teacher, Praxagoras, Herophilus postulated that the arteries carried only air or pneuma.[47] In typical Aristotelian fashion he conceived four purposeful forces which controlled all bodily activities. One was a nutritional force, located in the liver, while the second resided in the heart and produced heat. The third was a so-called sensitive force located in the nerves, followed by Herophilus' fourth thinking quality residing in the brain. Incidentally, Herophilus was the first physician to introduce into medicine the waterclock, a device with which to count the frequency of the pulse. Also ascribed to the anatomist are the first descriptions of a number of anatomical structures, such as the duodenum, the salivary glands and the vitreous humor in the eye.

The younger physician *Erasistratus* was more of a physiologist than an anatomist.[48] Born on the Aegean island of Chios, he had studied at Athens where he came into contact with the doctrines of Democritus, Strato and the Stoics. In Alexandria, Erasistratus carried out some anatomical studies. He described in detail the structure of the heart, the cerebral meninges, the epiglottis and prostate; but the bulk of his work was of a physiological character.

According to Erasistratus, the body consisted of atoms which were moved and vitalized by the actions of the innate heat. The various bodily tissues, with the exception of the brain, were composed of three elements, tightly intertwined together: arteries, veins and nerves. The mission of the arteries was to conduct the pneuma, while the veins distributed the nourishing blood and the nerves received sensory impressions and imparted motion to the limbs.

The blood originated from the digested food with the aid of the liver, and was projected to the right cardiac ventricle. In Erasistratus' view, the heart was composed of only two chambers since the auricles were considered to be merely extensions of the great vessels. From the right ventricle a vein with arterial characteristics (pulmonary artery) and the vena cava originated, whereas the left ventricle issued forth the aorta and the pulmonary veins.

Erasistratus believed that the left ventricle of the heart contained only pneuma, which was supposed to flow through the entire bloodless arterial system in the form of *pneuma zotikon,* being responsible for the body's vegetative functions. In addition, the left ventricle projected through the vessels of the neck and head *pneuma psychikon* to the dura mater, enabling the hollow nerves to feel and promote motion. By contrast, the veins conducted blood to the heart and the periphery for proper nutrition and growth of all organs. The lungs received their blood supply through the right cardiac ventricle and the pulmonary artery.

The Alexandrian physiologist speculated that the arteries and veins were connected through a series of anastomoses which were closed under normal physiological conditions. Only when a vessel of the arterial system was opened in a living organism would blood issue forth from the vascular gap, attracted from the veins through the arterio-venous connections. Erasistratus explained this phenomenon as a result of a "horror vacui" in the arteries, the fear-of-vacuum concept which he borrowed from Strato's physics.

The body's pneuma originated in the atmospheric air and was incorporated into the organism during the act of inspiration. From the lungs the pneuma was then conducted to the left ventricle of the heart by the pulmonary veins. Once in the left cardiac ventricle, this air was divided into two categories: the psychic pneuma went to the brain whereas the pneuma zotikon flowed to the rest of the organism.[49]

From the right ventricular chamber of the heart the blood was directed to the lungs during systole, flowing through the pulmonary artery without experiencing any reflux, thanks to the patency of the semilunar valves. During cardiac diastole the blood would again enter the right ventricle from the vena cava through the opened tricuspid valves. Erasistratus was the first to describe accurately the valvular mechanisms in the heart. Because of this knowledge, the Alexandrian physician came close to a correct understanding of the blood circulation. However, he continued to follow the opinions of Praxagoras of Cos, insisting that the arteries contained only air. It should be noted that Praxagoras separated the arteries from the veins for the first time because of their differences in pulsation. Finally, in the scheme of Erasistratus, the rest of the body was supplied with blood by the veins which originated in the stomach and liver, carrying the nutrients to the periphery.

In later years Erasistratus revised his theories and postulated that the hollow nerves actually conducted the psychic pneuma, beginning in the dura mater. The differentiation between motor and sensitive nerves has also been attributed to him. Muscular contraction was explained as an inflow of pneuma into the actual muscles, expanding them while shortening their fibres and placing them under tension.[50] Erasistratus, affected with a severe ailment, committed suicide by drinking a cup of hemlock.

Herophilus and Erasistratus were the founders of the most famous medical school in antiquity. For centuries thereafter, many excellent anatomists lived in Alexandria, following the traditions of these notable pioneers.[51] The last one among them was *Marinus of Tyre* (fl. 60-130 A.D.) who is reputed to have lived in Rome about 130 A.D. Marinus wrote a twenty volume, systematic treatise of anatomy, now lost, but extensively used by Galen. He observed the secretory glands in the intestines, distinguished seven cranial pairs of nerves, and described the central artery of the retina, thus carrying on the work of his Alexandrian predecessors. Hence, anatomy in Hellenistic times made rather impressive advances while physiology remained still largely speculative, due to the lack of physico-chemical knowledge and ignorance regarding the experimental approach.[52,53]

3.
GALEN OF PERGAMON AND ROMAN PHYSIOLOGY

In the two Punic wars Rome destroyed the power of Carthage and became the leading nation in the Western Mediterranean after the peace of 201 B.C. The conquest of the Greek settlements in southern Italy and Sicily brought about a growing contact with Hellenic language, literature and science. In 146 B.C. the city of Corinth was destroyed and Greece was annexed into the Macedonian province of the Roman Empire. Somewhat later, the kingdom of Pergamon was made the Asiatic province of the Romans. Most of the empire wrought by the conquests of Alexander the Great fell into the hands of the bourgeoning Mediterranean power, and Rome became the heir of the Hellenistic culture.

Primitive Roman medicine lacked any originality, both at the theoretical and practical levels.[54] Moreover, the first Greek physicians coming to Rome to improve their fortunes, did not exert any decisive influence on the development of the native medicine. Only from the year 46 B.C. onward did Greek medicine achieve wide recognition in Rome, as a result of Caesar's decree bestowing Roman citizenship on the free Greek physicians established there.[55]

The first prominent Greek healer on Roman soil was *Asclepiades of Prusa* (ca. 124-40 B.C.). Until recently, medical history has viewed this physician as being a quack, eager for publicity. However, according to M. Wellmann, Asclepiades about 91 B.C. developed a brilliant career in Rome, both as a philosopher and a physician. Among his patients supposedly were the great orator and poet Cicero, and the generals Crassus and Mark Anthony.

Asclepiades' medical ideas were based on the atomism of Democritus and the Epicureans. Molecules and the void were, therefore, for him the two essential forms of reality, also forming the living organism. Between the smallest bodily particles, Asclepiades postulated the existence of pores and passages in which pneuma and fluids could freely circulate. Thus, the normal relationship between the pores and their contents brought about health, and the humors played only a secondary role in the mechanisms of disease. Under these circumstances, Asclepiades became the founder of a theory of disease concerned with the solid parts of the body, accepted by other famous physicians such as *Themison* of Laodicae (fl. 31 B.C.-14 A.D.), *Thessalos* of Tralles (fl. 50-68 A.D.), *Soranus* of Ephesus (ca. 98-138 A.D.)[56] and others. For the development of physiology, however, the Methodist school did not contribute anything beyond the formulations of the Pneumatic physicians. The latter elevated the pneuma to the category of a goal-oriented, vital principle pervading the entire organism.[57]

In the second century A.D. the ancient world developed its final, broadly based medical theory concerned with the nature of man and the functions of his organism. This system, propounded by the physician and philosopher *Galen* of Pergamon (ca. 130-201 A.D.), dominated Western medicine unchallenged until the later Middle Ages.[58]

In Asia Minor, on the heights of the mainland, existed the city of Pergamon. The recently excavated ruins still reflect the magnificence of this wealthy settlement with its famous Asclepeion. In this city Galen was born about 130 A.D.,[59] the son of the architect and mathematician Nikon.[60] After initial studies in his home town, Galen received an extensive and varied philosophical and medical education in the cities of Smyrna, Corinth and Alexandria. In the latter metropolis, Galen acquired a thorough anatomical knowledge from Marinus, although this learning was based on animal structures with the possible exception of the skeletal system. After 157 A.D., Galen returned to Pergamon and there became the physician for the gladiators performing in the Asclepeion. He remained in his native city for five years, traveling to Rome in 162 A.D.

Once in the capital of the great Empire, Galen became in a few years a much sought after practitioner and scholar, in spite of the considerable opposition he encountered there. The rather bitter polemics which he carried out with his medical rivals revealed both his ruthlessness and boastfulness. After a short absence, Galen returned to Rome in 168 A.D. as the personal physician of the emperor Marcus Aurelius and his son Commodus. He died in Pergamon in about 201 A.D.

Galen's entire writings comprise, according to his own estimate, about four hundred treatises of medical, philosophical or miscellaneous content. Their style reveals an author of great confidence and pride. A certain amount of Galen's medical works were commentaries on the Hippocratic writings, which he valued highly enough to adopt the humoral doctrine, the concept of elementary qualities and the idea of a healing force in nature.

Galen was familiar with the Platonic and Aristotelean philosophies, and his tendency to view physiology from a teleological standpoint was probably taken from the tradition of Aristotle. On the other hand, Galen's acceptance of three different types of soul and pneuma was clearly based on Plato's ideas.[61]

The Stoic philosophy of the first centuries of our era also influenced Galen's thought. Without his extraordinary vitality, industry and capacity for assimilation, Galen could not have mastered the tremendous literary and medical, as well as experimental and pedagogical work.[62] His most interesting works were *De Anatomicis Administrationibus* (On Anatomical Procedures),[63] *De Usu Partium Corporis Humani* (On the Usefulness of the Parts of the Body) and *On Natural Faculties.*[65] Galen's undeniable merit lay in the fact that he based his systematic medical theory on anatomical and physiological facts. Thus, his concept of disease and ultimately his practical and therapeutical derivations had a similar basis.

Galen's anatomical knowledge exceeded that of his predecessors.[66] Even though his anatomy was based primarily on dissections of animals such as monkeys, pigs, bears, cattle and dogs,[67] perhaps he was of the opinion that man and the animals did not differ substantially from one another in their anatomical structures. Many serious errors occurred because of this faulty assumption. Galen's description of the human skeleton was essentially correct except for the idea of a divided mandible, and a simian sternum composed of seven sections.

Galen coined the words *epiphysis* and *symphysis,* and revealed a good anatomical knowledge of the nervous system. Nerves, tendons and ligaments were clearly distinguished by him. The nerves, except for the optic pair, were not hollow structures according to Galen, although he granted the existence of them in invisible canals. He described the existence of seven pairs of cerebral nerves. Regarding muscular anatomy, he discovered a series of new muscles, like the M. interossei, M. popliteus, M. plantaris and platysma. Moreover, he accurately portrayed the eyeball with its six muscles, the orbital septum and the lacrimal sac. Although many gaps remained, it must be conceded that Galen established a solid foundation in anatomy. Unfortunately, he was the last distinguished anatomist for over a thousand years.

Galen's physiological concepts can be divided into those which form a general theory of vitality, and the ideas related to the special physiology of certain organs. For this purpose we will examine Galen's book *On the Usefulness of the Parts.* We are also grateful for the new and substantial additions to our knowledge concerning Galen which have come from the works of R.E. Siegel.[68]

According to Galen, living matter consisted of varying mixtures of the four elements: fire, water, air and earth, and the four fundamental qualities: hot, wet, cold and dry. Health was achieved through a balanced blending of these elements and qualities.

Matter only came alive through the action of organizing spiritual entities. Galen considered the human soul to be an immaterial principle in some of his writings, while in others the soul was merely some material, identified with the pneuma. In accordance with Plato, he distinguished three varieties of pneuma: first, the *spiritus animalis* or animal spirit, located in the brain, which was the sentient and thinking soul also responsible for bodily motions. Secondly, Galen conceived of a *spiritus vitalis,* dwelling in the heart and responsible for the pulse, circulation of the blood and distribution of vital heat. Thirdly, there was a *spiritus naturalis* or natural spirit, existing in the liver and in charge of nutrition, blood formation and general metabolism, although it played only a very secondary role in the overall Galenic scheme.[69] Finally, each of these souls was conceived as being a fine material entity, the carrier of certain forces, while each organ in turn had its own functional capacity: a *dynamis* or "faculty." According to Galen, the bodily organs exhibited four especially important faculties: attraction, retention, alteration and expulsion.[70]

Galen declared that all vital functions followed strict physiological laws, displaying great expediency at the same time and pursuing a high purpose. He was convinced that "Nature was concerned with avoiding excesses and deficiencies or exerting undue care."

The blood was composed of the four cardinal humors: red blood, yellow bile, black bile and phlegm. The predominance of one or the other of these humors determined the so-called temperaments: sanguine, choleric, melancholic and phlegmatic when the blood, yellow bile, black bile or phlegm was in excess respectively. For Galen, the blood filled both the arteries and veins. He demonstrated the existence of blood in the left cardiac ventricle in a vivisection experiment, when he punctured the left chamber of the heart in an animal.

The formation of blood occurred in the liver thanks to the assistance furnished by the vital heat which was implanted in that organ. The constitutive elements of the blood came from the stomach to the liver through the portal vein system. For Galen those remaining or excessive particles which were not used to make blood were attracted by the gallbladder and spleen where they became respectively the yellow and black bile.

In Galen's view only thin and pure blood arrived at the right heart, carried there by the vena cava. For him the veins were vessels merely connected with the right side of the heart, while the arteries were hooked up with the left heart. He considered the heart to be like a muscle and described in detail the two ventricles with their valvular mechanisms, while the auricles were not mentioned, obviously because they were considered to be simply extensions of the large incoming veins.

Galen postulated that the cardiac septum had multiple pores, thus allowing communication between the two ventricles.[71,72] He also described the functional autonomy of the excised heart, as he had observed it during his vivisection experiments.

In the Galenic scheme, the movement of the blood was not solely initiated by the heart. The task of the left ventricle was to communicate a pulsatile force to the arteries and absorb pneuma as well as some fine blood from the lungs through the pulmonary veins. But, in the left ventricular chamber blood was oozing through the septal pores from the right side, being mixed with pneuma and warmed up by the action of the innate heat residing there. Then, this especially prepared ebullient blood was driven into the arterial system during the cardiac systole, allowing simultaneously the ascension of excessive vapours and waste fumes through the pulmonary veins back into the lungs (see Fig. No 3).

On the other hand the right ventricle of the heart attracted blood from the vena cava and projected a certain portion of it to the lungs through the pulmonary artery. The rest flowed across the septal pores to the left ventricle as already indicated. In Galen's view, the blood flowed mostly towards the periphery, both in the arteries and the veins. Inside the liver, however, the blood circulated in two opposite directions: part of it was diverted through the vena cava towards the right side of the heart, while the rest went inside the venous system to all the other bodily organs.

Under certain circumstances, Galen envisioned a reversal of this blood flow. Although he conceived of only a little mingling between the arterial and venous blood at the level of the lungs and in peripheral anastomoses, he postulated the presence of small amounts of pneuma in the veins. Therefore, Galen had no reason to propose the existence of a significant circular blood-

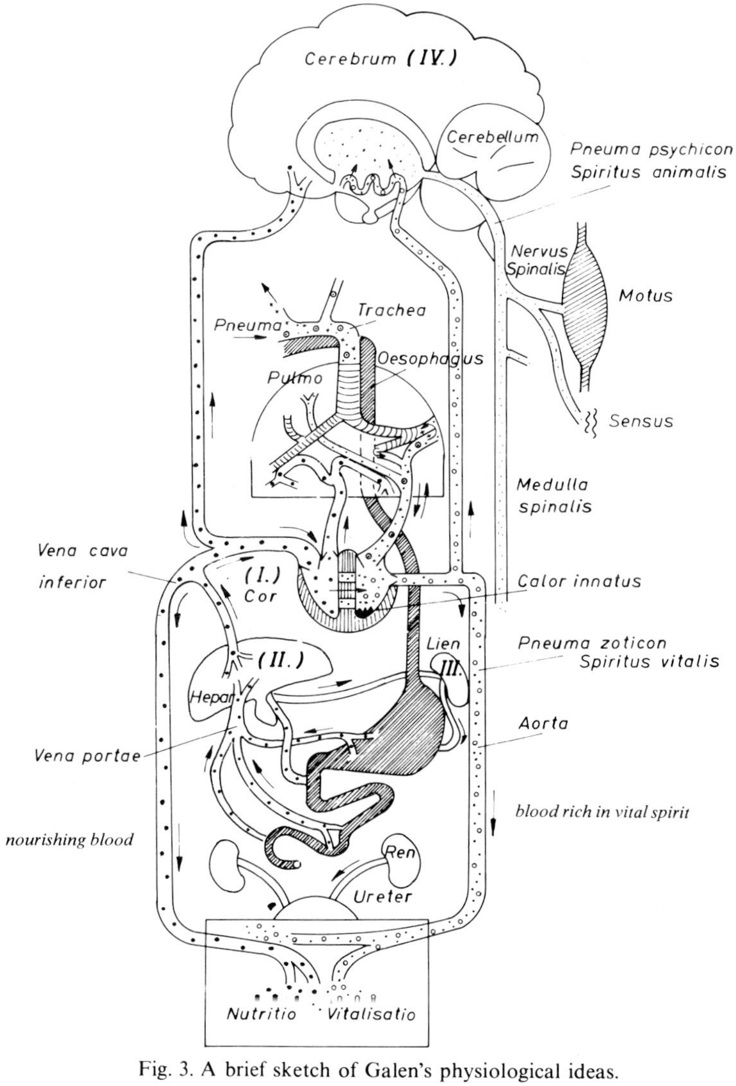

Fig. 3. A brief sketch of Galen's physiological ideas.

flow in the body. In his scheme, only a few drops of blood actually circulated through the vessels. Regarding vascular structures, he described in more detail than any of his predecessors, the great vessels, the ductus arteriosus and the point of entry of the right coronary vein into the right auricle of the heart.

For Galen the purpose of the respiration was to provide the organism with air necessary to sustain life, and cool the fire burning in the left ventricle of the heart. He was familiar with the mechanics of breathing, the role of the diaphragm and the intercostal muscles, whose innervation he discovered.

While the followers of Hippocrates had assumed a process of food disintegration in the stomach, Erasistratus, Asclepiades and others had envisioned merely a mechanical trituration and separation achieved mainly through gastric contractions.[73] Galen accepted both interpretations, adding other supplementary factors such as the vital heat and the pneuma physikon. As a result of vivisection experiments, he recognized that the digestion actually began in the mouth, and that the process of swallowing was due to esophageal movements and an attractive faculty inherent in the stomach. The decomposed food was, in turn, absorbed, thanks to a faculty of attraction residing in the veins and the liver.

Nutrition, in Galen's view, was absolutely necessary in order to replace the living substance of the body which was continuously lost. The kidneys excreted water and other waste products, taken from the arterial blood by an attractive renal faculty. In sharp contrast with the views of Asclepiades who considered the ureters to be spermatic ducts, Galen experimentally ligated these structures and demonstrated the subsequent absence of urine in the bladder.[74]

Before Galen's time apparently only the Alexandrian anatomist Marinus and the Macedonian Lykus had successfully examined the muscular system. Galen, in turn, devoted great interest to this subject and wrote two treatises *On the Movements of Muscles.* In the Galenic view, only voluntary muscles were genuinely considered to belong to the muscular system, whereas the contractile structures such as the heart, gut, uterus and bladder were only called "muscle-like," and endowed with an autonomous capacity to move.

The skeletal muscles consisted of tendinous fibres and a fleshy, insensitive substance. Their movements depended on the nerves which pierced the fleshy portions. For Galen, the capacity of contraction appeared to be inherent in the muscle itself, and was triggered into action by the pneuma psychikon arriving through the nerves from the brain.[75,76]

Although the first worthwhile knowledge regarding the nervous system had already been gathered by the Alexandrian physicians, Galen clearly was the first to stress the functional unity of the brain, spinal cord and peripheral nerves. The cerebrum was viewed as the place of origin for all the nerves, sensory perceptions and voluntary movements, as well as the seat of thought.

The spinal cord arose from the brain and finally branched out into the various peripheral nerves. Galen attempted to elucidate the importance of the different spinal levels by performing numerous serial sections of the cord in his vivisection experiments. The motor nerves originated in the outer layers of the brain, the sensory ones coming from the inner and softer areas of the organ. Galen believed that the various membranes covering the brain and spinal cord were functionally unimportant and instead postulated that the cerebral activity was solely conveyed by the pneuma psychikon, leaving unanswered the question concerning the nature of this spirit.

According to Galen, this variety of pneuma actually had two origins: the atmospheric air absorbed by the brain from the nostrils, and the pneuma zotikon carried from the left heart by the arteries of the head. In the brain itself, the pneuma psychikon was obtained from its two sources at the level of the choroid plexus in the lateral ventricles where it was also stored. In Galen's view the nervous substance, which made up the nerves, contained invisible passages which allowed the transmission of sensory impressions and the flow of pneuma psychikon to the peripheral muscles.

In Galenic physiology the crystalline lens was considered to be the main organ of vision, rather than the corpus vitreus or the retina. The latter was destined to nourish the vitreous body and perceive those changes occurring at the level of the lens. In turn, the cornea was seen as a continuation of the sclera, allowing the light to penetrate towards the lens. Galen conceived both optic nerves as hollow structures, communicating at the level of the chiasma, a fact which explained the consensual pupillary reaction.

Unfortunately, the various refractory conditions in the eye were virtually unknown in antiquity. Galen offered two interpretations of the visual process: a pneumatic and a geometric theory. According to the former, the pneuma flowed from the eye to the contemplated object and then returned to its point of departure in the form of an image, which then passed through the optic nerve to the lateral ventricles of the brain (thalamus). Faulty vision was also explained by Galen as the result of pneumatic anomalies. Like his predecessors, he imagined that the rays sometimes emerged from the eyes, while on other occasions these beams could also come from the object itself. Geometrically, the rays formed a cone with its apex located at the level of the crystalline lens.[77] Both eyes received slightly different pictures which were then fused into a new and single total image. Double vision could occur when one of the eyeballs was displaced.

This concludes a somewhat detailed description of Galen's physiological conceptions. They constitute, in a sense, a real summary of the total knowledge of physiology available at the end of antiquity, which remained basically unchallenged until modern times.

Finally, a judgment regarding Galen's place in the history of physiology remains to be rendered. In contrast with other interpretations, I view, together with Buess[78] and Siegel, Galen's physiology as the most important accomplishment in this field of endeavor during all of antiquity. He seems to have carried out numerous and ingeniously conceived animal experiments concerning the heart, nervous system, ureters, etc., dispelling the erroneous concept that the left cardiac ventricle and arterial system contained only air. Moreover, Galen studied the functions of the muscles in a series of dissections, recognizing the essential nature of the muscular contraction and the differences between voluntary and involuntary muscles. Finally, he described the communications between the cerebral ventricles, demonstrating also the functional relationships between the brain, spinal cord and peripheral nerves.

Contrary to the Aristotelean view which had assigned to the vital heat the responsibility for the respiratory movements, Galen recognized that the thoracic motions were caused by a series of muscular contractions. He also abandoned the concept of the brain as a glandular structure, furnishing further significant explanations about the phenomenon of vision.

Galen has often been denounced as a speculative thinker because of his teleological explanations, but, in my opinion, this aspect has been exaggerated. The question about the significance of a certain biological relationship remains always a fruitful heuristic procedure suitable for the investigation of organic beings. Surprisingly, Galen was on the right path with his causal as well as coordinated interpretation of the phenomena. Unfortunately, he did not realize that an explanation regarding the purpose and goal of a given physiological event was only a partial answer, but such an insight was too much to expect. Thus, I feel justified in vindicating Galen's reputation.

Not only was Galen the first methodically operating experimental physiologist of antiquity, but for more than a thousand years the last one who enriched this subject with original thoughts and results. Those speculative elements of the doctrine dealing with the pneuma occupied a large place in the Galenic system. Originating in early antiquity and reinforced by the Stoic philosophy, these ideas were useful to Galen, although they lamentably obscured many important facts. Therefore, some fateful errors survived and lingered on. Among them was the idea of pores in the interventricular septum of the heart, and several notions acquired from animal anatomy and transferred without proof to the human frame.

Nevertheless, in the total picture, Galen's positive accomplishments for physiology substantially predominate over the isolated errors. I believe he did not exaggerate when he said about himself: "Hippocrates accomplished something and above all, established a path. But I have paved the way and made it passable, just as the emperor Trajan did with the military roads of the Roman Empire".[79]

REFERENCES

[1]W. Nestle, *Vom Mythos zum Logos.* Die Selbstentfaltung des griechischen Denkens von Homer bis auf die Sophistik und Sokrates. Stuttgart, 1940.

[2]Benjamin Farrington, *Greek Science,* Baltimore, 1963.

[3]J.B. de C.M. Saunders, *The Transitions from Ancient Egyptian to Greek Medicine,* Lawrence, Kansas, 1963.

[4]H. Frankfort, J.A. Wilson, T. Jacobsen, *Before Philosophy,* Baltimore, 1949.

[5]A. Esser, "Die Medizin im Gang der abendländisch-morgenländischen Geistesgeschichte," *Ärztl. Forschung* 1 (1947) p. 317.

[6]John Burnet, *Early Greek Philosophy,* 4th ed., Cleveland, 1957.

[7]H. Erhard, "Alkmaion, der erste Experimentalbiologe", *Arch. Gesch. Med.* 34, (1941), p. 77.

[8]Joseph Schumacher, *Antike Medizin. Die naturphilos. Grundlagen der Medizin in der griechischen Antike,* Bd. 1, Berlin, 1940.

[9]Charles Singer, *A Short History of Anatomy and Physiology from the Greeks to Harvey,* 2nd ed., New York, 1957.

[10]R.O. Steuer, *WHDW, Aetiological Principle of Pyaemia in Ancient Egyptian Medicine,* Baltimore, 1948.

[11]J. Stannard, "The Presocratic origin of explanatory method", *The Philosophical Quarterly,* 15, (1965) pp. 3-16.

[12]R.A. Horne, "Atomism in ancient medical history", *Medical History,* 7 (1963), pp. 317-329.

[13]G.E.R. Lloyd, "Experiment in early Greek philosophy and medicine", *Proceedings Cambridge Philological Society,* 190 (1964), pp. 50-72.

[14]J. Longrigg, "Philosophy and medicine", *Harvard Studies in Classical Philology* 67 (1963), pp. 147-175.

[15]L. Edelstein, "The relation of ancient philosophy to medicine", *Bull. Hist. Med. 26* (1952), pp. 299-316.

[16]F. Kudlien, "Early Greek primitive medicine", *Clio Medica* 3 (1968), pp. 305-336.

[17]*Hippocrates,* An English translation by William H.S. Jones (The Loeb Classical Library), 4 vols., Cambridge, Mass., 1948-53.
Hippocrates Medical Works, A new translation from the original Greek made for English readers by the collaboration of John Chadwick and W.N. Mann, Oxford, 1950.

[18]M. Pohlenz, *Hippokrates und die Begründung der wissenschaftlichen Medizin,* Berlin, 1938.

[19]L. Edelstein, "The genuine works of Hippocrates", *Bull. Hist. Med.* 7 (1939), pp. 236-248.

[20]E. Schöner, "Das Viererschema in der antiken Humoralpathologie," Beiheft 4, *Sudhoffs Archiv,* Wiesbaden, 1964.

[21]K. Deichgräber, *Hippokrates, Entstehung und Aufbau des menschlichen Körpers,* Leipzig and Berlin, 1935.

[22]R. Fuchs, "Geschichte der Heilkunde bei den Griechen," *Hdb. d. Gesch. d. Med.* 1, Jena, 1902.

[23]G. Senn, "Über Herkunft u. Stil der Beschreibungen von Experimenten im Corpus Hippocraticum," *Arch. Gesch. Med.* 22, (1929), p. 217.

[24]T. Meyer-Steineg, "Die Vivisektion in der antiken Medizin." *Internat. Mschr. Wiss., Kunst. u. Techn.* 6, (1911), p. 1941.

[25]P. Diepgen, "Die Hippokratiker und die Lehre vom Blutkreislauf," *Münch. med. Wschr.*, (1939), p. 299.
P. Diepgen, "Haben die Hippokratiker den Blutkreislauf gekannt?", *Klin. Wschr.*, (1937), p. 1820.
P. Diepgen, "Hippokrates im neuen Licht," *Hippokrates*, (1938), p. 1237.

[26]G. Sticker, "Zur Geschichte des Blutkreislaufs," *Münch. med. Wschr.*, (1938), p. 258 and . 282.

[27]R. Kapferer, "Der Blutkreislauf, seine Darstellung in den hippokratischen Schriften," *Hippokrates*, (1937) p. 697; (1938) p. 251; (1939) p. 1331; (1940) p. 84 and p. 646.

[28]H. Diller, "Die Lehre vom Blutkreislauf, eine verschollene Entdeckung der Hippokratiker?", *Arch. Gesch. Med.* 31, (1938), p. 201.

[29]L. Aschoff, "Über die Entdeckung des Blutkreislaufes," *Freiburger Forsch. z. Med. Gesch.* H. 1, Freiburg, 1938.

[30]F.R. Hurlbutt, "Peri Kardies. A treatise on the heart from the Hippocratic corpus: introduction and translation," *Bull. Hist. Med.* 7 (1939), pp. 1104-1113.

[31]*Plato's Cosmology*. The Timaeus of Plato translated with a running commentary by Francis M. Cornford, New York, 1957.

[32]T.J. Tracy, *Physiological Theory and the Notion of the Mean in Plato and Aristotle*, (Studies in Philosophy 17), The Hague, 1969.

[33]W. Jaeger, *Diokles von Karystos. Die griech. Med. u. die Schule des Aristoteles*, Berlin, 1938.

[34]W. Jaeger, *Aristoteles. Grundlagen einer Gesch. s. Entw.*, Berlin, 1923.

[35]A. Volprecht, *"Die physiologischen Anschauungen des Aristoteles,"* (Diss. Greifswald, 1895).

[36]*The Works of Aristotle*, eds. J.A. Smith and W.D. Ross, London, 1928-52.

[37]C.W. Sievert, *"Die Physiologie bei Aristoteles,"* (Diss. Münster, 1948).

[38]Thomas E. Lones, *Aristotle's Researches in Natural Science*, London, 1912.

[39]David Ross, *Aristotle*, 5th rev. ed., London, 1949.

[40]E. Clarke, "Aristotelian concepts of the form and function of the brain," *Bull. Hist. Med.* 37 (1963), pp. 1-14.

[41]Max Neuburger, *Die Lehre von der Heilkraft der Natur im Wandel der Zeiten*, Stuttgart, 1926.

[42]Edward A. Parsons, *The Alexandrian Library*, glory of the Hellenic World, its rise, antiquities and destruction, Amsterdam, N.Y., 1952.

[43]G.C. Pournaropoulos, "The Greek medical school of Alexandria," *Communications to the 17th International Congress of the History of Medicine* I (1960), pp. 308-316.

[44]*The Fragments of Praxagoras of Cos and His School*, coll. ed. & trans. F. Steckerl, Leiden, 1958.

[45]L. Glesinger, "Zur Frage der angeblichen Vivisektionen am Menschen in Alexandria," *Communications to the 17th International Congress of the History of Medicine* I (1960), pp. 287-293.

[46]J.F. Dobson, "Herophilus of Alexandria," *Proceedings of the Royal Society of Medicine* 18 (1925), pp. 19-32.

[47]E. Műhsam, "Zur Lehre vom Bau und der Bedeutung des menschlichen Herzens im klass. Altertum," *Janus* 15 (1910), p. 797.

[48]J.F. Dobson, "Erasistratus," *Proceedings of the Royal Society of Medicine* 20 (1927), pp. 825-832.

[49]L. G. Wilson, "Erasistratus, Galen, and the pneuma," *Bull. Hist. Med.* 33 (1959), pp. 293-314.

[50]Th. Meyer-Steineg, "Die Vivisektion in der antiken Medizin." *Internat. Mschr. Wiss., Kunst u. Techn.* 6, (1911), p. 1491.

[51]O. Temkin, "Studies on late Alexandrian medicine," *Bull. Hist. Med.* 3 (1935), pp. 405-430.

[52]I.M. Lonie, "Erasistratus, the Erasistrateans and Aristotle," *Bull. Hist. Med.* 38 (1964), pp. 426-443.

[53]H. Schöne, "Das Experiment in der Biologie und Physik der Griechen." *Verh. d. 55. Vers. dtsch. Schulmänner in Erlangen,* Leipzig und Berlin, 1926, p. 8.

[54]John Scarborough, *Roman Medicine,* London, 1969.

[55]Sir Thomas Clifford Allbutt, *Greek Medicine in Rome,* The Fitzpatrick lectures on the history of medicine delivered at the Royal College of Physicians of London in 1909-10, London, 1921.

[56]*Soranus' Gynecology,* translated with an introduction by Owsei Temkin, Baltimore, 1956.

[57]M. Drabkin, "A select bibliography of Greek and Roman medicine," *Bull. Hist. Med.* 11 (1942), pp. 399-408.

[58]Rudolph E. Siegel, *Galen's System of Physiology and Medicine,* an analysis of his doctrines and observations on bloodflow, respiration, humors and internal diseases, Basel and New York, 1968.

[59]J. Ilberg, "Wann ist Galenos geboren?", *Arch. Gesch. Med.* 23, (1930), p. 289.

[60]George Sarton, *Galen of Pergamon,* Lawrence, Kansas, 1954.

[61]E. Hauke, *"Galen: Dass die Vermögen der Seele eine Folge der Mischung des Körpers sind."* (Diss. Berlin, 1937).

[62]Th. Meyer-Steineg, "Studien zur Physiologie des Galenos," *Arch. Gesch. Med.* 5, (1912), p. 173; 6 (1913), p. 417.
Th. Meyer-Steineg, *Ein Tag im Leben des Galen,* Jena, 1913.

[63]*Galen on anatomical procedures,* transl. of the surviving books with introduction and notes by Charles Singer, London and New York, 1956.
Galen on anatomical procedures, the later books, transl. by W.L.H. Duckworth, ed. by M.C. Lyons and B. Towers, Cambridge, England, 1962.

[64]Galen, *On the Usefulness of the Parts of the Body* (De usu partium). Transl. from the Greek with an introduction and commentary by Margaret Tallmadge May, 2 vols., Ithaca, New York, 1968.

[65]Galen, *On the Natural Faculties,* English transl. by Arthur J. Brock, Cambridge, Mass., 1947.

[66]E. Hintzsche, "Über das anat. Wissen Galens und seiner Vorläufer," *Ciba Z.* 8 (1944), p. 3411.

[67]R. Töply, V. Ritter, "Geschichte der Anatomie," in *Puschmanns Hdb. d. Gesch. d. Med.,* Bd. 2, Jena, 1903, p. 155.

[68]Rudolph E. Siegel, *Galen on Sense Perception,* his doctrines, observations and experiments on vision, hearing, smell, taste, touch and pain, Basel and New York, 1969.

[69]O. Temkin, "On Galen's pneumatology," *Gesnerus* 8 (1951), pp. 182-184.

[70]Lester S. King, *The Growth of Medical Thought,* Chicago, Ill., 1963, pp. 43-85.

[71]D. Fleming, "Galen on the motions of the blood in the heart and lungs," *Isis* 46 (1955), pp. 14-21.

[72]J. Bylebyl and W. Pagel, "The chequered career of Galen's doctrine on the pulmonary veins," *Medical History* 15 (1971), pp. 211-229.

[73]C.E.A. Winslow, R.R. Bellinger, "Hippocratic and Galenic concepts of metabolism," *Bull. Hist. Med.* 17 (1945), pp. 127-137.

[74]V.A. Triolo, "An interpretative analysis of Galenic renal physiology," *Clio Medica* I (1966), pp. 113-128.

[75]E. Bastholm, *The History of Muscle Physiology*, Copenhagen, 1950.

[76]C.M. Goss, "On movement of muscles by Galen of Pergamon," *Amer. J. Anat.* 123 (1968), pp. 1-26.

[77]J. Hirschberg, "Galen und seine zweite Anatomie des Auges," *Berl. klin. Wschr.* (1919), p. 635.

[78]H. Buess, "Das Vermächtnis der Antike: Galenos von Pergamon," *Schweiz. Med. Wschr.* 87 (1957), p. 173.

[79]E. Holländer, *Anekdoten aus der med. Weltgeschichte*, Stuttgart, 1925.

CHAPTER II:
Physiology During the Middle Ages

1.
EARLY MEDIEVAL TIMES—
ISLAMIC PHYSIOLOGY—SALERNO

Early medieval times

Ancient medicine and physiology culminated as well as ended with Galen. The primary reason for this development was the decline of the culture which had prevailed during antiquity. External as well as internal circumstances contributed to the decay: wars, high taxes, diminished trade, unemployment, depopulation of the countryside and loss of patriotism. Likewise, the intellectual climate of the ensuing centuries became unfavorable both for science and medicine. Following the deposition of the ancient pantheon of gods, oriental influences in the form of sun worship, the cult of Mithra, astrology and magic began to pervade the universal outlook. Both Neoplatonism and Christianity were ill-suited to allow concern for the natural, earthly appearances. Thus, stagnation and even decay were the necessary sequels of this new attitude. In the ensuing political chaos and the wars of survival from the fourth to the eighth century, not only numerous manuscripts but also the scientific tradition were lost on Italian soil.

Somewhat more favorable was the fate accorded to Constantinople, which became the capital of the Eastern Roman Empire in 395 A.D. Not only was Byzantium spared from the multiple migrations of the Barbarous nations, but it experienced a period of high cultural achievement, especially under the reign of the emperor Justinian (527-565).[1]

Prominent among the Byzantine physicians were *Oribasios of Pergamon* (ca. 325-395 A.D.), *Aetios of Amida* (ca. 502-550)[2] and *Alexander of Tralles* (525-605).[3] None of them acquired any new physiological knowledge, but they all were instrumental in transmitting important portions of the extant ancient literature to posterity. During the Arabic conquest of the city of Alexandria in the year 641, *Paul of Aegina* (625-690) lived in that city. His writings were instrumental in transmitting the learning of antiquity to the Islamic culture.[4,5]

Islamic physiology

The transmission of classical medical knowledge took a second and rather odd route. In Syria, the schools of the cities of Edessa and Nisibis came into being under the influence of Christianity towards the end of the third century. Together with theology, these institutions also preserved medical traditions and promoted translations from the classical Greek writings. The school at Edessa soon came under the influence of the Nestorians.[6] After the council of Ephesos, however, this religious sect was accused of heresy, and its members were cruelly expelled from Edessa by an imperial order issued in 489 by Zeno.

Thus, the Nestorians were forced to flee to the Persian Empire, taking with them their Greek cultural heritage, including the extant classical medical literature. In Jondishapur they founded a famous medical school under the auspices of the reigning Sassanian dynasty, and proceeded to translate their numerous medical writings from the Greek and Syriac languages into Persian.[7] Among the authors whose works were translated at Jondishapur, were the writers of the Hippocratic Corpus, Galen, Dioscorides, Ruphus, Oribasius and others.

When the Arabs conquered Persia in the seventh century, all these texts came into their possession, and they were subsequently not only preserved, but translated, commented on and incorporated into Islamic medicine. Therefore, a considerable amount of Graeco-Islamic medical knowledge became available in Mesopotamia and Persia during the eighth century, a wealth of material which reached its culmination between the ninth and the eleventh centuries.

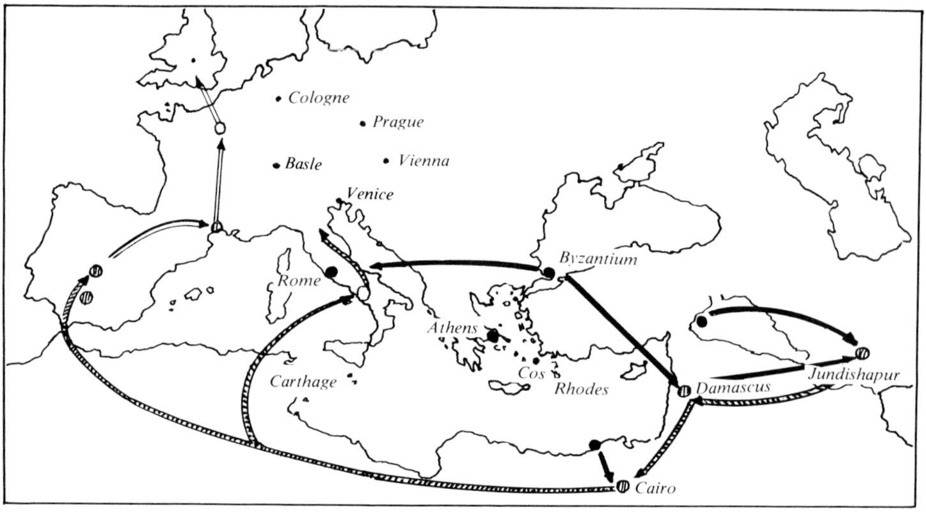

Fig. 4. *Main pathways of transmission of ancient medical knowledge to the Middle Ages.* From the Byzantine Empire (dark arrows) through the Islamic cultural centers (grey arrows) to the Christian West (light arrows).

Hence, the Islamic world came into possession of almost the entire Greek scientific literature in useful Arabic translation by the year 900.[8] By contrast, the medieval Latin West, which had lost this knowledge to a great extent, often recovered those accomplishments which had once originated on its own soil through the Islamic culture (see Fig. No. 4). We must also remember that, following the death of the prophet Mohammed in 632, the adherents of Islam extended their power and beliefs through Syria, Mesopotamia, Persia, Egypt and northern Africa after prolonged and successful wars. In 641, the ancient cultural center of Alexandria fell into Islamic hands, while between 710 and 712 almost the entire Iberian peninsula was occupied. A Moorish state and the caliphate of Cordova — in Granada — survived until 1492.[9]

The Islamites deserve posterity's highest esteem because they did not senselessly destroy the cultures during their conquests. Rather, Islam tolerated and consciously preserved these cultural heritages and incorporated them into its own civilization. Even today, the great accomplishments of the golden age of Moslem culture are still preserved in cities like Cordova, Granada, Seville and Toledo.

Islam also produced some very remarkable physicians.[10,11] Among them was *Hunain Ibn Ishak* (ca. 809-873), a Nestorian,[12] one of the earliest translators of Greek works into Arabic — he was called Johannitius by the Christians.[13] Another great physician of the medieval Muslim world was the Persian *Abu Bekr Mohammed Ibn Zakariah ar Rasi* or *Rhazes* (ca. 850-923). For a long period of time he lived as a physician and teacher of medicine at Bagdad. His impressive collection of Graeco-Arabic medical writings appeared in the form of an encyclopedia, the *Al-Havi* or *Continens Medicinae,* a monumental work still reprinted in Latin during the early sixteenth century.[14]

Fifty years after Rhazes' death, another great Persian physician was born: *Abu Ali Al Hussein Ibn Abdallah Ibn Sina.*[15] Called simply Ibn Sina or *Avicenna* (ca. 980-1037), this prolific author exhibited a universal genius, with an intellectual capacity of the highest rank. His great medical compilation, the *El Quanûn* or *Canon Medicinae,* was a veritable code of medical principles which had attained the certainty of laws.[17] The extensive work represented a thorough systematization of all the medical knowledge available since antiquity, an attractive compendium known for its clarity, logical formulation and substantial content. The medieval West was greatly impressed by Avicenna's Canon. Thus, it is understandable that its author became the unquestionable authority for the adherents of scholastic medicine.[18] Finally, two other distinguished Islamic authors should be mentioned: the physician *Ibn Zuhr* or *Avenzoar* (ca. 1091-1161)[19] and *Averroes* (1126-1198),[20] the latter a philosopher who was greatly influenced by the Aristotelean philosophy.

Islamic physiology and medicine were primarily Greek, although modified in part by the prevailing religious dogmas, the climate and race, and translated into Arabic language. Neither in anatomy nor physiology did the Muslims progress beyond the medicine prevalent in antiquity.[21] Anatomy in particular made no further advances because the Islamic religion forbade the dissection and depiction of the human body.[22]

The theory of elements, qualities and pneuma dominated the general conception of vitality. The physiology attributed to certain bodily organs was based in part on Aristotelean and Galenic models. As Aristotle before him, Johannitius selected the heart as the center of sensations, while the brain was merely seen as a cooling agent. The Islamic physician *Ibn an Nafis* (1210-1290) has the distinction of being the first in history to postulate the existence of a pulmonary circulation from the right to the left ventricle of the heart.[23] We will return to this subject in more detail later.

Also lacking in originality were the Islamic conceptions of blood formation, respiration, digestion and muscular motion. In opposition to Galen, Islamic physicians viewed the brain as a source of sensation and motion. It contained three chambers, considered to be the seat of the special actions performed by the soul. In general, those functions were *Imaginatio, Cogitatio* and *Memoria,* but their cerebral localization was not always clear from the writings. As a rule, the first ventricle contained the imagination, the second was the seat of thought and judgment, while the third harbored the memory. Rhazes, on the other hand, mentioned the existence of four cerebral cavities. He assigned the *Phantasia* to both lateral ventricles, *Cogitatio* to the third ventricle and *Memoria* to the fourth.

Many Muslim authors wrote extensively on the physiology of the senses, but essentially no new contributions emerged from their endeavors except for some original work on vision.[24] In several treatises written in Arabic and concerned with ophthalmology, new knowledge regarding the physiology of vision was expressed.[25] Three theories had been prevalent in antiquity: Pythagoras claimed that the rays departed from the eye and reached their objects, while Plato postulated an encounter between the light coming from the eyes and that emanating from the object. Finally, Aristotle's version was that the objects emitted some vibrations which then reached the eye.

In a work called *Opticae Thesaurus Alhazeni,* the Islamic mathematician *Ibn al-Haitham* or *Alhazen* (ca. 950-1030) attempted to confirm the Aristotelean viewpoint, combining it with some notions regarding light reflection, refraction and the question of magnification.[26] This book enjoyed for many centuries the same reputation accorded to Euclid's optics, in the thirteenth century version of the *Thuringo-Polonius* written by the Silesian physicist and philosopher Witelo (ca. 1220-1278).[27] Even Kepler called his work on dioptrics *Ad Vitellionem Paralipomena* in 1604.

According to Alhazen, the convergent rays emanating from a given object struck the eye after following a rectilinear path. There, they formed an image on the anterior surface of the crystalline lens. For Alhazen, the lens was the center of vision, representing the retina in a way, since each point on the lens corresponded to a point of the object. The unity of the images was seen as the result of a crossing of the optic nerves. The rays emanating from a given object, on which we directed our eyes, coincided with these directional lines which divided the ocular fields into certain segments. Therefore, we could only see an object clearly when the axis of both eyes were directly focused upon it.

Regarding the physical aspects of reflection and refraction, Alhazen went considerably further than Euclid.[28] In contrast with Aristotle, he also declared, from experiences with colors, that light needed a certain time for its propagation.

In a later work, written about 1296, a Syrian ophthalmologist, *Salah-ad-Din* (fl. 1296), represented the theory of vision for the first time in a geometric fashion. He also provided further proof for the spherical shape of the bundle representing the visual rays which penetrated the eye. Furthermore, the Islamic physicians were the first to mention the pupillary contraction which occurs after greater light exposure, while Galen had merely spoken about the consensual reaction of the pupils. The connection between ocular optics and the sensation of vision was still conceived as a "visual spirit," an especially subtle substance related to the psychic pneuma, to which all disturbances of the eyesight were attributed.[29]

The Medical School at Salerno[30]

While a medicine essentially based on the Greek model flourished in Islam, the healing art sank to very low levels in the Christian West, as seen both in Germany and Italy. Nevertheless, certain fragments of Hippocratic and Galenic medicine, as well as writings of Soranus, had been saved from destruction by Christian monastic orders.

In the year 529, Benedict of Nursia had founded the order of the Benedictines, and established the monastery of Monte Cassino, south of Rome. Following the example of their leader, Cassiodorus, who was sympathetic to scientific pursuits, the order proclaimed that the practice of medicine was to be a duty for all of its members. Thus, valuable manuscripts were collected at Monte Cassino, which became the fountainhead for a northward stream of monks who participated in the conversion of Germany to Christianity. These missionaries, endowed with medical knowledge, practiced their art in many of the newly erected abbeys of the northern lands.

It now appears that the necessary impulse for the creation of the oldest medieval medical school in the West also came from the monastery of Monte Cassino. The new institution, founded around the year 900, was located at Salerno, a town south of Naples.[31]

At Salerno, the cleric and physician *Alfanus* (ca. 1015-1085)[32] translated the work of bishop Nemesius (fl. around 500) into Latin entitled *On the Nature of Man.* This manuscript contained, among other topics, some discussion of the physiological problems of motion, respiration, nutrition, generation and movement of the blood.[33]

We are also greatly indebted to the Arabic author *Constantinus Africanus* (ca. 1010-1087),[34] who translated various Islamic texts into Latin at Monte Cassino, thus acquainting the West with these valuable writings. Africanus' merit was due to his role as a highly successful intermediary between the Muslim culture based on Graeco-Roman models, and the Latin West. In addition to Africanus, many other physicians contributed to the establishment of Salerno as a famous medical center.[35]

Another fortunate circumstance increased the contact between the Western and Islamic cultures. In 1085, Alfonso VI of Castile conquered the city of Toledo from the Muslims, and a great treasure of valuable manuscripts from a variety of Arabic and Graeco-Roman sources fell into the hands of the victors. Many of these writings were subsequently translated into Latin by *Gerard of Cremona* (1114-1187) and made available for study to the Western scholars. Therefore, a great amount of knowledge from antiquity was rediscovered, a fact which had considerable influence on the development of medicine.[36]

During the twelfth century a second important medical school developed at Montpellier, in southern France. Favored geographically because of its close relationship with the Muslim and Hebrew medical schools established on Spanish soil, the city received some valuable elements of Islamic and Greek learning.[37] In 1220 the Montpellier school received from Pope Honorius II the necessary academic rights and statutes to conduct examinations. Its most illustrious period began with Bernard of Gordon (d. 1318) and the famous physician *Arnaldo of Villanova* (ca. 1235-1311).[38] Villanova taught at Montpellier during the decade 1289-1299 and possessed a comprehensive knowledge of the natural sciences, medicine and philosophy.[39] Details about the anatomical contributions of another teacher, the great French surgeon *Henry of Mondeville* (ca. 1260-1320), will follow later.

The accomplishments of Graeco-Islamic medicine had in the meantime penetrated beyond the Alps. There is evidence that bishop Bruno of Hildesheim (d. 1161) already possessed eighteen works of Constantinus Africanus in the middle of the twelfth century. Perhaps some of this information had become available also at the Benedictine monastery of Rupertsberg at Bingen where the learned Hildegard von Bingen (1098-1179) wrote two historically important works. Entitled *Physica* and *Causae et Curae,* these treatises typically reflected the medical conceptions of the period.[40] Hildegard's writings have recently been re-edited and commented upon.

Hildegard's knowledge in anatomy and physiology appears to be scant, reflecting the fact that the clarity and maturity of the classical Galenic conceptions had not as yet been regained.[41]

2.
SCHOLASTICISM—RENAISSANCE—HUMANISM AND THE END OF THE MIDDLE AGES

Scholasticism

The political and intellectual viewpoints inherited from antiquity did not suddenly vanish in the Latin West. On the contrary, most people during the early Middle Ages considered themselves to be the proud successors of the classical traditions. Unfortunately, those early centuries exhibited a staunch conservatism and, except for religion, little creativity. The famous works of the Roman philosopher *Boethius* (ca. 475-525) transmitted some of the science and wisdom of antiquity to the West. Above all, Boethius translated fragments from Aristotle, which he published together with some commentaries.

The leading and cultivated strata of society was exceedingly thin in the early Middle Ages, even down to Carolingian times. The feudal lords and the aristocracy held all the available positions of power in the state, the church and in society at large. Their attitude and style of living were conservative. Within their ranks the clerics became the true upholders of culture, a situation which remained virtually unchanged until the emergence of the Hohenstaufen dynasty.

Then from the twelfth century onward, the political evolution allowed the growth of another social group: the lower nobility. This sector of society gradually achieved a position of power within the medieval framework, which was inspired by the new ideals of chivalry. The enthusiasm displayed by these people during the various crusades testified to their vitality, devotion to religious sentiment and practice of the Christian ideals of life.

This was the period of early Gothic architecture, in which men like the Italian friar and Saint *Francis of Assissi* (ca. 1182-1226), the philosopher *Thomas Aquinas* (ca. 1225-1274), the Italian poet *Dante Alighieri* (1265-1321) and the German philosopher and saint *Albertus Magnus* (ca. 1193-1280) lived, a time in the history of science which has been termed "scholastic."

The term "scholastic" refers to something pertaining to "schools" and was originally used to designate the studies and doctrines of systematized theology. Therefore, the real "scholastics" were those teachers who gave instruction in the seven liberal arts derived from antiquity. In the medieval "scolae" or schools, these men taught grammar, dialectics, rhetoric, arithmetic, geometry, music and astronomy, and in addition lectured on theology and medicine.[42] The church regulated and indeed supervised this type of education and most higher schools and universities received their teaching privileges directly from the Pope.

The early scholastic period of the eleventh and twelfth centuries was primarily concerned with bringing the older philosophical writings into accordance with the utterances made by the Fathers of the Church. For this purpose, a "scholastic," dialectical method of argumentation was used. An abstract and detached impression of the natural appearances replaced the previously metaphorical and comparative understanding of the natural things.

The so-called high scholastic period fell approximately in the thirteenth century and was characterized by its greater attention to sciences other than theology. The primary reason for this important shift lay in the gradual availability of numerous translations of the writings of Aristotle. The rediscovery of ancient texts also brought to light many medical, physical and mathematical works hitherto unknown in the West.

Thus, a rich and potentially fruitful field of endeavor opened up for those men practicing the scholastic method. Their task now was to purify the corrupted texts and achieve a necessary conformity among the various rediscovered versions of any given work. The peak of this scholastic period can be dated by the separation which gradually took place in the universities between the theological and philosophical faculties (Montpellier 1120, Bologna about 1119, Padua 1222, Paris 1200, Oxford 1249, Cambridge 1284, Prague 1348, Vienna 1365, Heidelberg 1386 and Cologne 1388).

Spurred on by the newly discovered knowledge, scholars began with great zeal to shift their attention towards all spheres of reality, and the interest in the natural world strongly increased. Both natural science and medicine were greatly stimulated. However, the renewed attention to nature was motivated more by the wish to recognize the greatness and magnificence of the Divine Master in the works of His creation. Thus, the need to discover new truths or unknown relationships was largely absent in scholastic times. The opinion prevailed that the writings of Galen, Hippocrates, Avicenna, Euclid and others contained everything worth knowing, just as the Bible and the works of the Church's Fathers included all articles of faith. This viewpoint should be stressed, since we can only do justice to scholasticism if we take it into consideration and view the entire period as merely a transitional evolutionary step in the ascent of Western science.[43]

The historical period under consideration provided only a few works which were relevant to the history of physiology. In the thirteenth century three notable men lived at the famous University of Paris: the English Franciscan and "doctor mirabilis" *Roger Bacon* (1214-1294),[44] the German Dominican and "doctor universalis" *Albertus Magnus* and finally the "doctor angelicus" *Thomas Aquinas.*[45] In addition, the following scholars, who summarized in their works the contemporary knowledge of the natural sciences, should be mentioned: the Silesian physicist *Witelo,* the English physician *Bartholomew* (fl. ca. 1250), the French encyclopedist *Vincent of Beauvais* (ca. 1190-1264) and the English encyclopedist *Thomas of Cantimpré* (ca. 1204-1271).

Most scholastics drew heavily, though not slavishly, from the old authoritative writers. Those items concerning physiology and vital functions which were reported, had been taken from Aristotle, Pliny, Hippocrates, Galen, Nemesius, Isidor of Seville, and the Arabic writings of Rhazes, Avicenna, Alhazen and others. Above all, the scholastic authors attempted to achieve conformity and concordance between the various extant versions.

From a combination of Galenic and Aristotelean concepts originated the idea of the heart as *membrum principale,* the very source of the entelechy and main seat of action for the *anima.* The other bodily functions were grouped around this central position of the heart according to purely teleological points of view.[46]

The writings of Roger Bacon, Witelo and Albertus Magnus also contained some original observations and thoughts, and even showed the beginning of simple, experimental procedures. By contrast, Thomas of Cantimpré furnished an anatomy and physiology based essentially on the ideas of Aristotle and Galen. His descriptions were contained in the first book of a twenty volume encyclopedia entitled *De Natura Rerum,* which was translated into German by Konrad of Megenberg (1309-1374) under the title *Buch der Natur.*[47]

Fig. 5. Albertus Magnus (1193-1280).

Albertus Magnus, count of Bollstaedt in Swabia, (see Fig. No. 5), had the great merit of having furnished the West with numerous hitherto unknown writings of Aristotle, especially a series of works concerned with the natural sciences. Although he had a high esteem for the Greek philosopher, Albertus recognized that Aristotle, like other authors, had not been immune to error: "Aristotle multum erravit in ista regione."

Albertus' own writings, which were very extensive, contained abundant personal observations. At one point he wrote: "The method based on syllogisms is not viable in the special inquiries which have to be made in the natural sciences," while in another place he declared: "The natural sciences have not as their goal the simple narration of facts, but rather, should attempt to fathom the causes of the natural phenomena." In the latter dictum, Albertus already expressed clearly the principle for genuinely scientific procedures.[48]

The most important sources for Albertus' physiological concepts were his commentary in the work *Parva Naturalia* and the treatise *De Animalibus Libri XXVI.* He was also acquainted with some of Galen's works. The knowledge of the human nature in both Albertus Magnus and Thomas of Cantimpré was based on the ancient doctrine of the elements, qualities and humors. Health was dependent on proper nutrition as well as climatic and cosmic influences. The specific physiology of the various bodily organs was Aristotelean or Galenic in character,[49] although in this respect we find that the knowledge of Albertus was much more extensive than that of other scholastic authors. For example, Albertus had an exact knowledge of the nervous system while Thomas of Cantimpré could not differentiate properly the nerves from the tendons, which his translator, Konrad of Megenberg, called the "pantarteries."

Following Aristotle, Albertus designated the heart as the most important organ for the formation of blood instead of Galen's liver, and he described three cardiac chambers. Regarding cerebral physiology, Albertus declared that although the brain was the organ of sensation and motion, these attributes were actually only derived from the centrally located heart. In the true Aristotelean tradition, Albertus considered the heart to be the source of the vital force, heat and spirit.

By contrast, the brain by its very nature was cold, and also possessed three cavities: the first one contained the olfactory organ, the second a "virtus cogitativa" and "virtus imaginativa," while the third and posterior chamber harbored the organs of "memoria" and "reminiscentia." According to Albertus, the brain was insensitive and contained no blood because of a lack of vessels. Contrary to Thomas of Cantimpré's work, Albertus had a surprisingly accurate knowledge of the function and distribution of the cerebral and spinal nerves.[50]

In the scholastic era some original contributions were made both by Witelo and R. Bacon in physical optics and the theory of vision, although these authors followed the ideas of Alhazen to a great extent. Roger Bacon belonged to the group of most notable medieval physicists. After his move from Oxford to Paris in 1240, he became acquainted with the works of Aristotle and the Islamic physicians. His independent reasoning and experimentation soon brought him into conflict with his monastic superiors, and he suffered in virtual isolation and imprisonment for many years.[51]

Roger Bacon was far ahead of his time in his ideas on scientific method. He wrote: "Natural science is an experimental science. All research is based on the study of nature. This declaration should not be viewed as a merely theoretical conclusion, but rather as a call for genuine action, observation and comparison. The natural sciences can only be studied on the objects of nature instead of carrying out philosophical debates about the writings of ancient authors concerned with natural science."

However, regarding the content of his works, Roger Bacon did not advance significantly beyond the Islamic writers. He closely followed Alhazen's ideas concerning the path of the light. The lens was the focal point of the pyramidally converging rays, while a visual spirit mediated the sensation. For sharp vision, both visual axes had to intersect in the contemplated object. Bacon was also well aware of residual colored images.[52]

Witelo[53] who also called himself "Thuringopolonius" because of his maternal Polish ancestry and his father's Thuringian origin, wrote around 1270 a work called *Perspectiva.* Basically, this book represented a systematized replica of Alhazen's optical knowledge. Witelo proposed that the lens was the site of optical sensation, completed in the optic commissure. He was aware of double images of any given object both in front and behind the point of fixation, and like Aristotle, he believed that vision occurred through rays which reached the eye from the objects.

Compared with the achievements of the classical and Islamic authors, the scholastic period only made negligible physiological discoveries. Nevertheless, the various writings helped to improve the general level of knowledge in the Latin West, allowing that part of the world to achieve again a level equal to that of classical antiquity. Unfortunately, the scholastic method could not provide new insights for medical practice nor improve the theoretical bases of medicine such as anatomy, physiology and pathology.

With his *Lectiones* and *Disputationes,* the Italian physician *Thaddeo Alderotti* (ca. 1223-1303) has been recognized as the founder of the scholastic method of teaching in medicine.[54] During the same period, and after more than a thousand years of omission, the first autopsies were carried out in northern Italy. At first, these dissections were only isolated occurrences, as in Cremona in 1286 and Bologna in 1302, the latter performed by *Bartholomeo da Varignana.*

Therefore, the Bolognese anatomist *Mondino de Luzzi* (ca. 1275-1326), also called Mondinus, was able to base his *Anathomia* (1316)[55] on personal dissections which he had been able to carry out.[56] However, these autopsies did not advance anatomical knowledge as one would have expected. The reason for this was the method used to carry out these dissections. In typical scholastic fashion, the physicians restricted their actions to merely pointing out those anatomical descriptions in the cadaver already made by the classical authors. Nevertheless, one can detect a gradual turn towards reality, nature and observation in these autopsies. To be sure, this shift in attitude originated in the scholastic period and then slowly, without sharp boundaries, passed on into the Renaissance.

Renaissance and Humanism

The ideas, models and self-consciousness of any given historical period are an expression of experiences which left a strong imprint on the people of that time. For example, the invention of the compass at the beginning of the fourteenth century made it possible for enterprising navigators to venture into the open sea, far away from safe coastal waters.

As a result of the Crusades and the founding of Tyrannis on Italian soil under Frederick II, an active commercial relationship already existed between the West and the Orient. Now, with the advances of navigation, seaways were discovered and promising colonies founded for trading purposes. A new economy based on money gradually replaced barter, and rich, commercial cities began to flourish, such as Venice, which in 1422 boasted 190,000 inhabitants.

Not only in Italy, but also in northern Europe, many cities evolved and expanded as a result of these trade routes. With the urban development a new social group came to the fore upholding the cultural traditions: the urban middle class. Its members became prosperous and demanded new luxuries. A feeling of confidence permeated their ranks, finding their greatest expression in the appearance of the famous Italian tyrants, "condottieri" and princes such as *Francesco Sforza* (1401-1466), the Duke of Milan, and *Cesare Borgia* (ca. 1476-1507).

The feeling of self-confidence was coupled with a strong individualism and joy in life. Moreover, citizens, princes and priests revealed by their actions a shift towards worldly affairs, a move away from the tutelage of theological and scientific authorities which had dominated the scholastic era.

The separation from traditional values and the abandonment of the old order began first in Italy. Chronologically, the development coincided with a new and conscious return to the art and literature of classical antiquity. Writers such as Virgil, Livi, Plutarch and Homer were again greatly admired and imitated while archeological diggings unearthed such famous sculptures as the Apollo of Belvedere and the Laocoon from the rubble of centuries. Aristotle was abandoned in favour of Plato and under the influence of *Cosimo Medici* (1339-1464) of Florence, a Platonic Academy was founded.

The new attitude toward classical antiquity did not stress continuity and further elaboration of the past. Instead, the emphasis was on creating a

Renaissance, a veritable rebirth of the true Graeco-Latin tradition. This scholarly and art loving *humanism* first appeared in Italy and was associated with the names of three famous poets: *Dante Alighieri, Petrarch* (1304-1374) and *Giovanni Boccaccio* (1313-1375). In 1352, Petrarch composed his famous satire on scholastic medicine.

About 1450, many merchants and scholars fled the city of Constantinople in the face of the advancing Turks, taking with them a wealth of Greek manuscripts which further stimulated the humanist movement. Its adherents even attempted to speak and write a flawless and pure classical Greek.

Together with the breaking away from scholastic traditions, a new approach towards nature emerged at about this time.[57] Petrarch was the first to climb a high mountain just to admire the natural beauty of the surrounding countryside. The first collections of animals were started and gardens containing a great variety of plants were established. Moreover, the natural sciences began to shed their close connections with Aristotle. The development of the impetus theory by the French mathematician *Nicholas d'Oresme* (ca. 1323-1382) in particular led scholars away from the Aristotelean "motus naturalis."[58]

The question of whether the nature of a substance was based more on qualitative or rather quantitative attributes led to many disagreements, while other authors vigorously debated the alternatives of merely describing or actually explaining the natural events. Notable mathematicians, physicists and technicians were sponsored by the ruling families of the northern Italian principalities, such as the Sforza, Borgia and Medici. Artists such as the painters *Donato Bramante* (1444-1514), *Michelangelo* (1475-1564) and *Leonardo da Vinci* (1452-1519) furnished the Renaissance with a new art style. Finally, the invention of the printing press by the German *Johannes Gutenberg* (ca. 1400-1468) provided after 1450 the possibility of diffusing classical knowledge and culture on a scale hitherto unknown.

The limits between the scholastic period and the Renaissance are somewhat flexible. Roger Bacon actually belongs chronologically to the era of scholasticism, but in his departure from established traditions he should be considered also as an early Renaissance figure. The same is true of the English philosopher *William of Occam* (ca. 1285-1349), who stressed that all knowledge had to be grounded in internal as well as external experiences. In turn, the German philosopher and mathematician *Nicholas of Cusa* (ca. 1401-1464) combined his metaphysical and theological speculations with the notion of mathematical infinity. He thus expressed the transition from a static viewpoint (geometry and arithmetic) to a more dynamic conceptual approach followed later by Kepler, Newton and Leibnitz.

Medicine meanwhile remained strongly traditional, although some doubts on the scope and perfection of the old medical doctrines were expressed. In 1495, a hitherto unknown or little observed disease, syphilis, made its appearance in Europe, while another condition which had not been described in antiquity spread across the lands in 1529: the sweating sickness. Some febrile illnesses were recognized as being more than merely a fever. The Italian physician and humanist *Girolamo Fracastoro* (1478-1553) distinguished the spotted fever as specific disease entity, and at the same time clearly realized that several epidemic diseases were actually caused and transmitted by contagion.[60]

In spite of the movement away from tradition and authority, Renaissance physiology did not achieve many new insights into the various organic functions. This fact was not surprising, since the precondition for such a physiological development was a more detailed and accurate knowledge of human anatomy, as well as a better understanding of physical and chemical phenomena.

Therefore, the Renaissance scholar who gathered the most advanced physiological knowledge of the times had simultaneously achieved by himself a considerable familiarity with the structures of the human body. This person was the Italian genius *Leonardo da Vinci* who represented the ideal of the "uomo universale," the universal man of the Renaissance.[61] Leonardo was an excellent painter, sculptor, musician, poet, scientist and technician, all in one person, and his thirst for knowledge thrust him into the field of human anatomy, which he studied by carrying out his own dissections. Leonardo examined animal and human organs, comparing their structures, although not in any systematic fashion.[62] These inquiries were carried out in the years 1490, 1503-1507 and 1510-1515 and he planned to publish a complete text of anatomy with the help of the Italian anatomist *Marc Antonio della Torre* (1473-1506) of Pavia. Unfortunately, della Torre died prematurely, and the project was cancelled, although most of Leonardo's sketches have been preserved.[63]

The content of these drawings covered the whole field of anatomy. At the same time, the accompanying text and Leonardo's notebooks contained his views regarding the function of certain organs. These sketches and notes, only recently discovered,went significantly beyond the physiological concepts held in antiquity, but being unavailable, they scarcely influenced the development of contemporary physiology. Thus, they remain merely as invaluable documents of the new Renaissance outlook regarding nature.[64]

Leonardo's anatomical achievements have received detailed attention[65] whereas the physiological concepts of the Italian genius were recently subjected to an extensive analysis[66] which replaced the previous, almost inaccessible works of F. Bottazzi.[67] For reasons of space, only some of Leonardo's most important physiological concepts, characteristic of his approach, will be mentioned here.

For Leonardo, the strong contraction of the heart caused whirls of blood which, because of their friction against the irregular cardiac walls, were heated up and thus produced a vital spirit. The sigmoid valves in the heart were closed by the spiraling spinnings of the blood in their cusps and some reflux movement. (See Fig. No. 6.) Then, when the blood collided with the valves and closed them, a certain tone was created at the level of the great vessels and transmitted throughout the arterial system.

All veins and arteries originated from the heart instead of the liver. Leonardo believed that the lungs were of a spongy consistency and that the finest bronchial branches had blind endings. He did not believe that the

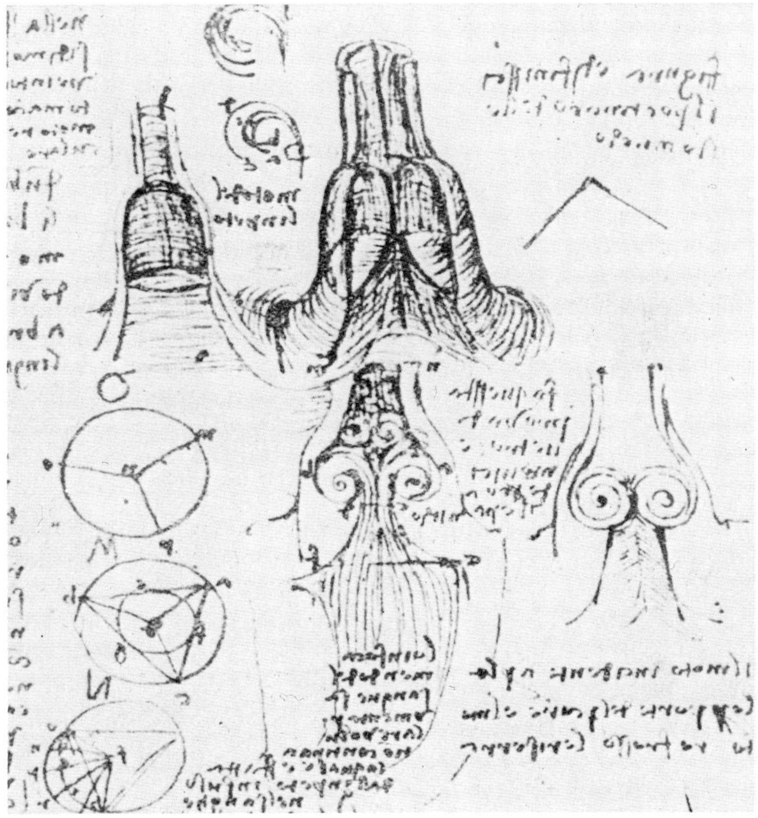

Fig. 6. *Drawings of Leonardo da Vinci (1452-1519) regarding the mechanical function of the semilunar cusps in the valves of the heart* (from his *Quaderni d'Anatomia* IV, fol. 11v.)

air could pass from the lungs into the vessels and the heart, since an isolated lung kept its air when artificially inflated. Rather, concluded Leonardo, like Galen before him, only the freshness and coolness of the air was communicated from the finer bronchial tree to the vascular branches. On the other hand, the lungs passively followed the movements of the thorax, performed by the costal muscles. Both the thoracic and diaphragmatic motions were compared with the action of a pair of bellows. Leonardo also postulated that the abdominal muscles could move the diaphragm and thus force the lungs to expel the air.

The muscles consisted of fine fibres connected with a nerve. Through the hollow center of the nerve arrived the stimulus necessary for the contraction. Leonardo didn't believe that air was responsible for the swelling exhibited by the contracted muscle, and he declared that the muscle attracted the one on which it was inserted through the tendon, in a motion which closely followed the direction of the muscular fibres themselves.

The Italian author was well acquainted with the gross anatomy of the nervous system, describing the cerebral ventricles and the functional connections between the brain, spinal cord and nerves. Regarding vision, he thought, like Alhazen before him, that the rays entered the eye from the object, traveling in straight lines until they reached the pupil, where they crossed. However, the reason that we perceive the objects in an upright position was due to a second crossing which Leonardo postulated at the level of the eye's lens. He felt that ocular refraction was similar to the phenomena occurring during the passage of light from the air into the water, and he was also aware of the pupillary reaction to light. Leonardo realized that a needle point which was smaller than the diameter of the pupil could not be perceived, while the size of brightly illuminated objects increased against a darker background (simultaneous contrast). He also described the ocular spectrum.

Leonardo postulated the origin of the voice in the larynx itself, and he deduced the physical conditions necessary for its production. He described the role of the glottis, the buccal cavity, the tongue and the palate for the formation of speech.

From this short selection of Leonardo's attempts to explain some of the bodily actions it is clear that teleological considerations were displaced by questions concerned with the specific causes for certain phenomena. For this purpose the Italian author successfully used mechanical models, such as in the mechanism of valvular motion and the bellowing action of the chest. Moreover, the attempt to explain the older concept of vital heat formation in the heart through parietal friction was truly remarkable.

However, his idea of a blood circulation remained obscure. It is highly improbable that Da Vinci was aware of the pulmonary circulation, in spite of the fact that there are several suggestive passages in his writings which could be interpreted in that direction. The functions of hearing and vision were explained in purely physical terms, although it is fair to say that many optical phenomena had already been interpreted in a somewhat similar fashion before his time.

In all, the new spirit of the Renaissance was noticeable for the first time in Leonardo's biological conceptions. It was a conscious attempt to express vital actions as the results of natural, *physical forces*. Understandably, his explanations were predominantly mechanical, since contemporary applied mechanics was especially popular in irrigation techniques and warfare technology — two fields also developed by Leonardo. The unfortunate fate which befell his notebooks and sketches — they were lost and only became available again in the nineteenth century — prevented him from exerting any lasting influence with his new insights. Most of the more than 750 anatomical drawings became available after 1898. Nevertheless, Leonardo's approach, the use of physical principles and experimental analysis, was soon used by other authors in whose hands the method became a highly rewarding tool for the attainment of further knowledge.[68]

REFERENCES

[1]Norman Hepburn Baynes, *The Byzantine Empire,* London and New York, 1962.

[2]See biography by F. Kudlien in *Dict. Scientif. Biog.,* I, New York, 1970, pp. 68-69.

[3]His biography, written by F. Kudlien, appeared in *Dict. Scientif. Biog.,* I, p. 121.

[4]Paul of Aegina, *The Seven Books of Paulus Aegineta.* Transl. from the Greek by F. Adams, 3 vols., London, 1844-47.

[5]O. Temkin, "Byzantine Medicine: tradition and empiricism," *Dumbarton Oaks Papers* 16 (1962), pp. 97-115.

[6]P. Reznikoff, "Role of the Nestorians in the preservation of Greek medicine," *New York Journal of Medicine* 67 (1967), pp. 3263-3268.

[7]I.A. Brody, "The School of Medicine at Jundishapur: the birthplace of Arabic medicine," *Trans. & St. Coll. Phys. Phil.* 23 (1955), pp. 29-37.

[8]See: Seyyed H. Nasr, *Science and Civilization in Islam,* pref. by G. de Santillana, Cambridge, Mass., 1968.

[9]Bernard Lewis, *The Arabs in History,* 4th ed., London, 1966.

[10]Edward G. Browne, *Arabian Medicine,* The Fitzpatrick Lectures 1919-1920, Cambridge, England, 1921.

[11]Donald Campbell, *Arabian Medicine and Its Influence on the Middle Ages,* 2 vols., London, 1926.

[12]For a biographical sketch see: G. Strohmaier in *Encyclopedia of Islam,* new ed., Leiden, 1965-71, 3, pp. 578-581.

[13]M. Meyerhof, "New light on Hunain ibn Ishâq and his period," *Isis* 8 (1926), pp. 685-724.

[14]G.S.A. Ranking, "The life and work of Rhazes," *Proc. XVIIth Int. Congr. Med.,* London, 1914, pp. 237-268.

[15]M. Meyerhof, "Thirty-three clinical observations by Rhazes (circa 900 A.D.)," *Iris* 23 (1935), pp. 321-356.

[16]See A.M. Goichon's biography in the *Encyclopedia of Islam,* III, pp. 941-47.

[17]Avicenna, *A Treatise on the Canon of Medicine of Avicenna,* incorporating a translation of the first book by O. Cameron Gruner, London, 1930.

[18]S.M. Afnan, *Avicenna, His Life and Works,* London, 1958.

[19]See biography by R. Arnaldez in *Encyclopedia of Islam,* III, pp. 976-979.

[20]His biography, also by R. Arnaldez, appeared in *Ibid,* III, pp. 909-920.

[21]H.M. Leon, "Physiology and medicine under the Khalifs," *Islamic Culture,* 1 (1927), pp. 388-405 & pp. 560-566; 2 (1928), pp. 291-308.

[22]G. Rath, "Eigenes und Übernommenes in Avicennas Anatomie." Ärztl. Forschung 3, (1949), p. 1.

[23]L.G. Wilson, "The problem of the discovery of the pulmonary circulation," *J. Hist. Med.,* 17 (1962), pp. 229-244.

[24]B.S. Eastwood, "Averroe's view of the retina — a reappraisal," *J. Hist. Med.* 24 (1969), pp. 321-341.

[25]J. Hirschberg — J. Lippert, *Geschichte der Augenheilkunde bei den Arabern,* in Grafe-Saemischs Hdb. d. gesamten Augenheilkunde 13, Leipzig 1908.

[26]D.C. Lindberg, "Alhazen's theory of vision and its reception in the West," *Isis* 58 (1967), pp. 321-341.

[27]L.M. Sadi, "Ibn-al-Haitham (Alhazen), medieval scientist," *Univ. Mich. Med. Bull.* 22 (1956), pp. 249-273.

[28]E. Wiedemann, "Zu Ibn al Haitams Optik." *Arch. Gesch. d. Naturwiss. und Technik* 3 (1910), p. 1.

[29]M. Meyerhof — C. Prüfer, "Die Lehre vom Sehen bei Hunain ibn Ishâq." *Arch. Gesch. Med.* 6 (1912), p. 21.

[30]P.O. Kristeller, "The School of Salerno," *Bull. Hist. Med.* 17 (1945), pp. 138-194.

[31]H.P. Bayon, "The masters of Salerno: the origins of professional medical practice," in *Science, Medicine and History* I, pp. 203-219.

[32]R. Creutz, "Der Frühsalernitaner Alfanus und sein bislang unbekannter *Liber de pulsibus*," *Sudhoffs Arch* 29 (1937), pp. 57-83.

[33]Pascal Parentes, transl., *The Regimen of Health of the Medical School of Salerno,* New York, 1967.

[34]See article by M. McVaugh in *Dict. Scientif. Biogr.*, 3 (1971), pp. 393-395.

[35]K. Sudhoff, "Constantin der Afrikaner und die Med. Schule von Salerno," *Arch. Gesch. Med.* 23 (1930), p. 293.
K. Sudhoff, "Die erste Tieranatomie von Salerno und ein neuer salernitanischer Anatomietext." *Arch. Gesch. Math. Naturwiss. u. Technik* 10 (1927), p. 136.

[36]W.D. Sharpe, ed. & transl. "Isidore of Seville: the medical writings," *Trans. Amer. Phil. Soc.* 54 (1964), pt. 2.

[37]S. Cooper, "The medical school of Montpellier in the 14th century," *Ann. Med. Hist.* 2 (1930), pp. 164-195.

[38]See article by M. McVaugh in *Dict. Scientif. Biogr.* I (1970), pp. 289-291.

[39]P. Diepgen, *Arnald von Villanova u. die Medizin des Mittelalters. Lynchos.* Stockholm, 1939.

[40]R. Creutz u. J. Steudel, *Einführung in die Geschichte der Medizin in Einzeldarstellungen.* Iserlohn, 1948.

[41]Hildegard von Bingen, *Das Buch von dem Grund und Wesen und der Heilung der Krankheiten,* ed. H. Schipperges, Salzburg, 1957.

[42]Frederick B. Artz, *The Mind of the Middle Ages A.D. 200-1500,* 3rd revised ed., New York, 1962.

[43]Lynn Thorndike, *A History of Magic and Experimental Science,* New York, 1934-47, Vols. 3 & 4.

[44]See A.C. Crombie and J.C. North in *Dict. Scientif. Biogr.* I, pp. 377-385.

[45]V.R. Larkin, "St. Thomas Aquinas on the movement of the heart," *J. Hist. Med.* 15 (1960), pp. 22-30.

[46]M. Schmid, "Der Weg zu Harvey," *Sitz. Ber. Physik-Med. Sozietät,* Erlangen 79 (1958), pp. 66-101.

[47]K.V. Megenberg, *Das Buch der Natur,* ed. by F. Pfeiffer, Stuttgart, 1861.

[48]See article by W.A. Wallace in *Dict. Scientif. Biogr.* I (1970), pp. 99-103.

[49]W. Pagel, "Harvey, foetal irritability — and Albertus Magnus," *Med. Hist.* 10 (1966), pp. 409-411.

[50]H. Balss, *Albert Magnus als Biologe, Werk und Ursprung,* Stuttgart, 1947.

[51]Stewart C. Easton, *Roger Bacon and His Search for a Universal Science,* New York, 1952.

[52]Roger Bacon, *The Opus Majus,* transl. R.B. Burke, 2 vols., Philadelphia, 1928.

[53]Cl. Baeumker, "Vitello ein Philosoph und Naturforscher d. 13. Jhs.," *Beitr. z. Gesch. d. Philosophie d. Mittelalters* 2, Münster, 1908.

[54]Vern L. Bullough in *Dict. Scientif. Biogr.,* I, 1970, p. 107.

[55]Charles Singer, ed. and transl., *The "Fasciculo di Medicina"* (Venice, 1493), 2 vols., Florence, 1925.

[56]E. Hintzsche, "Wege der Überlieferung der Galenschen Anatomie," *Ciba-Z* 8 (1944), p. 3421.

[57]Wallace K. Ferguson, *Europe in Transition (1300-1520),* Boston, 1962.

[58]A.C. Crombie, *Medieval and Early Modern Science,* 2 vols., New York, 1959.

[59]Paul Oskar Kristeller, *Renaissance Thought;* the classic, scholastic and humanistic strains, revised and enlarged ed., New York, 1961.

[60]G. Fracastoro, *De Contagione,* transl. by W.C. Wright, New York, 1930.

[61]Charles D. O'Malley & J.B. de C.M. Saunders, *Leonardo da Vinci,* New York, 1952.

[62]J. McMurrich, *Leonardo da Vinci, the Anatomist,* Baltimore, 1930.

[63]S. Esche, *Leonardo da Vinci. Das Anatomische Werk,* Basel, 1955.

[64]Leonardo da Vinci, *Notebooks,* arranged and rendered into English by E. McCurdy, New York, 1958.

[65]K.D. Keele, "Leonardo da Vinci's influence on Renaissance anatomy," *Med. Hist.* 8 (1964), pp. 360-370.

[66]Kenneth D. Keele, *Leonardo da Vinci on Movement of the Heart and Blood,* London, 1952.

[67]F. Bottazzi, in *Arch. per l'Antropologia* 1902; *Arch. ital. Anat.* 1905, 1907; *Riv. d'Italia* 1907; *Sapere* 1938; *Rass. Clin Sci* 1939; *Progressi di terapia* 1939.

[68]Charles D. O'Malley, (ed.), *Leonardo's Legacy: An International Symposium,* Berkeley, 1969.

CHAPTER III

Foundation and Development of Physiology in the Sixteenth and Seventeenth Centuries

1. THE RENAISSANCE OF ANATOMY IN THE SIXTEENTH CENTURY

The entire sixteenth century appeared to be in a state of great inner turmoil. More distinctly than in any other historical period, two great intellectual currents, *tradition* and *reformation,* coexisted, displacing each other temporarily in their fight for supremacy. The contest was not limited merely to the religious realm, but also involved the sciences, arts and economics; all fields of endeavor in which some daring innovators emerged during this age.[1]

Likewise, the occurrences and affairs of everyday life suffered great transformations. A booming commerce, encompassing the southern and central regions of Europe, created prosperous cities and contributed towards the development of a broadly based and selfconscious *middle class.* An erudite education no longer remained a clerical privilege, thanks to the invention of printing, and even the layman began to take active part in the debates of the age.

The century became an era of awakening and stirring *individualities,* contractors, merchants, artists or scientists who left their personal imprint on their works and creations. In 1493, Paracelsus' birth date, the great German painter *Albrecht Dűrer* (1471-1528) executed the first self-portrait in the history of European painting, and at last man appeared to discover his own originality.

The outlook which, during the Middle Ages, had been directed towards the heavens in the face of a theocentric cosmology, now focused joyfully on the earthly realm. Large and small natural objects elicited the fond attention of the artist, and likewise were the subjects of new interpretations by scholars and scientists. The German painter *Albrecht Altdorfer* (ca. 1480-1538) selected a whispering forest and Dűrer a violet as subjects for their pictorial art.

Although this Renaissance *ideal of secularism* penetrated all aspects of life, it also encountered some stiff opposition in certain quarters. In 1483, the year in which the German theologian *Martin Luther* (1483-1546) was born, the Spanish inquisitor general *Tomás de Torquemada* (1420-1498) became the leader of the infamous Inquisition operating in his country. The Church attempted to suppress the internal as well as external criticism to which its actions and dogma were increasingly subjected. Thus, the Spanish physician and theologian *Michael Servetus* (1511-1553), recognized as a rediscoverer of the pulmonary circulation, was burned at the stake in Geneva for heresy. Even in Italy, the humanists gradually lost their influence and recognition. The ensuing *Counter-Reformation* fought vigorously against these men because of their freely expressed religious disbelief.

Nevertheless, some great humanists such as the German scholars *Johann Reuchlin* (1455-1522) and *Ulrich von Hutten* (1488-1523), and the Dutch theologian *Desiderius Erasmus* (ca. 1466-1536) of Rotterdam, flourished in northern Europe. At various universities, the followers of Galen and Hippocrates still exerted great influence. Carefully compiled and philologically correct editions of Galen, Avicenna, Aristotle and other authors were witnesses to the staunch conservatism still running high among scholars. No less than four hundred and eighty-one writings of Galen were edited during the Renaissance period.

The year 1543 truly constitutes a watershed in the evolution of Western thought. With the death of the Polish astronomer *Nicolas Copernicus* in 1543, the canon of Frauenburg, his great work entitled *Opus de Revolutionibus Coelestibus,* was published. Although presented as a hypothesis, Copernicus demonstrated in his book, on the basis of more accurate astronomical measurements and calculations, that the earth was not stationary, but moved around the sun.[2]

In the same fateful year 1543, another work appeared which was as revolutionary for anatomy as the Copernican treatise was for astronomy. The great treatise was *De Humani Corporis Fabrica,* written by the Flemish physician and anatomist *Andreas Vesalius* (1514-1564),[3] a text of human anatomy based on careful personal dissections and preparations. The various bodily structures were described in detail and reproduced in drawings, just as Vesalius himself had seen them, not merely depicting what could be inferred from reading Galen's anatomical works.

However, the work of Vesalius was not without precursors, and it also found a host of important followers. Gradually, during both the sixteenth and seventeenth centuries, a growing understanding of the human frame occurred throughout Europe, the first and indeed indispensable precondition for genuine progress in physiology.[4]

The Salernitan instructions for dissection which originated around 1150 illustrated the expanding need to further medical knowledge through anatomical investigations, even though these proposed dissections were to be carried out on the pig. In turn, the first human examinations in Cremona and Bologna have already been mentioned. Although the interest in anatomical studies was experiencing a notable resurgence, Bologna in 1318 was the scene of a trial against four persons who had exhumed a cadaver for dissection.

The contemporary teachers of anatomy used several charts which can be traced back to Alexandrian sources according to the research of K. Sudhoff.[5] These depictions had been transmitted in part through Western monastic medicine and also with some Persian manuscripts. The already mentioned *Anathomia* of Mondinus was the most popular textbook of anatomy well into the sixteenth century. Another Italian anatomist and physician, *Alessandro Achillini* (1463-1512) of Bologna, supplemented Mondino's work describing among other things the anvil and malleus, two of the ossicles of the internal ear.[6] He also depicted the cerebral ventricles, the infundibulum and the fornix in the brain.

Likewise, the Italian physician *Berengario da Carpi* (1470-1530)[7] published a commentary on Mondino's anatomy in 1521. In this work, Berengario essentially advocated Aristotelean and Galenic ideas, although some critical comments were also included.[8] At about the same time, Leonardo da Vinci carried out his anatomical observations already discussed.

North of the Alps, anatomy was even more strongly enchained by tradition than in Italy and Montpellier. The first public dissection at Vienna was carried out in 1404, and in Basel during the year 1531. In those cities, as in Paris, the anatomists were mainly Galenists who, in their loyalty to the authorities and doctrines of Galen, were not inclined to doubt the accuracy of his works.

Therefore, the accomplishment of the young Andreas Vesalius should be valued highly.[9] He was born in Brussels into a family which had originally come from Wesel — one of Andreas' ancestors had adopted the name "Vesalius" in memory of his homeland. During his student years at Paris, Vesalius excelled in the empirical knowledge of anatomy over all his fellow students, and soon even surpassed that of his teachers. He did not miss any opportunities to dissect animals or examine the bones of executed criminals.

Although Vesalius had studied at Paris under the direction of the Galenist anatomists *Gűnther von Andernach* (1505-1574) and *Jacques Dubois Sylvius* (1478-1555),[10] he soon freed himself from the shackles of tradition. After more studies at Louvain and Padua, Vesalius became professor of anatomy at the latter Italian university. While teaching there, he avidly dissected human cadavers and prepared their organs. The first results of his untiring labors were a series of anatomical plates, which were published in 1538 under the title *Tabulae anatomicae sex*. These drawings still contained a number of inaccuracies based on assumptions taken from Galenic anatomy, such as the liver with five lobes.[11]

However, Vesalius gradually became more cautious and doubtful about the anatomical knowledge which he had inherited from the ancients, recognizing that Galen's descriptions had been based mostly on animal dissections. Further and intensive anatomical work produced Vesalius' main contribution, the *De humani corporis fabrica libri septem,* published at Basel in 1543.[12] The brilliant illustrations accompanying the book had been drawn by the Flemish artist *Jan Stephan of Calcar* (1499-1546).[13] The text was based on Vesalius' own observations and departed in several hundred instances from the Galenic descriptions, thus making the book a landmark in the evolution of anatomy. The impact of the Vesalian publication truly established anatomy as the basis for medicine and set the standards for the anatomical method. The tremendous progress made is symbolized by the contrast existing between a skeletal depiction of the fourteenth century and that of the *Fabrica* (see Figs. No. 7 and 8).

The reaction to the Vesalian innovations was prompt. The initial response was largely negative, exemplified when the Italian anatomist *Realdo Colombo* (1516-1559), a student and successor of Vesalius at Padua, attacked his former teacher. Soon the Parisian anatomists followed suit, not without some reason, since Vesalius had kept several erroneous views along with his many corrections. This fact reflected the increasingly more accurate anatomical observations being carried out which soon made the Vesalian mistakes obvious.

In 1551, Vesalius' old teacher Sylvius composed a defamatory work in which he violently attacked his former student as "an ignorant, slanderous liar, inexperienced in all things, ungrateful and godless, a monster of ignorance who with his pestilential breath was attempting to poison Europe, and whose errors were so numerous that merely listing them would be an endless task." Even the capable Italian anatomist *Gabriello Fallopio* (1523-1562) discovered some errors in

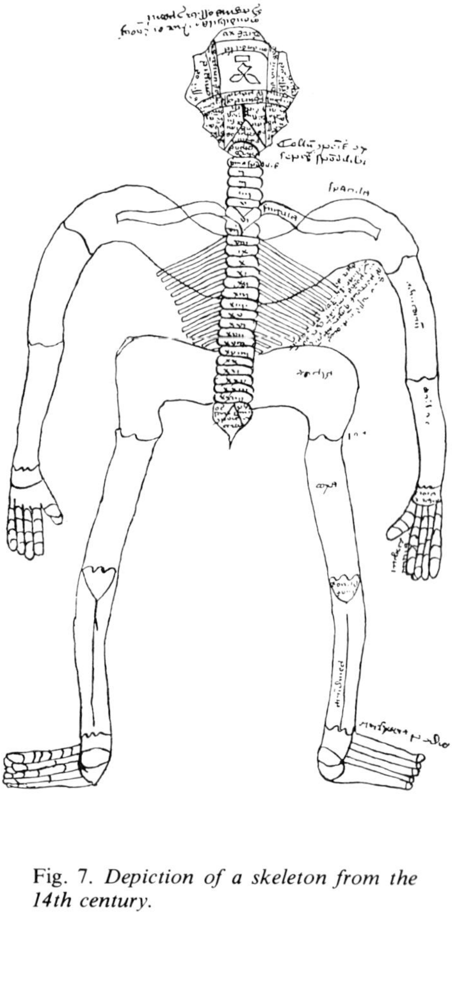

Fig. 7. *Depiction of a skeleton from the 14th century.*

Vesalius' work. Incidentally, Falloppio described for the first time the aqueduct of the vestibule, the osseous labyrinth, the chorda tympani and many other structures.[14]

In 1544, Vesalius became a personal physician to the emperor Charles V and he abandoned his anatomical dissections. He died in 1564 while returning from a pilgrimage to Jerusalem.

Following Vesalius, the sixteenth century also witnessed the activities of a series of notable anatomists, like the Italian *Bartholommeo Eustachio* (ca. 1524-1574).[15] Eustachio discovered the auditory tube or tuba pharyngo-tympanica which now carries his name, and some details pertaining to the organ of hearing.

We cannot go into details about the numerous anatomical discoveries subsequently made. Suffice it to mention some of the prominent authors of the period, such as the Italian physician *Guido Guidi* (1500-1569), the French anatomist *Charles Estienne* (1504-1564) at Paris, the Swiss anatomists *Felix Platter* (1536-1614) and *Gaspard Bauhin* (1560-1624) in Basel, the Italian surgeon and student of Fallopio, *Hieronymus Fabricius ab Aquapendente* (1537-1619) as well as the Flemish anatomist *Andriaan van den Spieghel-Spigelius* (1578-1625), both in Padua. The careful and painstaking work of the various anatomists provided the bases for the physiological progress which occurred in the next centuries.[16,17]

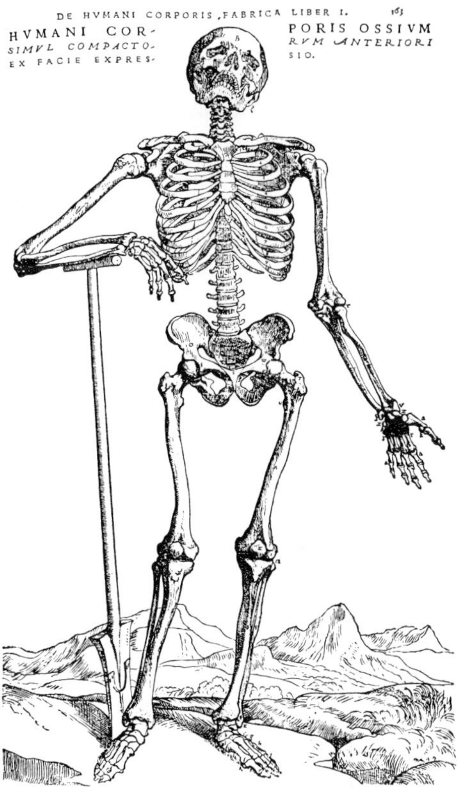

Fig. 8. *Depiction of a skeleton from Andreas Vesalius' De Humani Corporis Fabrica,* Basle, 1543.

The knowledge of morphology, however, only seldom contributes towards an immediate solution of the existing physiological questions. Thus, it is not surprising to note that the physiological conceptions of the sixteenth century did not develop far beyond the Galenic doctrines. In this sense, the following quotation from Vesalius' *Fabrica* is noteworthy, inasmuch as it reflects the degree to which the willful anatomist was nevertheless bound by the contemporary views. Vesalius wrote: "The septum of the ventricles, composed as I have said of the thickest substance of the heart, abounds on both sides with little pits impressed in it. Of these pits, none, so far at least as can be perceived by the senses, penetrate through from the right into the left ventricle, so that we are driven to wonder at the handiwork of the Almighty, by means of which the blood sweats from the right into the left ventricle through passages which escape human vision."[18]

In the last chapter of the *Fabrica,* Vesalius described some important physiological observations: the lungs collapsed when the chest was perforated, and there was a loss of voice when the recurrent nerve was severed. Using a pair of bellows, it was possible to keep an animal with an opened thorax alive and its heart beating.[19]

On the basis of carefully executed anatomical studies, and combining this knowledge with the observation of organic functions in living animals, physiology was to experience meaningful progress only in the seventeenth century. These necessary new insights were achieved less by vivisection experiments than by the interpretation of the morphological observations, since the experimental method was still employed on a small scale.

The contemporary anatomical and physiological views or *sex res naturales,* were subjected to further rationalization and systematization by the French physician and astronomer *Jean Fernel* (1497-1558)[20] in the middle of the sixteenth century. In his work *De naturali parte medicinae* (1542), Fernel seemed to have followed more Aristotle than Galen. His physiology was based on the elements, qualities and humors while his anatomy distinguished between simple and complex organic structures. The seven books of Fernel's work were divided into 1) anatomy 2) elements 3) temperaments 4) spirits and innate heat 5) the capacity of the soul 6) functions and humors, and finally 7) human conception and the semen.

For Fernel, human nature was composed of soul and body. The soul was the sole reason for the existence of the body, producing and directing all the vital appearances. For such a task, the soul needed instruments, the bodily organs, which were activated by three independent and different spirits. The accompanying illustration (see Fig. No. 9) is an attempt to visualize the structure of Fernel's physiology. It was a systematized summary of the total knowledge available at the time, predominately from classical and Islamic sources, and represents a truly remarkable level of achievement.[21]

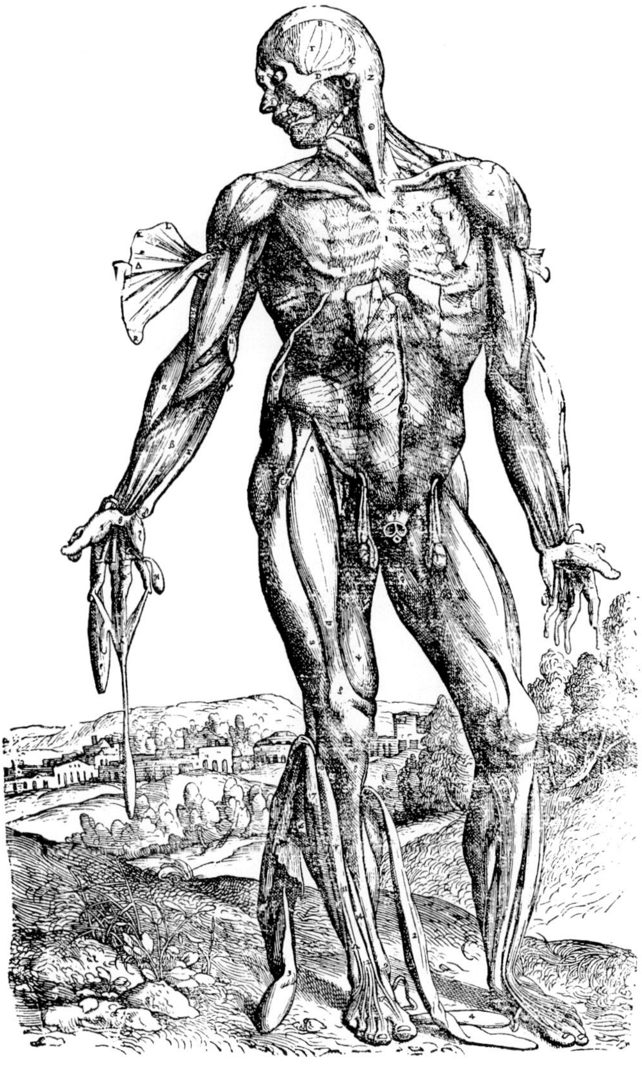

The *Fabrica* of VESALIUS is adorned with a series of large muscular figures of great artistic and scientific merit. They exhibit the muscles in a state of contraction and are, in fact, living anatomy. VESALIUS, unlike GALEN, was fully aware of the difference between nerves and tendons, and he showed how nerves and distributed to all the muscles. As regards the physiology of the vascular system, however, VESALIUS did not advance very far from the GALENIC standpoint though he expresses doubts as to the permeability of the interventricular septum.

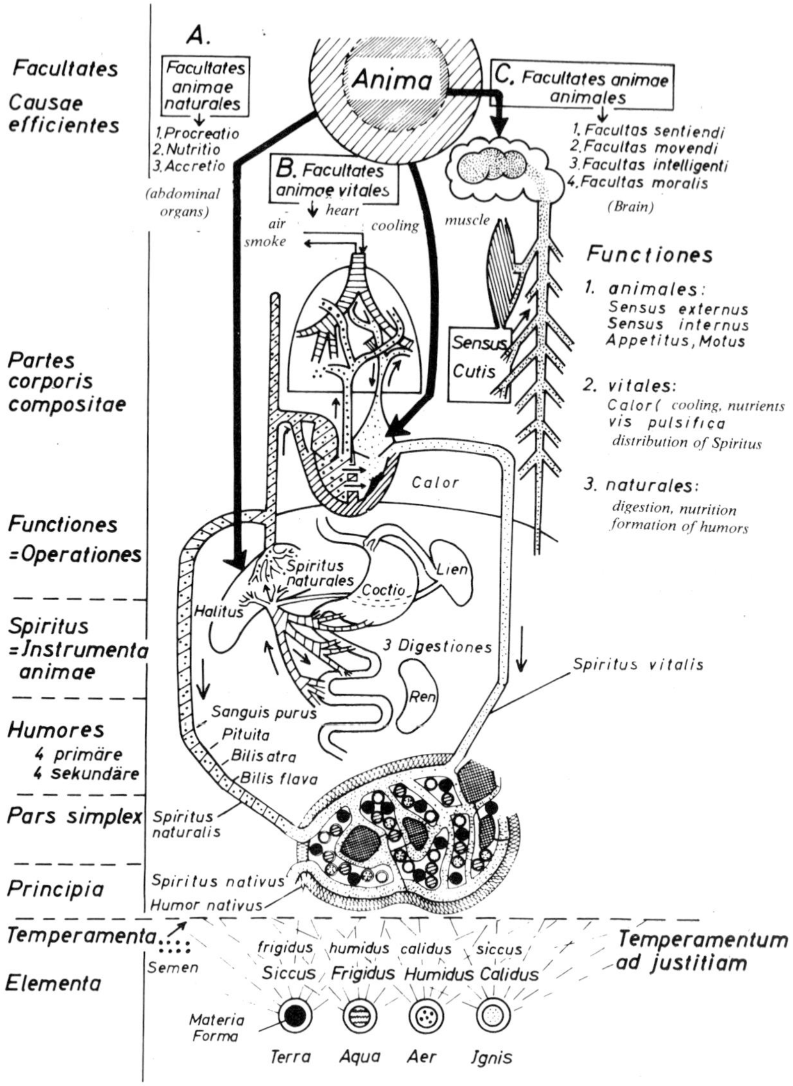

Fig. 9. Schematic representation of Jean Fernel's physiology according to his *De Naturali Parte Medicinae,* 1542.
The anima is responsible for the three fundamental faculties. The entire scheme with its humors, faculties and spirits is essentially Galenic and the philosophical bases Aristotelean.

2.
THE APPLICATION OF CHEMICAL PRINCIPLES IN PHYSIOLOGY BY THE IATROCHEMISTS

At the beginning of the sixteenth century the development of chemistry tended to shift from general and conceptual interests to more practical and technical concerns. The Middle Ages had been engaged with a passion in theoretical speculations regarding the nature of the elements, their differences and possible transformations. This pre-chemical evolutionary step is known as alchemy, and its main body of knowledge was based on ancient traditions derived in part from the ancient Egyptian priesthood.[22]

Certainly, alchemy was more natural philosophy than chemistry in the present meaning of the discipline. It attracted the greatest minds of the Middle Ages, such as Albertus Magnus, Roger Bacon, Arnaldo of Villanova and others. Alchemy was concerned with a comprehensive knowledge of the inner essences of nature on the theoretical level — *alchimia speculativa* — just like astrology and magic. The alchemical method thus was an attempt to grasp the fundamental characteristics of the entire cosmos and its components through the use of analogies. The goal was to find the unity within the plurality, the hierarchical chain of beings with their spiritual components from simple matter to the celestial bodies.[23]

On the practical level, however, alchemy dealt with laboratory procedures — *alchimia practica* — and the problems of the metals, their origin, composition, transformation and inherent forces. The central issue was, of course, the question of transmutation — transformation into other metals — and the production of gold.[24,25]

The infusion with astrological elements has been ascribed to the sect of the Ismailis, a powerful branch of Shia Islam, which integrated into their doctrines such sciences as astrology and alchemy. Their works were among the most influential in all of Islam, especially the esoteric numerical symbolism and study of nature. The sect celebrated Hermes Trismegistus, a name given by Neoplatonists and others to the ancient Egyptian god Thoth, a divinity of knowledge, after whom alchemical writings were often called "hermetic."[26]

From this tradition originated the alchemical idea that the metals possessed souls like the planets and could be related to the various signs of the zodiac, thus receiving a series of astrological symbols. In this fashion, gold was connected with the sign for the sun, silver with the moon, iron with Mars and mercury with the planet of the same name. Moreover, the chemical processes were symbolized as veritable "chemical weddings," as great or small magistries, and the entire art viewed as something mysterious. "The alchemist searches for all the natural phenomena in his retort: life and death, generation and decay, transformation and perfection, the metamorphosis of plants and animals, and the allegories related to egg and birth, the Virgin and the King."[27]

The Islamic traditions exerted an especially strong influence on the alchemical development. The chemical knowledge available in the tenth century was primarily based on the writings of the Islamic alchemist *Jabir Ibn Haiyan* (fl. ca. 776) better known by his Latinised name *Geber*. In turn, the later Middle Ages added much new experience to this already existing body of knowledge. Alcohol, mineral acids, niter, sulphur, mercury, arsenic, as well as combinations of mercury with gold, silver, lead, tin and copper were also known.

The processes of distillation, sublimation, precipitation and melting were often practiced. Alchemical compounds were even recommended for the treatment of diseases, and Arnaldo of Villanova used an "aurum potabile" for human rejuvenation, and mercury for curative purposes.[28]

Another alchemist, probably a Benedictine monk only known under the pseudonym of *Basil Valentine* (fl. 1413), promulgated the use of antimony in medical practice. Like most alchemists, Basil combined a sober empirical knowledge with magical ideas. He prepared mercury and niter compounds, sugar of lead, and described the preparation of hydrochloric acid from vitriol and common salt. On the theoretical level, this author viewed sulphur as the principle of activity or combustion, while the mercury was passive and volatile. Salt was an intermediary, non-combustible principle, and all three of them symbols of essential chemical properties found in the various elements.

In the sixteenth century, the Swiss physician and chemist *Theophrastus Bombastus von Hohenheim* (1493-1541) better known as *Paracelsus,* developed further the relationship between medicine and alchemy,[29] discovering new drugs as well as techniques for the preparation of effective remedies. Paracelsus introduced a series of essences, tinctures and extracts into therapeutics.

During the seventeenth century, alchemy gradually became an extinct science. This decline was to a great extent the result of alchemy's nefarious combination with astrology, macro-microcosmic analogies, the doctrine of signatures, and the often deliberate deceptions associated with the making of gold. Therefore, the inception of modern chemistry as a science from the alchemical background, can be traced back to the seventeenth century,[30] as reflected in the works of the English chemist *Robert Boyle* (1627-1691).

Beginning with Paracelsus, alchemical-chemical explanations were increasingly used to explain vital phenomena, although the interpretations appeared to be tentative and the language still vague. Because of the obscure meaning of the neoplatonic and magical Paracelsian ideas, their author was for a long time considered to be a quack rather than a great physician and thinker of the period.[31]

The influence of Paracelsian concepts was obvious in the work of another notable physician, *Johannes Baptista van Helmont* (1579-1644), as well as other chemical writers. Unfortunately, since these ideas regarding medicine and the natural sciences were submerged in a neoplatonic framework, they were rather shortlived. Many alchemical and Paracelsian viewpoints can be found in the literature of the sixteenth century German Rosicrucians such as *Valentin Weigel* (1533-1588), *Jakob Böhme* (1575-1624) and *Johann V. Andreae* (1586-1654). Therefore, as the Enlightenment began, Paracelsus continued to be quoted only in the occult sciences until his rediscovery in the twentieth century. We are grateful for the recent attempt to present the Paracelsian ideas within the context of the Swiss physician's own time.[32]

Paracelsus (Fig. No. 10) was born in the Swiss town of Einsiedeln in 1493, the descendant of a Swabian noble family, the Bombasts of Hohenheim. He died in 1541 at the age of forty-eight in Salzburg, barely two years before the publication of Vesalius' revolutionary work on anatomy.[33] During his years of restless travel, Paracelsus wrote his great works, like the *Paragranum* (1530), the *Grosse Wundarznei* (1536), *Defensiones* and *Labyrinthus medicorum* (1538).[34]

Fig. 10. Theophrastus von Hohenheim called "Paracelsus" (1493-1541).

The importance of Paracelsus in the development of physiology lay in his approach. By viewing the vital actions as chemical phenomena, he broke with the past abruptly and thus paved the way for the subsequent chemical interpretations of digestion, metabolism and other functions. His radical departure from the previous medical views centered primarily on the elimination of the classical doctrine of the humors and elements.

For Paracelsus, the fundamental principles of the body were not the ancient four elements, but three substances: sulphur, mercury and salt. Each of them was a symbolic expression for an essential chemical quality: sulphur representing combustibility, mercury the volatility and salt the residual. The changing mixture of these three substances brought about the various characteristic inorganic compounds of living bodies. According to Paracelsus, the vital forces and actions within the organism were directed by an alchemical force called the "Archeus." Although dependent upon the material substrate, this chemical director developed its own activity in the form of a force which shaped matter.[35]

Therefore, Paracelsus urged the physicians to imitate the "Archeus." He recommended the use of mercury, lead, copper, gold, sulphur, silver, antimony and other metals. Moreover, he also anticipated various physiological relationships which were confirmed later on through adequate research. Paracelsus already rejected before Borelli's time the classical doctrine of the heart as the center for the vital heat. Instead, he postulated the existence of an "ignis digestionis" — a digestive fire — taking place throughout the body and responsible for its temperature, an anticipation of the metabolic activities occurring in living beings.

Just as the stomach seemed able to extract only those elements necessary for nutrition from the food, Paracelsus envisioned that every organ could likewise incorporate the needed substances and dispose of waste in the form of tartar. For this scheme the ancient "anatomia localis" was no longer useful, and had to be replaced by a new "anatomia essentialis" and fire. Paracelsus declared that "medicus igne, non cultro necessario anatomicus," since there were no finished products.

Everything for Paracelsus was in a state of flux, on the road towards greater perfection. The actions which nature was capable of performing spontaneously, could be reproduced by those who understood the natural workings with the aid of alchemy. The initiated was able to aid the impure matter to achieve its ultimate perfection — "ultima materia" — through the art of chemical separation.[36] Thus, alchemy also meant for Paracelsus the theoretical attempt to comprehend the inner mechanisms of nature and the living body, and likewise the practical effort to free the prevailing natural chemical forces.

Hence, Paracelsus was truly a transitional figure at the dawn of modern times. One aspect of his knowledge was deeply rooted in the Middle Ages, neoplatonism and the magical realm of alchemical thought. The other Paracelsian tradition expressed awareness of empirical facts and natural appearances, trying to couch them in certain terms and concepts. Yet, Paracelsus' opposition to all historical aspects and his blanket indictment of tradition without a critical appraisal of the past doctrines, had largely negative effects. Moreover, the Swiss rebel neglected anatomical considerations within his medical theory and concept of vitality, thus thwarting largely the impact which the Paracelsian ideas might have had for the subsequent development of medicine and physiology.[37]

Almost one hundred years later, the famous German physician *Daniel Sennert* (1572-1637) was more successful in relating chemistry to medicine. Sennert, an adherent of Aristotle and atomism, was opposed to the Paracelsian ideas. Like many authors in the seventeenth century, he was very interested in the theoretical and practical aspects of chemistry. Also to be mentioned are the German chemist *Johann R. Glauber* (1604-1670) who discovered the salt which still carries his name, and the German physician *Hadrian Mynsicht* (1603-1638), who first crystallized the tartar emetic.

The new chemical approach gradually led to the discovery of new substances and improved concepts regarding the nature of the elements, their combinations and transformations.[38]

The real founder of iatrochemistry was undoubtedly *Johann B. van Helmont* (see Fig. No. 11). Like Paracelsus before him, whom he considered his intellectual predecessor, van Helmont did not concentrate on certain isolated questions. Instead he attempted to create a comprehensive system of nature, man, health, and disease in predominantly chemical terms.[39]

Fig. 11. Johannes Baptista van Helmont (1577-1644).

The characteristic conflicts of the baroque period were reflected in the work of van Helmont.[40] A sober and open approach to nature coexisted with a profound yearning for religious and philosophical relationships, a combination of scientific and speculative trends typical for the whole era.[41] His most important chemical discovery was the clear distinction between "gas" and steam, and the identification of carbonic acid — "gas sylvestre" — which formed during the fermentation of wine and the combination of potash with acids.

Van Helmont used scales to weigh the elements used in his chemical experiments, and he postulated the continued existence of these substances in the various chemical combinations. He also recommended the measurement of specific gravity of the urine.

According to van Helmont, every organ possessed its own "archeus," a material, actively directing principle which conferred functional specificity to the organ in question.[42] He interpreted the digestive process as a phenomenon of chemical decomposition, and for this purpose envisioned the existence throughout the body of five "gas" elements, which, according to van Helmont, acted as a "fermentum." These "ferments" were invisible, spiritual agencies, capable of transforming the various substances, especially the food in the stomach and intestinal tract.

For van Helmont, the "ferments" generated a series of six fermentations in the body in order to transform the ingested foodstuffs into "living flesh." The first of such reactions occurred in the stomach through the action of a specific ferment and the presence of an acid, followed by a second alkaline process in the duodenum, where the food was neutralized and received a third ferment with the arrival of the bile. Van Helmont postulated the existence of two more ferments, elaborated by the liver, which transformed the absorbed chyle into blood. One of them was supposedly located in the left ventricle of the heart, warm as light, which changed the venous into arterial blood, a substance rich in vital spirit. For this conversion van Helmont believed that the heart needed to be constantly in motion. The sixth and final fermentation took place in the various tissues, where the blood was assimilated into flesh. Moreover, a certain amount of vital blood gave rise in the brain to the "archeus."

Van Helmont stressed that it was not the bodily heat which digested the food, producing its own breakdown and fermentation as Galen had proposed.[43] Instead, the temperature of the organism was the result of fermentative processes. In this scheme, which strove to abolish the old Hippocratic model, pulse and respiration contributed to the diffusion of the heat throughout the body, rather than exerting a cooling action. However, Galen had already considered that breathing and pulse furnished innate heat throughout the body. Van Helmont argued that no smoky wastes escaped from the heart to the lungs, and no air entered the blood from the lungs.[44] Thus, van Helmont increased the doubts about the prevailing classical physiology with his chemical conceptions, and he furnished physiology with several new and useful ideas.[45]

Fig. 12. Frans de le Boë-Sylvius (1614-1672).

Less original but more important for the diffusion of iatrochemical ideas was the work of the German physicist and chemist *Frans de le Boë* (1614-1672) — better known as *Franciscus Sylvius* (see Fig. No. 12). The child of a refugee family of Hugenots, Sylvius was born in 1614 at Hanau, and studied medicine at Basel. After his graduation, he practiced at Hanau, Paris and then Leyden, where after 1658 he became a greatly sought after practitioner and teacher of medicine, dying in 1672 as a victim of typhus.

Sylvius was not a religious or neoplatonic thinker, but rather a naturalist, important also as a physician, anatomist (fossa Sylvii), botanist and chemist. In contrast to van Helmont, who had virtually ignored the discovery, Sylvius became a strong advocate of Harvey's circulation of the blood, and a forceful supporter of the Cartesian mechanical philosophy.[46]

For Sylvius all *vital actions* could be explained in chemical terms as a series of fermentations. Hence, the food was disintegrated in the digestive system in three separate, fermentative steps. The first one began in the mouth, where Sylvius for the first time recognized the role played by the saliva. Exposed to the action of the so-called "triumvirate of fluids" — acid saliva, acid pancreatic juice and alkaline bile — the foodstuffs were broken down with production of gas which escaped during the ensuing ebullition.

In the heart, Sylvius envisioned the presence of a vital fire which was kept burning by chemical reactions. The blood which circulated through the lungs was cooled there with the incorporation of "sal nitrum" from the air. Upon its arrival in the left ventricle of the heart, the blood was agitated and heated again, and thus entered the arteries in that state, in order to furnish the bodily organs with proper nutrition and warmth.

The action of the nerves was mediated by the animal spirits which flowed through the hollow structure of the nerves to the periphery. Sylvius considered health to be the result of a proper mixture of acid and alkaline elements, while disease was an expression of chemical disturbances with the production of pungent acid and alkaline substances. Therefore, according to Sylvius, therapy had to be directed at these offending compounds, treating the acid ones (acrimoniae) with alkaline drugs and those noxious alkalis with proper acid medications.[47]

This short summary of the doctrines propounded by the most distinguished representatives of the iatrochemical school shows the fruitfulness of the new chemical views, especially for the explanation of the digestive phenomena. Yet, the essentially correct observations were included into speculative generalizations, giving rise to frameworks in which empiricism and speculation were inextricably intertwined.

Nevertheless, the approach facilitated the recognition of specific chemical activities occurring in the organism. The ideas about fermentation, the existence of acids and alkalis in the body, were constantly repeated in the contemporary medical literature. Sylvius also stimulated specifically one of his students, the Dutch anatomist *Regnier de Graaf* (1641-1673), who in 1664 discovered the pancreatic juice and was able to isolate it from an artificial pancreatic fistula (see Fig. No. 13). The reason for ascribing to this juice an erroneous acid character was understandable under the prevailing circumstances. At the time, the distinction between acids and alkalis was made by checking their taste or testing them with alkaline substances such as potash, and then observing whether the compound exhibited any ebullition.

The strongly speculative ideas of the iatrochemists came under attack early, especially by the German physician and chemist *Johannes Bohn* (1640-1718), the anatomical prosector at Leipzig.[48] In his work *Exercitationes Physiologicae* (1668-1677), Bohn accepted the need for chemical explanations, especially concerning the digestive processes. He postulated, for example, that the saliva prepared the foodstuffs for their gastric digestion. For Bohn, the saliva's watery portion extracted salty particles from the food, while the oily and spiritual salivary components combined with the sulphurous parts of the nutriment. The entire digestive process consisted of the disintegration of the ingested material into fine atoms.

Fig. 13. Pancreatic fistula of R. de Graaf.

For Bohn the pancreatic juice did not appear to boil when combined with alkali and was thus not acid. Moreover, he denied that the bile was of alkaline nature and secreted from the walls of the gallbladder. He correctly deduced that the liver was the real source of bile and also refuted the existence of a nervous fluid circulating in the nerves.

Bohn's ideas reflected the influence of both the English physician *Thomas Willis* (1621-1675) and the Italian anatomist *Marcello Malpighi* (1628-1694). He was one of the most important experimental physiologists before Haller, expressing more sympathy for iatrophysics and atomism than iatrochemistry. Bohn's critique, however, was still only tentative because of a lack of proper methods of investigation.

3.
THE DISCOVERY OF THE CIRCULATION OF THE BLOOD
(Ibn an-Nafis, Servetus, Colombo, Caesalpino, Harvey and De Waal)

During the seventeenth century, physiology was substantially enriched by a series of great discoveries. Among them, William Harvey's explanation of the blood circulation was undoubtedly the most important and momentous one. His new theory interpreted the movement of the blood in a closed circle through the heart and vessels, and it served as a magic key for the understanding of numerous, hitherto unexplained observations of the heart, arteries and veins.

There is something peculiar about Harvey's discovery. One might say that it could have been made many decades prior, since the gross anatomical characteristics of the cardiac chambers and valves, as well as the great vessels and venous valves were previously known. Even the opportunities to observe a living, beating heart in cattle and other animals before slaughter had not been missed before the advent of Harvey. Moreover, the pulmonary circulation had already been recognized.

Thus, what was needed to achieve the "discovery" was essentially a new idea together with a renewed and persistent attention and observation of living nature. In addition, courage was needed in order to avoid assimilating Galen's ideas without proof, because "no author can be older than nature herself." How and to what extent Harvey gathered the unassailable proof for his circulatory theory will be described in detail, but the discovery had important antecedents. A large number of recent and older accounts depicting the background for Harvey's innovation have become available.[49] Due to limitations of space, I shall only provide a short survey of the development of the problem of circulation, and refer for the details to the various monographs.[50,51]

From their relatively advanced knowledge of the heart and large vessels, the Hippocratic physicians came to the conclusion that the blood prepared in the liver was sent across these vessels to the various organs. In spite of several suggestive passages, from which certain authors concluded that the Hippocratic school was aware of the circulation,[52] one can definitely affirm that these interpretations are erroneous.[53,54]

Irrespective of the textual interpretation, it would be very strange indeed if posterity, from classical times onwards, should have misinterpreted these dubious passages continually. After all, the Hippocratic writings were subjected to careful analyses and studies for many centuries.

In Plato, whose physiological conceptions coincided to a great extent with the ideas expressed in the Hippocratic work entitled *Περι Χαρδιης, there is no allusion to a theory of circulation. The same can be said for the views of Aristotle and the Alexandrian physicians, who did not entertain any theories of a circular movement of the blood.[55]

The notions of Galen concerning the movements of the blood and the role of the heart have already been described in detail. According to the Galenic conception, all blood-carrying veins originated in the liver while the blood and spirit-carrying arteries had their beginnings in the heart. In this scheme, the blood prepared in the liver flowed peripherally in small amounts through the veins into the various bodily organs, where it was consumed. A portion of this blood circulated from the liver to the right ventricle of the heart, where it became divided into two portions. One part went to the lungs

**On the Heart*

through the pulmonary artery while the other one flowed across the inter-septal pores of the heart into the left ventricle. Subsequently, this blood, mixed with pneuma and heated, departed from the left ventricle and flowed to the aorta, from where it reached the lungs and the periphery of the body. In turn, the arteries communicated with the veins through a series of anastomoses, which allowed some pneuma to enter the veins while the arteries received some blood.

For nearly fifteen hundred years the basic Galenic conception of circulation maintained its almost unquestioned validity. Although the Galenic ideas did undergo change especially among Islamic writers, the fundamental scheme could be found virtually unchanged in the Islamic, Scholastic and Renaissance physiology, including the writings of Vesalius. But there was one exception to this statement of fact. In 1924, a young Egyptian physician named Moyyi el Din el Tatavi, discovered a hitherto forgotten thirteenth century commentary written by the Islamic physician *Ibn an-Nafis* (ca. 1205-1288) on the *Canon* and anatomy of Avicenna. In this text, Ibn an-Nafis expressed some unmistakable doubts about Galen's ideas of the blood circulation.[56]

Wrote Ibn an-Nafis: "When the blood in this (right) chamber has been refined, it must reach the left one where the (vital) spirit originates. Now, there is no communication between these two cavities, in spite of what some have hitherto believed, not even invisible pores, like Galen had postulated. The pores of the heart are solid (impermeable) and its substance thick. Therefore, the attenuated blood reaches the lung through the vena arteriosa (pulmonary artery), where it spreads across the pulmonary substance and mixes with the air. Subsequently, the finer parts of the blood, mixed with air, flow into the arteria venosa (pulmonary vein) in order to reach the left cardiac chamber."

From this passage it is clear that Ibn an-Nafis had distinctly recognized the smaller, pulmonary circulation, in deliberate opposition to the views held by both Avicenna and Galen.[57] Thus, the Islamic physician seems to have been the first to postulate a circulation through the lungs. Unfortunately, the innovation was apparently not noticed and therefore wasn't incorporated into the available medical literature. In these texts, only the virtually unchanged Galenic doctrine of circulation was consistently reproduced.[58]

Like the Hippocratic writings, a controversy regarding Leonardo da Vinci's possible knowledge of the pulmonary circulation has also emerged. Some authors have offered arguments in favor of this hypothesis[59] while other scholars have considered it highly improbable or absolutely false.[60] Another historian has admitted that Leonardo was truly at the threshold of this circulatory discovery but never went beyond mere suggestions.[61] To be sure, Leonardo accurately described the action of the cardiac valves in several passages of his writings. But the question concerning the whereabouts of the blood in the lungs — obvious to us today — may not have impressed Leonardo, who was still laboring under the impact of the traditional teachings.

Even Andreas Vesalius, who had observed the impenetrable cardiac septum with his own eyes, did not conclude from this evidence the real path of the blood through the lungs. Instead, the great anatomist admired the power of God which conducted the blood from the right to the left ventricle across invisible channels. Whether the Vesalian statement was a sarcastic answer which concealed a deeper knowledge of the problem is highly doubtful.[62]

The merit for having again postulated the existence of a smaller circulation through the lungs has been hitherto ascribed to the Spanish physician *Michael Servetus*. (See Fig. No. 14.) Born in Villaneuva, Aragon, Servetus showed a burning interest for theological and scientific questions which carried him throughout Europe. He studied anatomy with J. Sylvius and Günther von Andernach at Paris, probably somewhat later than Vesalius. In time, Servetus' considerable medical reputation allowed him to become temporarily the personal physician to the archbishop of Vienna. But his true inclination was directed towards existing theological problems, and the purpose of his anatomical work was merely to demonstrate, through the wonders of creation, the greatness of God, and glorify Him for the conception of the human body.[63]

Fig. 14. Michael Servetus (1511-1553).

Servetus' theological writings showed him to be an intellectually independent thinker. In his book *De Trinitatis Erroribus* (1531), he refused to accept the doctrine of the Divine Trinity, while in another work, *Restitutio Christianismi* (1553), he expressed further heretical views which brought him soon into conflict with the Inquisition. Servetus committed the unfortunate mistake of sending one of his books to the French theologian and reformer *John Calvin* (1509-1564) at Geneva, with whom he had previously corresponded. Calvin promptly denounced Servetus to the Inquisition at Lyons. The anatomist fled and in his flight passed through Geneva where he was recognized while attending church. He was promptly arrested and jailed.

Servetus suffered greatly during his imprisonment, but did not recant from his previously expressed theological views, an attitude which led to his execution on October 27, 1553. He was burned at the stake in Geneva together with copies of his *Restitutio*. Almost all the remaining copies were also collected and burned by the Inquisition.[64]

In the anatomical considerations of this work, we read the following: "The passage of blood does not occur, as commonly believed, through the cardiac septum, but is carried instead by a longer route across the lungs, in a most remarkable and ingenious fashion. The blood is elaborated in the lungs, where it acquires a light yellow color, and from the arterial vein (pulmonary artery) it flows to the venous artery (pulmonary vein). This view is confirmed on account of the remarkable size of the arterial vein, which would not exhibit such dimensions and strength in order to supply solely the nourishment to the lungs."

Thus, Servetus clearly expressed the opinion that the blood could not flow through the septum, but rather found its way across the lungs to the left portion of the heart. To be sure, this belief was expressed on purely theoretical grounds as in the case of Ibn an-Nafis, without the support of any new evidence.[65,66]

A few years after Servetus' *Restitutio,* a posthumous work by the Italian anatomist *Matteo R. Colombo* entitled *De Re Anatomica* was published in 1559.[67] In it, Colombo had written the following: "The blood is directed from the artery-like vein to the lungs, where it is attenuated, and together with air this blood is brought back to the left cardiac chamber through the vein-like artery."

Colombo claimed to be the first author to have arrived at this conclusion. He supported his idea by referring to observations made in animals, which showed the pulmonary vein always filled with blood. This fact was in conflict with the prevailing idea that only air was carried to the left ventricle. Moreover, Colombo declared that the intraventricular septum had no pores to facilitate the passage of blood between the two cavities. He also determined that the arteries would pulsate during the systolic emptying of the heart.

On the other hand, Colombo believed that he had observed the warming of the blood to occur in the left ventricle of the heart: ". . .following a cut into the left cardiac cavity (of a living dog), you will encounter a danger: there is such a great amount of heat in this chamber that it will prove to be intolerable to the exploring hand." Apparently the theoretical expectations of Colombo were stronger than his factual perception of the phenomena.

Regarding the priority of the rediscovery of the pulmonary circulation, the only clearly ascertainable facts are as follows: Servetus had already sent his manuscript of the *Restitutio* to Calvin in 1546, while Colombo expressed his views between 1545 and 1548, according to accounts from his students.[68] It is probable that both authors, Servetus and Colombo, arrived at the same conclusion almost simultaneously but independently from each other, with

Colombo furnishing at least some experimental evidence for his ideas. The capstone for this physiological development was eventually furnished by Malpighi who in 1661 observed, with the help of the microscope, the unknown but necessary links between the arteries and veins of the lungs: the pulmonary capillaries.

Colombo's work received widespread attention and his theory of pulmonary circulation became well known. However, and in spite of the discovery, Colombo himself and succeeding anatomists remained true to Galen's interpretation, believing that the general circulation consisted of blood flowing peripherally in the veins. This phenomenon was even more remarkable in view of the fact that the existence of venous valves had already been pointed out by the Italian anatomist *Giambattista Canano* (1515-1579) to Vesalius in 1546.[69] Furthermore, these valves were also mentioned in a posthumous work of the Parisian anatomist J. Sylvius in 1555 and gradually became common knowledge among anatomists by 1564.[70]

In turn, Hieronymus Fabricius, who taught with Galileo at the University of Padua, devoted a whole treatise, entitled *De Venarum Ostiolis* (1603) to the valves of the veins. Unfortunately, Fabricius, unlike Harvey, did not draw any revolutionary conclusions from his observations. He thought that the purpose of the valves was to delay the flow of blood to the periphery and thus avoid the flooding of the upper and lower extremities.[71]

The first author to describe the venous distention at the distal end of a ligature was the Italian botanist and physician *Andrea Cesalpino* (1519-1603). Cesalpino, an enthusiastic Aristotelean, had already described accurately the valvular actions of the heart in his work *Questiones Peripatheticae* (1571). He also coined for the first time the concept of "circulatio," apparently in analogy to the circular path of the distillation procedure.[72] In 1593, Cesalpino stressed the continuous return of venous blood to the heart and the outflow of blood from the heart to the arteries. Therefore, Cesalpino has been called the inaugurator, but not the discoverer of the circulation of the blood.[73]

In spite of a variety of interpretations, a majority of contemporary historians believe, in contrast with J.P. Arcieri[74] that Cesalpino never went beyond certain vague notions regarding the general blood circulation.

–The great English physician *William Harvey* (1578-1657) has the distinction of having been the first to discern correctly the circulation of the blood through the lungs and body. His treatise, entitled *Exercitatio Anatomica de Motu Cordis et Sanguinis,* published in 1628, clearly reflects Harvey's full realization of the significance inherent in his discovery.[75]

William Harvey (see Fig. No. 15) was born on April 4, 1578 at Folkstone, on the southern coast of England. In 1588 he went to the Royal School at Canterbury and then transferred to Caius College, Cambridge. From 1600 to 1602, Harvey studied under Fabricius at Padua, returning home after his graduation there in order to begin his medical career in London. In that city he became the physician of the philosopher Francis Bacon (1561-1626) and in 1632, official physician to the ill-fated king Charles I (1600-1649) whom he accompanied on his various war-connected travels. Finally, Harvey returned to private life and his studies. His last important work was the treatise entitled *De Generatione Animalium,* published in 1651, with its revealing dictum "Ex ovo omnia" on the title page. He died on June 3, 1657 at Rochampton and was buried on June 26 in the family plot at Hempsted.[76,77]

There are reasons for assuming that Harvey had already developed his ideas regarding circulation around the year 1615, and that he expounded them in his lectures of 1618 before the published work first appeared at Frankfurt in 1628. He had certainly brought back from Italy suggestions regarding the problem, taken from the writings of Vesalius, Colombo, Fallopio and Fabricius. Nevertheless, the proof of a "circulation" was his very own accomplishment.[78]

Fig. 15. *William Harvey* explains to Charles I his ideas concerning the circulation of the blood.

Harvey, in his *De Motu Cordis,* first expressed doubts regarding the prevailing conceptions of the blood movement. He wrote in the introduction of that work: "Why, I ask, when we see that the structure of both ventricles is identical. . .should their uses be imagined to be different, when the action, motion and pulse of both are the same?. . .And when we observe that the passages and vessels are severally in relation to one another in point of size

. . .wherefore should the one be imagined destined to a private or particular purpose, that, to wit, of nourishing the lungs, the other to a public and general function?" Then, Harvey continued: "Why was nature reduced to the necessity of adding another ventricle for the sole purpose of nourishing the lungs?. . .how comes it that spirits and fuliginous vapours can pass hither and thither without admixture or confusion? Good God!, how should the mitral valves prevent regurgitation of air and not of blood? Why should the pulmonary vein be presumed to be made for many — three or four different usages? For they will have it that air passes through this vessel from the lungs into the left ventricle; that fuliginous vapours escape by it from the heart into the lungs, and that a portion of the spiritous or spiritualized blood is distributed by it to the lungs for their refreshment. . .If fumes or fuliginous vapours and air permeate this vessel, as they do the pulmonary bronchia, wherefore do we find neither air nor vapours when we divide the arteria venosa? Why do we always find this vessel full of sluggish blood, never of air?"

Harvey concluded that: "If the pulmonary vein were destined for the conveyance of air, it has the structure of a blood vessel here. Nature had rather need of annular tubes, such as those of the bronchia, in order that they might always remain open. . .Still less is that opinion to be tolerated which . . .supposes the blood to ooze through the septum of the heart from the right to the left ventricle by certain secret pores. . . .But, in faith, no such pores can be demonstrated, neither in fact, do any such exist."[79]

The English physician eliminated all these inconsistencies by using the results of his careful observations, made on the hearts of animals and their great vessels in his vivisection experiments. On close examination of these facts, Harvey recognized that the older conceptions were erroneous. For example, Vesalius' assumption that the heart extended during contraction and thus behaved like a cupping glass absorbing blood, was entirely false. Instead, the heart chambers became hard during systole and ejected blood.

Harvey pointed out that if an artery was punctured during the experiments, it spurted out the blood mainly during the cardiac systole. Thus, for him the pulse originated in the arteries themselves as a result of their systolic filling with blood which allowed a certain expansion of these elastic tubes to occur. Moreover, both auricles contracted before the ventricles, sending their blood into the ventricular cavities.

Likewise, Harvey postulated that during cardiac systole the blood was impelled into the aorta on the left side and towards the lungs on the right side. In the porous lung tissue, the blood flowed continuously from the pulmonary artery to the pulmonary vein, since the cardiac septum was solid and did not allow any blood to ooze through.

From the aorta, the flow of blood travelled across the arterial system to the bodily organs and returned in the veins, a veritable "circulation" of the blood, as Harvey termed it. He estimated for the first time the stroke volume for the left cardiac ventricle at about half an ounce or fifteen grams, concluding that the quantity of blood flowing through the aorta in one day was many times greater than the amount which the liver could manufacture from the ingested nourishments.

In the veins, Harvey determined that the blood flow occurred in only one direction: towards the heart. He demonstrated this afferent circulation by placing a moderately tight ligature used for venesection around the arm, and observing the gradual venous engorgement distally from the point of occlusion. In the venous circulation, the valves located in the various veins actually prevented the blood from flowing back towards the periphery, thus facilitating its movement towards the heart.

At the end of chapter fourteen in *De Motu Cordis* Harvey wrote: "...it is absolutely necessary to conclude that the blood in the animal body is impelled in a circle, and is in a state of ceaseless motion; that this is the act or function which the heart performs by means of its pulse, and that it is the sole and only end of the motion and contraction of the heart."

Harvey's discovery was of epoch-making significance for the subsequent development of physiological thought.[80] It was based almost entirely on anatomical and vivisection observations (for example, ligatures) without considering at all chemical or physical viewpoints. Hence, this seventeenth-century landmark in physiology is a classical example of the so-called "animated anatomy" and the success which is attainable in the explanation of organic functions through a sagacious use of the available anatomical data.

If one analyzes Harvey's work not in the light of his accomplishments, but according to his explicit assumptions, a different picture readily emerges. Under those circumstances, Harvey's leading ideas appear, to a great extent, to be bound by tradition, reflecting an aspect of his personality which was typically baroque in nature. Important for arriving at such a conclusion is Harvey's later work, *De Generatione Animalium,* where he clearly acknowledged his adherence to the Aristotelean ideas.[81] Moreover, Harvey followed Aristotle's methodology of comparative anatomical observations, an intensive look at the various natural events, and the examination of different animals through dissection, injection techniques and the use of the magnifying glass.[82]

Regarding theory, Harvey stressed the supremacy of the senses as the source of truth rather than reason which he considered subservient to the empirical knowledge already gathered.

In Harvey's case, however, the facts of observation did not provide the necessary key: the idea of a circulation. This concept was therefore derived from the principle of circular motion, the essential form which Aristotle conceived for his entire cosmological scheme. The circular movement of the blood in the organism mirrored in man, the microcosm, those planetary motions occurring in the universe or macrocosm. The circulating blood passed the heart, source of vitality and heat, and was thus regenerated and properly restored.[83]

The micro-macrocosm analogy was also prevalent among the contemporary neoplatonists, such as the English physician and alchemist *Robert Fludd* (1574-1637) who in 1629 was one of the first to accept Harvey's doctrine of circulation.[84] Even earlier, the Italian philosopher *Giordano Bruno* (1548-1600) had expressed in 1591 such an analogy about the circular motion of the blood.[85] Therefore, Harvey accomplished a surprising synthesis of observations, some calculations — all part of a methodology we could call modern — together with speculative and metaphysical considerations. Harvey's dynamic thought, concerned with movement instead of static conditions and relationships rather than isolated findings, reflected to a great extent the contemporary thought.[86]

Harvey's discovery was soon widely known,[87] and prominent figures such as Borelli, Descartes, van Helmont and F. Sylvius learned about its main postulates. Lively arguments for and against the Harveian scheme were also quick to appear. The controversies became more vehement in the years 1648-1651, involving primarily Harvey himself, the French physician *Jean Riolan, Jr.* (1580-1657) and the English physician *James Primerose* (ca. 1590-1659).

Of special significance for the eventual recognition of Harvey's discovery were the two letters written by the Dutch physician *Jan de Waal* (1604-1649) or *Walaeus* to the Danish anatomist *Thomas Bartholin* (1616-1680) in 1640. In these writings, de Waal furnished new and convincing experimental proof for the passage of blood from the arteries to the veins. If, in animals, the femoral artery was ligated experimentally at the level of the knee joint, and the femoral vein emptied and then also tied, de Waal pressed blood from the arterial to the venous system. Accordingly, he concluded that there were, in fact, vascular communications between the arteries and veins.

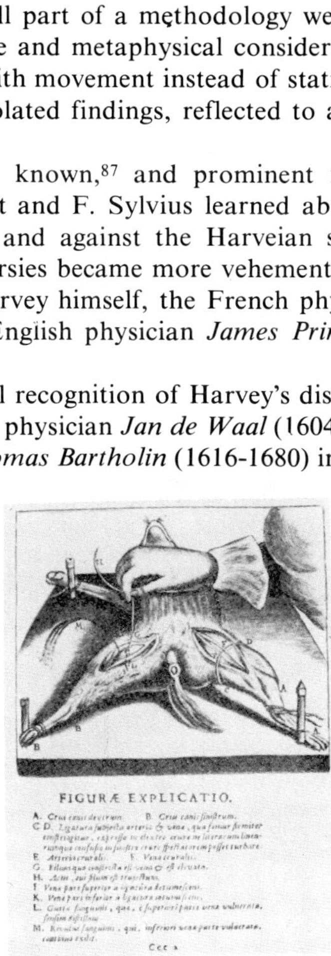

Fig. 16. *Experiments of J. de Waal* to confirm the circulation of the blood.

Moreover, de Waal performed other experiments in which he proved the unidirectional flow of the blood in both the arteries and veins. If the dissected femoral vein in a dog was pulled up on a string, the elevated proximal end of the vessel emptied while the distal gradually filled with blood and became distended. When the same experiment was performed on an artery, opposite effects were noted. (See Fig. No. 16.)

De Waal accurately described the portal circulation. He also wrote that the blood only squirted out of the arteries during systole, and denied the existence of spirits or air in the left ventricle of the heart which was solely filled with blood. He was convinced that the circulation merely occurred through forces of suction and thrust.

Hence, de Waal furnished a series of essentially new experimental proofs for Harvey's circulation of the blood, and became one of the most important supporters of the Harveian scheme. Nevertheless, the opposition to the new circulation was not easily squelched and continued to come from the entrenched Galenists. For example, Primerose attempted to refute de Waal's conclusions, but without lasting success. Finally, in 1661, fourteen years after Harvey's death, Malpighi made his anatomical discovery of the capillary vessels in the lungs. This event provided the most important piece of evidence for the correctness of Harvey's blood circulation.[88]

In the meantime, the knowledge concerning lymphatic vessels had increased considerably. In 1622, the Italian anatomist *Gasparo Aselli* (1581-1626)[89] of Cremona observed the lymphatics of the mesentery of a dog, but believed that these vessels emptied into the liver. Then, in 1651, the French physician *Jean Pecquet* (1622-1674) observed and described the ductus thoracicus or common lymph trunk, and the cisterna chyli in animals. Two years later, T. Bartholin reported for the first time the existence of the thoracic duct in man.

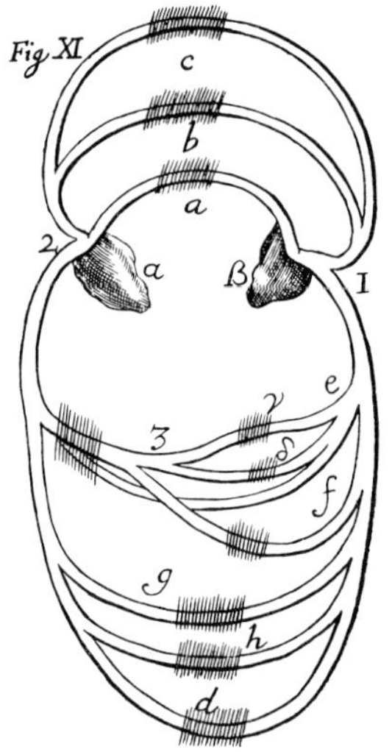

Fig. 17. Circulation Scheme of Bartholin (1655-1738).

The connection between the intestinal lymphatics, the thoracic duct and the venous system was only gradually recognized, and the existence of lymphatic vessels ascertained for the entire body. Whether the priority for these discoveries should be given to Bartholin or the Swedish naturalist *Olaf Rudbeck*[90] (1603-1702) is difficult to decide. Both authors made their findings independently of each other, although Rudbeck was more concerned with the anatomical aspects of the lymphatic system while Bartholin searched more for the physiological role played by the lymph.[91] (See Fig. No. 17.)

In any event, two crucial questions concerned with the movement of fluids, namely blood and lymph, had been markedly clarified around the middle of the seventeenth century. In truth, however, the discovery of the blood circulation was more than just another new finding, it was the starting point for modern physiology. Harvey's innovation forced a total re-orientation about almost all other organic functions, especially the cardiac activity, the vascular flow, the role played by the lungs and the liver, as well as the source of animal heat. Solutions for these problems were eagerly sought for during the remaining years of the seventeenth century, leading to the use of mechanical models for the explanation of physiological phenomena.[92]

4.
THE APPLICATION OF MECHANICAL PRINCIPLES FOR THE SOLUTION OF PHYSIOLOGICAL PROBLEMS: THE SEVENTEENTH-CENTURY IATROPHYSICISTS

The significant progress in anatomical knowledge attained during the sixteenth century provided the basis for the rapid advance of physiology in the next century. To be sure, a series of political, cultural and intellectual changes all contributed decisively to this advance. First in Spain, and later in France, the political absolutism with a new social structure made its appearance while Holland achieved its highest development and Germany was converted into the theater of the Thirty Years War. The spirit and vitality of the new age, called the Baroque, was in many ways different from the preceeding one, clearly turning away from the ideals and goals of the sixteenth century.[93]

The chronological boundaries of the Baroque are not rigid, but the period generally extends from the second half of the sixteenth century to about 1700. Historians of art consider the period from the death of Michelangelo in 1564 to the death of the painter Daniel Poeppelmann in 1637, while the historians of physics lengthened it from the appearance of Copernicus' *De Revolutionibus* (1543) until the death of Isaac Newton (1727).[94] Finally, in medicine the Baroque spirit clearly emerged only during the seventeenth century.

The Baroque period witnessed in Europe the Counter-Reformation movement, during which Catholicism significantly regained its former stature. To a certain extent, one can also talk about a conscious return to the medieval ideals since successful efforts were made to restore the older social order and place the Church in its previous position of authority.

This attitude was reflected in the great triumph of the Jesuit order which reached its apex of power in the period under discussion. The fanatic persecution of dissidents was reflected in the religious wars and the excesses committed during the witchcraft trials which overshadowed any previous expressions of intolerance. The last witch was burned as late as 1749 in Würzburg. In spite of this situation, magic, astrology, alchemy, witchcraft and demonology flourished as never before.

The seventeenth century was also a period of hedonistic pleasure and zest for life. The Baroque life style combined, like the contemporary architecture, sensuality and transcendentalism. An exuberant imagination together with a veritable intoxication of the senses, placed side by side animals and plants, angels and demons, God and the world, in plastic and ornamental schemes. The royal residences became the representatives of the Baroque culture, while the cities lost a great deal of their former independence and were integrated into new national states. Thus, the urban culture of the Renaissance faded away, and the center of the new life style shifted to the small societies living at the royal courts.

The Baroque spirit also expressed itself in science. The natural sciences changed their attention from the static to the dynamic aspects of the natural phenomena. Attention to metaphysical questions was balanced by a strong interest and concern with secular subjects, especially nature.[95] Among other accomplishments, Leibniz and Newton invented the calculus, Harvey discovered, as already discussed, the circulation of the blood, the Italian physicist *Galileo Galilei* (1564-1642) established the foundations of dynamic physics with his laws of motion.

The French philosophers *René Descartes* (1596-1650) and *Blaise Pascal* (1623-1662) were also mathematicians and physicists. *Johannes Kepler* (1571-1630), one of the most prominent mathematicians and astronomers of his day, clearly expressed the relationship between the occurrences of this world and the divine, natural order which they reflected. He wrote: "In truth, I must say that whenever I follow in my mind the marvelous order in which one thing is followed by another, it is like reading a divine declaration, expressed in essential worldly objects instead of letters: Man extend your reason to this world in order to understand these divine things" (from a 1604 calendar).

In the year 1620, the philosopher Sir *Francis Bacon,* chancellor to Queen Elizabeth I (1533-1603), published his work *Novum Organum.* In it, he stressed emphatically — although he was not the first to do so — the value of the inductive method and its systematic use in the natural sciences. In marked contrast with the religious restorations, the sciences had gradually begun to argue with Aristotle, Galen and other ancient authorities.[96] Galileo's life-long fight was directed against the doggedly defended theories of the Greek peripatetic philosopher, a fact which contributed to Galileo's conflicts with the Inquisition.

The scientific interest shifted to the research of individual problems, and the experimental method was now systematically employed. Simultaneously, there was an indiscriminate collecting of all kinds of observations, the emphasis being on the strange and the curious rather than ordinary events and objects. So-called curio cabinets were established, containing many specimens of anatomical deformities and monstrosities.[97]

The scientific activities developed mainly outside the still very conservative universities which depended on the State or the Church. For this purpose, various associations and *academies* were formed by interested people for the purpose of exchanging ideas and promoting scientific work. At Rostock, a *Societas Ereunetica* was created in 1622 under the motto: "Per inductionem et experimentum omnia," while the *Academia Naturae Curiosorum* originated in 1652 (later called *Kaiserlich Leopoldinisch Carolinische Deutsche Akademie der Naturforscher* or Royal Leopold and Caroline German Academy of Natural Scientists).

Another institution was the Italian *Academia del Cimento,* established in 1657 at Florence, among whose members were Galileo's last student, the mathematician *Vincenzo Viviani* (1622-1703), the mathematician and physicist *Giovanni A. Borelli* (1608-1679), the physician and poet *Francesco Redi* (1626-1667) and the already mentioned famous anatomist Marcello Malpighi. In London, the *Royal Society* was instituted in 1662, and in Paris the *Académie des Sciences* appeared in 1666. Journals such as the English *Philosophical Transactions,* the French *Journal des Savants* and the German *Acta Eruditorum,* provided fast information regarding new scientific discoveries, often in their pages in the form of letters from the authors.[98]

Hence, the seventeenth century exhibited sharp intellectual contrasts since medieval and modern elements were locked in a visible struggle for supremacy. An internal dichotomy, which often manifested itself in some of the individual scientific personalities of the day, lay at the bottom of the life style and thought of this eventful period.

In this period, physics and chemistry received an extraordinary impetus and therefore the seventeenth century has been justly called the age of *classical scientific work.* Since the time of Alhazen, Roger Bacon, Maurolycus and della Porta, the science of optics had made tremendous progress, especially through the work of Galileo, Kepler and Descartes. Conversely, the mechanics of solids and liquids were advanced by Galileo himself, while knowledge of the physical behavior of gases increased with the work of the German physicist *Otto von Güricke* (1602-1686), the Italian mathematician *Evangelista Toricelli* (1608-1647), the Dutch astronomer *Christiaan Huygens* (1629-1695) and the already mentioned English chemist Robert Boyle. With these men, the experimental method and the mathematical treatment of scientific problems became highly successful.[99]

As noted before, the gradual separation of alchemy and chemistry occurred. It was a century in which the spirit and method of exact scientific research finally came to the fore. The inductive method and experimental analysis were consciously employed, a prerequisite for the solution of physiological problems. Fruitful beginnings were made and great triumphs achieved in physiology during the seventeenth century, and it was indeed a classical period for physiological research. The details of many of these advances follow.

Under the influence of the success achieved in science and technology with physical methods, medicine likewise attempted to attain a theoretical basis founded on physics and especially mechanics during the sixteenth and seventeenth centuries.[100] This direction has been called *iatrophysics* or *iatromechanics,* as well as iatromathematics. Among its representatives were the Italian physicians *Santorio Santorio* (1561-1636) and *Giorgio Baglivi* (1668-1707), the Italian anatomist *Lorenzo Bellini* (1643-1704), the British physician *James Keill* (1673-1719) and Borelli. They were the most famous representatives of a great number of contemporary scholars who applied mathematico-physical principles to the solution of medical and biological questions.[101]

In this sense the term iatrophysicist had a broader meaning than was usually accepted. All of these authors wanted to measure and calculate the phenomena according to Galileo's dictum: "Measure all that is measurable, and make those things measurable, which have hitherto not been measured."

Mathematics took the place previously occupied by philosophical speculation. The inquiries were not only qualitative but now also quantitative. Furthermore, Galileo did not ask *why* the bodies would fall, but *how* they were falling. His epoch-making approach was not so much the insistence that experience had to be grounded on the observation of nature. More important, Galileo searched for the quantitative, objective and universal lawfulness — the laws of nature — through abstraction from the empirical reality of isolated phenomena. Because of his success in other fields, Galileo indirectly contributed much to the promotion and application of physical methods of measurement to organic functions, especially in Italy.[102]

One of the first scholars who advocated the use of measurements for the determination of the pulse, respiration and urinary excretion, was the German clergyman and philosopher *Nicholas of Cusa* (1401-1464).[103] The effort was carried out with even more zeal by Santorio, professor at Venice and later Padua.[104,105] Santorio invented an instrument to count the pulse — the "pulsilogium" or pulsimeter — consisting of a pendulum with a sphere of lead at the distal end. In addition, he used a thermometer to measure the body's temperature in accordance with the classic example of Heron.[105]

More interesting were Santorio's pursuits regarding metabolic questions. For more than thirty years, he compared his body weight and the differences between the ingested food and the excreta — namely stools and urine. To aid in the computations, Santorio used his now famous balance chair in which he often ate and slept (Fig. No. 18). During these experiments, he determined that the body lost continuously a certain amount of weight — 1¼ Kg in 24 hrs — in the form of "insensible perspiration." His results were published in the famous work *Ars de Statica Medicina* (Venice, 1614) which laid the foundations for the modern studies of metabolism.[107]

Fig. 18. Santorio's metabolic balance.

The physico-mechanical approach to physiology and medicine was significantly supported by the Cartesian philosophy. Descartes, in turn, had been encouraged to develop his mechanical views because of the great progress being made in the physical sciences. Born in La Haye near Tours,[108] Descartes became not only a great philosopher and thinker, but also an important mathematician and physicist.[109] He established the foundations for the method of analytical geometry and system of coordinates, and independently of the Dutch mathematician *Willebord Snell (Snellius)* (1591-1626), he discovered the laws of refraction.

In addition to his physico-mathematical activities, Descartes also carried out anatomical observations as evidenced by some of his records which have been preserved. His importance for the development of physiology lay in the fact that he was the first author ever to attempt to explain all bodily functions according to purely mechanical laws: *De Homine* (1662).[110]

According to Descartes, the human body was essentially a machine, functioning according to mechanical laws like a clockwork or a highly perfected automaton. This human machine was inhabited by a divinely created soul, which outside of commanding the voluntary motions, had no connection with the bodily operations.[111]

In the Cartesian scheme, the various organic functions were viewed separately as individual mechanisms, and thus a strange amalgamation of Galenic notions and physical concepts took place. Heat was the only principle of motion in the body, a quality produced by the cardiac fire and responsible for the circulation of the blood. Althought Descartes was familiar with Harvey's ideas about circulation and accepted them in general, he nevertheless postulated a highly personal theory regarding the blood's movements.[112]

Descartes envisioned that the heart's heating action brought the blood almost to the boiling point, allowing it to expand. At the same time, he felt that the air taken in at the level of the lungs caused a kind of effervescence which allowed the blood to flow into the arteries, distending them and thus creating the pulse. The cardiac valves, on the other hand, avoided the reflux of blood into the heart's chambers, and also regulated the evaporation of waste products. Finally, the direction of the blood vessels and the size of the pores in the various organs determined the blood flow and the rate of absorption of nutritious elements.

For Descartes, some small particles travelled with the blood to the brain and penetrated into the cerebral ventricles and the pineal gland through some fine openings. Once in these structures, the particles were converted into the animal spirits, a sort of fine air or liquid which could flow across the spongy nervous substance and enter the hollow nerves. In turn, this nervous principle or spirit circulated through the nerves to the body's muscles, where it was responsible for motion by inflating them.[113]

Descartes conceived of the pineal gland as the seat of the soul, as well as the center of imagination and perception (see Fig. No. 19). Each sensation was produced through a mechanical action at the peripheral endings of the nerves which in turn was transmitted to the brain.

Unquestionably, Descartes provided the essential impetus for the modern mechanistic thought in biology.[114] However, the Cartesian method of deducing from first principles the structure and laws of the cosmos provided a less modern emphasis on *ratio* (reason) instead of *observatio* (empiricism), pointing towards the beginnings of the Enlightenment.[115]

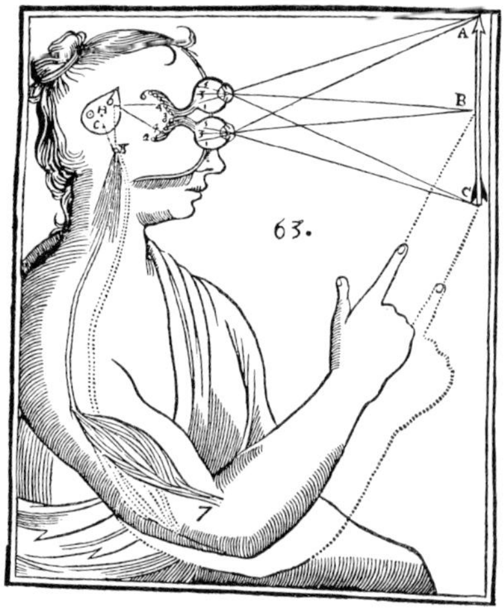

Fig. 19. Descartes: contact between the sensory organs and the soul at the level of the pineal gland.

On the other hand, Descartes was a representative of the baroque, partial to a dynamic interpretation of nature and using some Aristotelean concepts such as the ideas concerning cardiac action while rejecting others — doctrine of elements, qualities and faculties. Descartes uncritically accepted many portions of Galenic anatomy and physiology, abandoning the classical views only when they failed to harmonize with his mechanical views, as in the case of the blood circulation.

As in the rest of nature, the human machine functioned blindly according to predetermined and goal oriented laws of nature which had been instituted by God himself. This fundamental Cartesian idea encouraged the analysis of individual physiological processes, in order to discover their cause and effect relationship, with the expectation that such knowledge would make man the master and possessor of nature.

In 1649, Descartes went to Stockholm at the invitation of Queen Christina of Sweden (1626-1689), but he died the following year. The strongly speculative character of the Cartesian physiological conceptions coupled with his inadequate anatomical knowledge, were recognized by the contemporaries of Descartes. Among them was the Danish physician *Niels Stensen* (1638-1682) who severely criticized the Cartesian ideas during his sojourn at Paris in 1665. However, the mechanical philosophy found many adherents who attempted to follow in Descartes' footsteps.

The work of the famous physicist Giovanni A. Borelli of Pisa was also important for the development of physiology (see Fig. No. 20).[116] Under the direction of his teacher, the mathematician *Benedetto Castelli* (1577-1644), Borelli acquired a great amount of knowledge in the field of mathematics. He was one of the most notable members of the Florentine *Academia del Cimento,* and after 1672 lived in Rome under the patronage of Queen Christina of Sweden, who had assembled a group of notable scholars there.

Fig. 20. *Giovanni A. Borelli* (1608-1679).

In Rome, Borelli concluded his great work entitled *De Motu Animalium* in which he applied, on a grand scale, mechanical principles to physiological problems.[117] Although large parts of the treatise seemed to have been ready for publication in 1662, the book did not appear until 1680-81, a year after his death.

Borelli's work contained a series of investigations about the flight of birds, the swimming of fishes and the motion of human limbs. In addition, there were important descriptions regarding the mechanics of respiration, the phenomenon of muscular contraction, the work of the heart and the circulation of the blood. To be sure, many of Borelli's experiments, measurements and calculations contained numerous errors and deficiencies, but they should nevertheless be viewed as the initial and tentative attempts taken in the correct direction of physiological research.[118]

The main problem which attracted Borelli's attention was the question of muscular action on the skeleton and the mechanism of muscular contraction in general. Like Vesalius and Stensen before him, Borelli believed that the contractile element resided in the muscles themselves. These structures were composed of fine fibres covering the walls of small, rhomboid-shaped vesicles, arranged in long chains. (See Fig. No. 21.)

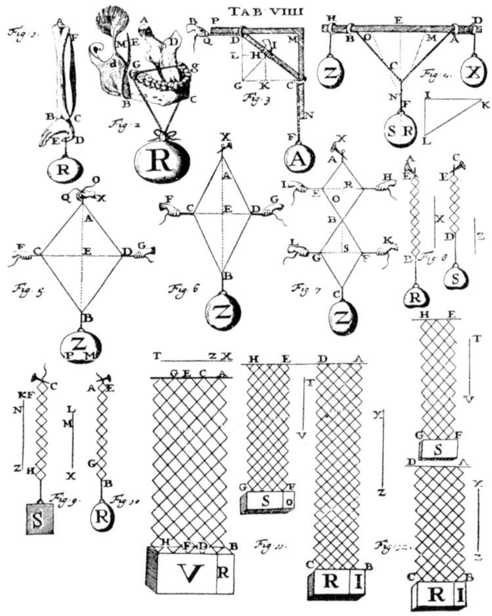

Fig. 21 *Borelli* — Geometrical and mathematical approach to the mechanical contraction of the muscles.

In accordance with the morphological disposition of the muscular fibres, Borelli presented a geometrical explanation of the muscular contraction: the rhomboid spaces all modified their diagonals, thus shortening the chains. For the contraction to occur, however, the author postulated the existence of a chemical process of fermentation. According to Borelli, the nerve endings within the muscle issued forth a certain "succus nerveus" which interacted with the blood present in the muscular fibres. The result was the production of an effervescence which inflated and shortened the fibres.

Borelli declared that the older theory of muscular contraction which had postulated an influx of air or animal spirits into the muscle, was no longer valid. He carried out some experiments to prove his contention by cutting through a muscle in the living animal which had been submerged in water. No air bubbles escaped to the water's surface when the muscle was contracted.

The same principles of muscular action were applied by Borelli to the activity of the heart. For him, this organ was automatically in motion, thanks to its muscular nature. During contraction, the cardiac fibres swelled and protruded into the chambers, an action which necessarily diminished the volume of these cavities and promoted the expulsion of blood. Moreover, the cardiac contraction, by impelling the blood into the arteries, forced these more elastic structures to stretch in order to receive the flow. Thanks to their muscular nature, the arteries responded to the distention with a contraction of their own, which forced the blood to circulate further towards the periphery.

Equally interesting was Borelli's attempt to calculate the work of the heart. According to the Italian physicist, the left ventricle could be balanced with a weight of 3000 lbs (1 Roman lb=327 gm). Another factor to be considered in the calculations was the resistance offered by the vessels to the flow of blood. Comparing the weight and performance of the heart muscle with that of the temporal and masseter muscles, Borelli concluded that the force of the heart during each contraction was about 135,000 lbs. Actually, a Dutch mathematician, *Daniel Bernoulli* (1700-1782),[119] was the first to make an approximately correct calculation of the cardiac action by considering it the product of force (weight of the blood) and the length of the impulse (height of the column of blood expulsed by the contracting heart).[120]

In any event, Borelli's pioneering efforts greatly stimulated subsequent scholars. He had completely discarded the existence of an immaterial principle in the nerves, postulating instead the flow of a liquid, the "succus nerveus", which mediated all nervous actions.

In another area, Borelli used the thermometer to determine whether the body heat really developed in the left cardiac ventricle. He discovered that the temperature of the heart was not any greater than that of other bodily organs.

For Borelli, like Leonardo da Vinci before him, the inflow and outflow of air to the lungs occurred in purely mechanical terms because of the negative pressure created by the expansion of the thoracic wall. The significance of respiration was not due to the cooling effect on the heart, an idea postulated since antiquity and accepted by Descartes. Instead, Borelli declared that the air penetrated through the small membranes located at the distal end of the bronchi — Malpighi had already described the lung's microscopical structure in 1661 — and was then absorbed by the blood. This blood contained certain elements necessary for life, a fact which Borelli adopted from the experiments conducted by Otto von Guericke and Robert

Boyle in 1660,[121] in which small animals perished in the rarified air. Borelli explained the life-sustaining action of breathing by postulating the incorporation of spiral and elastic particles from the air into the blood which conferred to the latter a certain "motus intestinus."

Finally, Borelli made an attempt to interpret the secretory functions of glands in mechanical terms. Adopting a theory developed by his pupil Lorenzo Bellini, he compared the kidneys to a sieve which filtered the various components of the blood, trapping those elements needed for urinary excretion.

Borelli, like most contemporary iatrophysicists, was an adherent of neoatomism, which had begun with the writings of the German physician *Daniel Sennert* in 1636, the physician *Sebastian Basso* (fl. 1621) in 1621 and the French philosopher *Pierre Gassendi* (1592-1655) in 1647. In 1637, Descartes developed his doctrine of the three types of corpuscles, followed by the work of *Henricus Regius* in 1641. A few years later, in 1654, the British physician *Walter Charleton* (1619-1707) introduced Gassendi's atomism into England in a work entitled *Physiologia Epicuro-Gassendo-Charletoniana,* and Robert Boyle established corpuscular chemistry in 1661.[122] Towards the end of the seventeenth century, Boyle was followed by J. Bohn of Leipzig and the German physician and systematist *Friedrich Hoffmann* (1660-1742) of Halle with their corpuscular interpretation of the blood, fermentation and vitality in general.

Viewed together, Borelli's mechanical interpretations of the organic functions were of great importance for the progress of physiological thought. His approach was widely imitated throughout Europe.[123] Moreover, the idea of applying physical measurements and using mechanical interpretations to explain physiological events, signaled the introduction of a new and valuable method in biology. Its early success can be largely ascribed to Borelli's endeavors, since similar previous efforts of Leonardo da Vinci had failed because they were never expressed in print.

The culmination of the iatromechanical current in Italy occurred with Giorgio Baglivi,[124] who was greatly influenced by both Borelli and Malpighi. Since 1692, Baglivi was professor of anatomy at the Papal School, the "Sapienza" in Rome.[125] The pride which Baglivi felt in the accomplishments of the iatrophysical school was reflected in the following passages of his work entitled *Praxis Medica* (1699): "Since physicians began to examine the structure and actions of a living body, not by physico-mechanical and chymical experiments but by geometrico-mechanical principles, they have not only discovered an infinite number of things that were unknown to former ages; but have made it out, that a human body, as to its natural actions is truly nothing else but a complex of chymico-mechanical motions, depending upon such principles as are purely mathematical."[126]

Baglivi was influenced by the Cartesian mechanical philosophy, and like

Descartes, also searched for the motive power of the human machine. In addition to cardiac action, he declared that the meninges and their extensions, the "fibrae membranacea" (membranous fibres of the small muscles) pulsated continuously, while another muscular force resided in the "fibrae motrices" (voluntary muscle fibres).

For Baglivi, the nervous condition was a sort of oscillation of fine particles, transmitted along the nerves. He observed the circulation of the blood in the mesenteric vessels of a frog, using a magnifying glass, and described a greater speed of flow in the center of the vascular lumen. In another experiment, the heart of an animal was removed from the body and the persistence of its automatic contractions observed. Wrote Baglivi: "Even when the severed heart was cut into several pieces, these independent parts continued to contract regularly, much to the amazement of the observers."

A student of Borelli and contemporary of Baglivi, Lorenzo Bellini, attempted to develop a theory of fever on the basis of these various iatromechanical conceptions. In addition, Bellini was the first author since Bartholomeo Eustachio to describe the renal tubules and minor calyces in his *Exercitatio Anatomica de Structura et Usu Renum* (Florence 1662).

In Northern Europe and England, iatrophysics and iatromechanics also received considerable attention. James Keill carried out exact measurements of the expired air, following the example of Borelli's approach. Keill estimated the number of air cells in the lungs to be over one and a half billion, and attempted to calculate their total surface. Another English physician *James Jurin* (1684-1750), performed similar experiments related to respiration, optics and cardiac activity.

Henceforth, the principle of iatrophysics, that is, the application of mechanical and physical concepts to physiological problems, was to remain a viable approach to obtain further knowledge in the field.[127] Stensen, Willis, Mayow, Lower, Hooke and many other scholars used this method to explain the various vital actions, although they gradually combined the mechanical explanations with some chemical interpretations. Their lives and work took place in Northern Europe, primarily in England, and thus merits a special section.

Finally, another great scholar should be mentioned here: Johann Kepler. The distinguished astronomer successfully transferred mathematical and physical knowledge to an important physiological problem: vision, thereby placing physiological optics on a sound and exact scientific basis. Witelo, Roger Bacon and the Italian mathematician *Francesco Maurolyco* (1494-1575) had surpassed the Aristotelean optics, and gone beyond the Islamic author Alhazen. The concepts about the direction of the rays in refraction, the reflexion occurring in mirrors, and the action of convex as well as concave lenses had progressed considerably.

Before the advent of Kepler, another Italian scholar, *Giambattista della Porta* (1538-1615), had done some valuable work in optics. Porta described in his work *Magia Naturalis* the already known principle of the "camera obscura," together with the laws of reflexion in a concave mirror, the effect of lens combinations and the phenomenon of vision. Regarding the latter subject, Porta did not go significantly beyond the ideas of his predecessors, and it is to the credit of Kepler that new insights were achieved in the field of vision.[128]

Kepler was born in 1571 at Weilderstadt, Würtemberg (see Fig. No. 22) and wrote his two great optical works at Prague: *Paralipomena in Vitellionem seu Astronomia pars Optica* (Frankfort, 1604) and *Dioptrice* (1610).[129] Both publications established the fundamental principles of the visual process which are still valid today. Kepler described for the first time the pupillary contraction which accompanies visual accommodation, and he corrected the older statements about the size of the visual angle.[130]

Fig. 22. *Johannes Kepler* (1571-1630).

In the fifth book of his *Paralipomena,* Kepler set forth for the first time, a correct account of how the light rays penetrated through the crystalline lens and reached the retina, in spite of the incomplete contemporary anatomical knowledge of the eye. Kepler also correctly postulated that the retina was the real organ of vision. On its surface, an *inverted image* of the contemplated object appeared, point by point. Moreover, Kepler explained the effect of glasses both in nearsightedness and presbyopia, depicting the directions which the rays of light followed through the lenses. He realized that the image of an object would remain out of focus until the rays could converge on the surface of the retina.

Kepler described his discoveries in the field of vision in the fourth section of the *Paralipomena,* the following way: "You, ingenious Porta, should have added to your description that the image on the crystalline lens was not clear. Moreover, vision cannot originate through a combination of the light with the elements of the lens, but only occurs when this light reaches the retina. Thus, the rays which are united at the level of the lens gradually diverge, forming point for point an inverted and sharp image at the level of the retina. If you, Porta, could have added these facts to your explanations you would have been able to explain completely the phenomenon of vision."[131]

In his life, Kepler experienced hard times. However, thanks to his great influence, he was able to prevent the burning of his mother, accused of being a witch. He died in 1630, during a voyage to Regensburg, and had to be buried outside of the city walls because he was a Protestant.[132]

A few years after the publication of Kepler's *Dioptrice,* another work appeared in 1619 written by the Swabian astronomer *Christoph Scheiner* (1575-1650) and entitled *Oculus sive fundamentum opticum.* For the first time, Scheiner described the fact that he had observed the sharp and inverted image of an object on the retina of an enucleated animal and human eye, thus confirming Kepler's thesis. Scheiner also described the lateral emergence of the ocular nerves from the eyeball and the eye's convergence.

Another observation, still known to students as the "Scheiner experiment", was made in Innsbruck in 1615. It demonstrated that the refractive conditions in the eye varied through accommodation for distant or near vision. For the experiment, a card is needed with two small holes in it, separated by a distance smaller than the pupillary diameter. The card is then placed close before the eyes, and the subject ordered to look through the holes in order to see two needles placed at some distance one after the other. If the person attempts to focus on the nearest needle, he will observe a double and inverted image of the second one.[133,134]

5.
FURTHER PHYSIOLOGICAL DEVELOPMENTS IN THE SEVENTEENTH CENTURY, ESPECIALLY IN GREAT BRITAIN

The British Isles made extraordinary contributions to the development of physiology during the seventeenth century. After the sea victory over Spain in 1588, England achieved the necessary political power to organize its world-wide trade and establish an extensive colonial empire in India and North America. National unity and increasing prosperity in the reign of Queen Elizabeth facilitated the flowering of intellectual life. The dramatist *William Shakespeare* (1564-1616), the philosopher *Thomas Hobbes* (1588-1679), and the poet *John Milton* (1608-1674) reflected the high cultural level of Elizabethan England.

Likewise, anatomy and especially physiology made enormous strides primarily through the application of physical and chemical principles. Thus, the physicians William Harvey, Francis Glisson, Thomas Willis, Richard Lower, William Croone, the naturalists Robert Hooke, John Mayow, Robert Boyle and others testify to the successful efforts of this period in solving a series of physiological questions through observations and experiments. Since Harvey's accomplishments have already been described, this section will deal with his eminent contemporaries.

The oldest of them was *Francis Glisson* (1597-1677), first a lecturer in Greek. Glisson later decided to study medicine, receiving his M.D. degree in 1634 at Cambridge, where he became Regius professor of medicine in 1636. A year prior he had been elected Fellow of the Royal College of Physicians in London whose presidency he held between 1667 and 1669.

Glisson was an excellent anatomist and pathologist, who examined carefully for the first time a glandular organ such as the liver before the advent of the microscope. The result of his research was the work *Anatomia Hepatis,* published in 1654, where Glisson described the liver's capsule which still carries his name, the hepatic vascular distribution and the formation of bile.

More important for physiology was the book *De Ventriculo et Intestinis* (London, 1677) which developed the idea of a *general* vital property of "irritabilitas"- irritability- from the previous concept of "irritatio" or "stimulatio" found in Galen, Caesalpino (1593) and Glisson's own writings.[135] The bodily tissues were formed of string-like fibres possessing "naturalis perceptio," "appetitus" and "motus." Thus, the living substance had its own dynamic principle, independent of the nervous system and the soul. This immanent activity was called a "substantia energetica" or "archeus." To be sure, Glisson spiritualized matter with these concepts, adopting in the process Paracelsian and Helmontian ideas, and preparing the way for the postulates of the philosopher Leibniz.

The muscular fibre, capable of perceiving stimuli without consciousness — "naturalis perceptio" — tended towards action — "appetitus" — and contraction — "motus." Likewise, the cardiac function was dependent on the irritability of the cardiac muscles fibres which were stimulated by the incoming blood. Glisson was not a mechanistic thinker like his contemporary, the physician Thomas Willis, who viewed the process of muscular contraction as a sort of explosion, resulting in the inflation of the passive fibres through the accumulation of a gaseous spirit. Indeed, Glisson rejected Willis' theory by carrying out a classical experiment which can be called the first known plethysmographic test in history. The arm of a well-developed muscular individual was totally submerged in a glass container with an upper funnel-shaped opening. During muscular contractions, Glisson observed that the water level actually decreased slightly in the receptacle, thus dispelling the notion that there was an increase in volume due to the inflation of the hollow muscles.

Finally, Glisson believed that all other components of the body, such as the bones, blood and other tissues, also had irritability, a property which we would designate today as reactivity. Glisson's doctrine of irritability was conceived on purely theoretical grounds, and not confirmed nor supported experimentally, like Haller's later concept. Moreover, in contrast to the Hallerian property, Glisson's irritability was a general characteristic of all living parts, not just muscular contractibility. Hence, it acquired relatively little significance and played no important role in the subsequent development of physiology.

Among the founders of the Royal Society was the physician *William Croone* (1633-1684),[136] who had developed another theory of muscular contraction, similar in many respects to that enunciated by Willis. Entitled *De Ratione Motus Musculorum,* Croone's work appeared in 1664 and was later reprinted as a supplement in Willis' *Cerebri Anatome.*

According to Croone, the muscular fibre consisted of small vesicles which were connected with each other through small openings. A nourishing liquid flowed from the arterial system into these muscular vesicles, where it encountered a second fluid of nervous origin. Croon postulated that the contact of both these humors produced the swelling and contraction of the muscles. It is probable that Willis was influenced by the ideas of Croone, since his work, *De Motu Musculari,* was published six years later, in 1670.[137]

Similar problems, especially the questions of muscular contraction, occupied the attention of the physician Thomas Willis (Fig. No. 23).[138] Born in Great Bedwyn, Wiltshire, on January 27, 1621, Willis was educated at Oxford and in his early years became very interested in physics, especially the Cartesian mechanical philosophy, the neoatomism of Gassendi and the iatrochemistry of van Helmont. This is reflected in Willis' first two publications of 1659, entitled *De Fermentatione* and *De Febribus.* According to his views, all bodies could be dissolved into spirits, sulphur, salt, water and earth, a theory already expressed by the Italian chemist Sebastian Basso (fl. 1621) in 1621. Within these elements existed a series of invisible particles of various forms and motions.

Fig. 23. *Thomas Willis* (1621-1675).

Willis conceived of fermentation as a transformation of substances brought about by internal movements of their component parts. In the organism this fermentation or inner motion occurred in the blood and its expression was the organic vitality. In contrast with Glisson, Willis was far more technical in his concepts. Elements of his thought can be found in other authors such as Borelli, Malpighi, Bohn, Craanen, St. Blancard, William Croone, Boerhaave and Friedrich Hoffmann. Among his collaborators were Richard Lower, Robert Hooke, John Mayow and John Locke.

Following Descartes, Willis considered the human organism to be a machine, and attempted to interpret in strict physical terms the functions of the so-called animal spirits. Therefore, he distinguished a flame-like spirit in the circulating blood and another similar to light in the nervous system. For Willis, all involuntary motions were directed by these two agents. The flame was responsible for heating the blood, and the heart was no longer viewed as the seat of the vital heat. Rather, the blood communicated its warmth to the cardiac organ during its passage through the chambers.

Thanks to a process of fermentation, the food provided all the components necessary for the formation of blood: spirit, sulphur, water, salt and earth. This activity occurred in the stomach and resulted in the formation of the chyle, a product which was subsequently transformed into blood at the level of the heart and also yielded the animal spirits in the brain.

Willis denied the Cartesian hypothesis that the nerves were hollow.[139] Since he postulated that these nerves were tightly stretched between the brain and the periphery, he conceived the flow of the animal spirits as occurring around the fibres of the nerve bundle, just as fluid would run around a tense violin string.

Willis also established a highly hypothetical doctrine of cerebral localizations for the most important functions of the mind and the body which persisted in medical theory for a long time. The corpus striatum was viewed as the organ of perception, while the reflection of these sensations in the corpus callosum allowed imagination to occur. For Willis, the cerebral convolutions were the seat of the memory, whereas the cerebellum and the pons were responsible for the involuntary motions of various organs, such as the heart beat, breathing and gastrointestinal peristalsis. For the first time, the ganglions of the sympathic nerves were described by Willis. Moreover, he clearly recognized the vagus nerve and assumed that its origin was in the cerebellum. Finally, the oliva was considered to be the center for speech.[140]

In connection with his iatrochemical views, Willis developed a highly speculative theory of muscular contraction in his work *De Motu Musculari* (London, 1670). The animal spirits were supposed to flow from the brain along the nervous fibres to the muscles and tendons. After arriving in the muscle, these spirits encountered sulphurous and nitrous blood particles, producing a kind of local explosion which dilated and shortened the hollow muscular fibres. Willis attempted to verify this theory experimentally by placing a ligature around the tendinous portions of a muscle and observing that the muscle was under those circumstances unable to contract. As mentioned before, Willis' theory of muscular inflation was quickly demolished by the experiments of Glisson.[141]

In addition to the two dissertations of 1659, Willis wrote another important work: *Cerebri Anatome* (London, 1664). With this book, he enriched considerably the anatomical knowledge of the nervous system, especially regarding the cerebral nerves and the vegetative system.

Willis' conception of the animal spirits was often copied in later times, especially by Friedrich Hoffmann. Willis distinguished a thick, slow flowing nervous fluid which had a nutritive function, from another fast traveling spirit inside the fluid behaving like a ray of light. Willis ascribed to the animal spirits both the centrifugal and centripetal nervous actions, and postulated that some transmissions could return to the periphery — the first clear exposition of reflex action.[142]

Moreover, Willis distinguished between an *anima rationalis* (rational human soul) and *anima brutorum* (animal soul). The latter, in turn, could be divided into the *anima sensitiva* responsible for the nervous functions, and the *anima vitalis* or flame of life which constituted the basis for vitality.

By contrast, Willis' collaborator and assistant, the physician *Richard Lower* (1631-1692) (Fig. No. 24), was more empirical in his assumptions and activities. Born in 1631 at Cornwall, Lower studied in London and later at Oxford, where he worked with Willis until 1666.[143]

Under the influence of Harvey's circulation of the blood, Lower performed his first rather spectacular experiment in 1665, by transfusing blood from one dog to another. The experience was later repeated with two sheep, a feat already performed by the English scholar *Francis Potter* (1594-1678) in 1650 and the physician *William Croone* in 1662. Two years later, Lower followed the example of the French physician *Jean B. Denis* (1643-1702) in attempting a blood transfusion from a sheep to a man. Similar experiments had been performed in animals by the famous physician and chemist Robert Boyle, who published his results in a work entitled *Considerations touching the Usefullness of the Experimental Natural Philosophy* (1663). In a related field, the great astronomer, ar-

Fig. 24. *Richard Lower* (1631-1691).

chitect and chemist *Christopher Wren* (1632-1723) had performed intravenous injections on dogs in 1656. For this purpose, he infused a warm solution of opium using a quill and the bladder of a pig.

One of Lower's most significant works was the *Tractatus de Corde* (London, 1669).[144] After describing the fascicular arrangement of the cardiac muscular fibres and the distribution of the cardiac nerves, Lower declared that the blood experienced a change of color in the lungs instead of the heart. Maintaining an artificial respiration in accordance with the method developed by Robert Hooke, Lower injected some darker venous blood into the pulmonary artery of an animal, obtaining a much brighter red blood from the pulmonary vein. He also observed that the upper layers of blood obtained from a person through bloodletting and collected in a bowl exhibited a light red color. He therefore concluded that this change of color was somehow related to the air and absolutely necessary for the maintenance of all vital functions. On a different subject, Lower observed that the ligature of the inferior vena cava gave rise to ascites.

In his efforts to understand the phenomena of respiratory physiology, Lower confirmed with his experiments some of the observations and conclusions of the natural philosopher *Robert Hooke* (1635-1703), temporarily an assistant of Robert Boyle. Hooke, with Boyle, had constructed an air pump which facilitated their studies on the relationship between the volume and pressure of various gases. In the year 1667, Hooke carried out an experiment already described by Vesalius, which consisted in maintaining the life of an animal whose chest had been opened. For this purpose, Hooke used his air pump in order to inflate artificially the collapsed lungs. In certain instances, the rhythmical pumping could even be stopped for a short time without compromising the animal.[145]

Thus, Hooke concluded that the essential aspect of respiration did not consist in the movement of the lungs nor the expedient passage of blood through the lesser circuit. Rather, what seemed important was the change in color which the venous blood experienced in its passage through the lungs supplied with air. Hooke postulated that a certain "menstruum" joined the blood from the air. If this phenomenon did not take place, as in cases when the artificial respiration was stopped, the animal went into convulsions and died shortly thereafter.

Hooke published his respiratory experiments in his work *Micrographia* (London, 1667), a book which also contained a host of microscopical observations and the first reference to "cells" in plants. As curator of the Royal Society in London between 1664-1702, Hooke greatly promoted and cultivated the experimental sciences.[146]

Further important discoveries came from the twenty-five year old collaborator of Lower, the physician *John Mayow* (1643-1679). Born in 1643 at London, Mayow studied both law and medicine. His first work entitled *Trac-*

tatus duo seorsim editi: quorum prior agit de Respiratione, alter de Rachitide, appeared in 1674 at Oxford. It contained a series of investigations concerned with respiration, including fetal breathing, as well as the phenomenon of muscular contraction.[147]

Among other topics, Mayow carefully described the mechanics of breathing already sketched by Borelli, noting that the lungs only passively followed the thoracic motions. He declared that respiration had nothing to do with cooling or promoting the flow of blood. Rather, the real need for breathing air was based on the transfer of some elastic nitrous particles, the "spiritus nitroaereus," from the air into the blood. (See Fig. No. 25.) The effects of niter in gunpowder and fireworks were already known, especially the capacity to burn without air under the water.[148]

Saltpetre was a substance which could maintain a fire, and Boyle's experiments had shown that it was also present in the air. Indeed, Boyle had shown that when he withdrew air from a glass container with the vacuum pump, the flame was extinguished and the small animal within it perished. Thus, an inescapable conclusion followed: the existence in the air of a nitrous component necessary for respiration as well as the process of internal organic combustion. Under those circumstances, declared Mayow, breathing merely accomplished the contact between the atmospheric air and the circulating blood, and therefore niter particles from the air were transferred to the blood, where they reacted with sulphurous and combustible particles. The body's temperature originated as a result of the chemical fermentation between the aerial niter and the inflammable salty portions of the blood. This reaction also accounted for the observable change in blood color and temperature.

Fig. 25. *Experiments of John Mayow* demonstrating that the animal uses a certain portion of air during respiration.

From Mayow's respiratory theories it is evident that, in advance of his times, he possessed a fairly clear idea about the atmospheric component we now recognize to be oxygen.[149] Indeed, he already ascribed metabolism, respiration, and the calcination of metals with formation of acids to a certain component in the air almost a hundred years before Lavoisier. However, Mayow's chemical interpretation of respiration elicited little attention for many years, until the discovery of oxygen and other gases.

It is difficult to decide whether Mayow's ideas were truly original, since Boyle, Hooke and even Lower had expressed somewhat similar views.[150] In addition, it should be pointed out that Mayow indiscriminately transferred his fundamental concepts to various medical questions.

In the course of our descriptions we have repeatedly alluded to the experiments of *Robert Boyle.*[151] Although he contributed more indirectly to physiology, his excellent works in chemistry and physics stimulated the practice of using experiments in order to ascertain the nature of biological phenomena. In his work entitled *Some Considerations Touching the Usefulness of Experimental Natural Philosophy* (Oxford, 1664), Boyle expressed the following ideas about the value of physiology in the investigation of nature: "I remember I have not yet told you, that the study of physiology is not only delightful, as it teaches us to know Nature, but also as it teaches us in many cases to Master and Command her."[152]

Among other accomplishments, Boyle determined the atmospheric pressure in 1660 through the use of a mercurial column. Simultaneously with Mariotte, he established the relationship between pressure and volume in gases. Boyle also improved with the help of Hooke the airpump of Otto von Guericke, and he gradually freed chemistry from its iatrochemical bindings. Moreover, he contributed to the knowledge of chemical substances and their combinations, and studied the passage of light through prisms, electrical phenomena and other topics.

Boyle rejected the Aristotelean substantial forms just like Descartes, Regius, Gassendi and others had done before him. He imagined that matter was composed of corpuscles, which exhibited a certain configuration, size and motion. In 1684, Boyle carried out the first analysis of blood, checking its properties: color, taste, temperature, combustibility and weight, as well as its components: serous and red portions, volatile and fixed salts, oil, mucus, reddening effect when shaken in the air, etc.

For physiology, Boyle was directly important on two counts. First, he strictly rejected the application of Paracelsian elements in chemistry, while at the same time criticizing the iatrochemical conceptions regarding acid and alkaline humors. In the process of refuting the latter views, Boyle introduced the use of the litmus

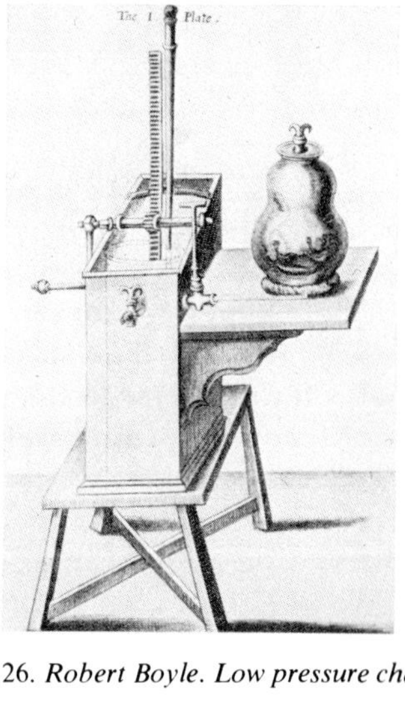

Fig. 26. *Robert Boyle. Low pressure chamber.*

paper.

Boyle's other contribution was related to the importance of respiration in physiology. In his now famous experiment with the air pump, he studied the effects of the vacuum on combustion and animal life. In the work entitled *New Experiments, Physicomechanical Touching the Spring and Weight of the Air and its Effects* (Oxford, 1660), Boyle demonstrated that the animals became rapidly asphyxiated as the flame extinguished in the rarified atmosphere of the testing chamber (see Fig. No. 26). He explained this result by declaring that the gases produced during combustion, and the elements which the blood had given up through expiration had all inhibited further burning and life.

To be sure, similar experiments had already been conducted by Otto von Guericke in 1650, and Mayow as well as Hooke had arrived at similar conclusions. Historians to date are still undecided whether Boyle was truly the first scholar to point out the significance of certain air particles for the respiratory continuity.[153] Finally, it is interesting to note that Robert Boyle was also the first to attempt the removal of gases present in the blood with the aid of his air pump. This experiment was carried out in 1670, but Boyle did not pursue the question further.

While the members of the Royal Society were engaged in solving the aforementioned problems, the Danish physician *Niels Stensen* or Nicolaus Stenonis — also called Steno — was similarly concerned with questions of muscular contraction.[154] Newer works by G. Scherz have greatly contributed to an understanding of Stensen's ergographic efforts.[155]

Born in 1638 at Copenhagen, the son of a goldsmith, Stensen (see Fig. No. 27) studied in his home town under the direction of the Danish anatomist *Thomas Bartholin* and later in Amsterdam with the Dutch physician *Gerhard Blasius* (ca. 1625-1692). A recently rediscovered manuscript entitled "Chaos" and dated 1659, is a work of the twenty-one year old Stensen. It reflects the author's remarkable versatility of interests and scientific maturity. Stensen was apparently acquainted with and appreciative of the works published by Descartes, Kircher, von Helmont, Sartorio, Kepler and many others.

Fig. 27. *Niels Stensen* (1638-1682).

While studying with Blasius,

Stensen discovered in 1660 the ductus parotideus, the excretory duct of the parotid gland which now carries his name. As a result, he became embroiled in a violent controversy with his teacher, Blasius, regarding the priority of this anatomical finding. Several other meritorious anatomical works followed this initial discovery. Stensen investigated the lacrimal glands of the eye, and then devoted his attention to the muscles.

In 1664, he published his *De musculis et glandulis observationum specimen* and three years later the work *Elementorum myologicae specimen seu musculorum descriptio geometrica.* In the first of these two publications, Stensen described the direction of the muscular fibres in the heart before Lower's work *De Corde.* Indeed, he demonstrated for the first time the muscular nature of the heart. For thousands of years, numerous authors had ascribed a series of mysterious functions to that central organ. In 1664, Stensen wrote: "They have considered the heart as the seat of vital heat, the throne of the spirit or the soul itself. They have revered the organ as the sun or a king, but if one looks at the heart with more care, only its muscular nature can be found."

Stensen left Copenhagen for Paris in 1664, and gave a famous lecture in the French capital entitled "Lecture on the anatomy of the brain". In his talk, Stensen severely criticized the shallowness and ineptitude of those contemporaries who, like Descartes and Willis, had dealt with cerebral physiology. He disagreed with their anatomical concepts and the physiological ideas deduced from the structures, pointing out the true difficulties associated with the procurement of proper anatomical brain preparations.

From Paris, Stensen went to Italy to the court of the Medicis at Florence. His contact with the Catholic faith led to theological studies and his conversion to Catholicism in 1667. At the same time his second treatise on muscles appeared, three years before Willis' *De Motu Musculari,* and it is probable that the English physician was acquainted with Stensen's work.

The most remarkable thing concerning Stensen's views of the muscular system was the strictly mechanical and geometrical treatment which he gave to the phenomenon of muscular contraction. The muscle consisted of long fibres and some transverse, connecting tissue elements. The fibres were arranged in parallelogrammatic

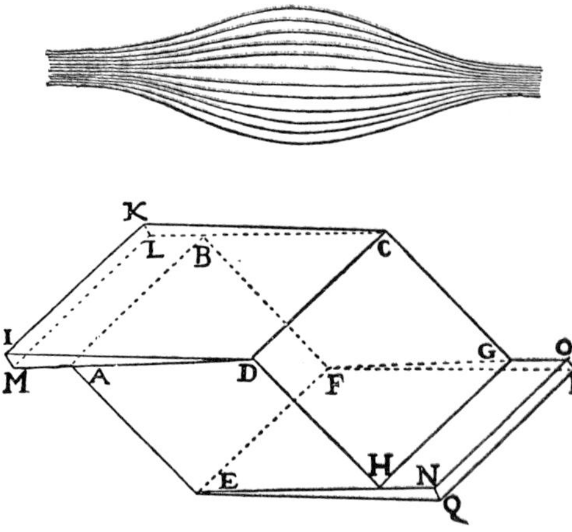

Fig. 28. *Stensen.* Geometrical theory of muscular contraction.

fashion instead of the hitherto accepted parallel disposition, and they originated at various angles in the tendinous portion. During contraction there was a purely geometrical change in each of these fibres. They decreased in total length while augmenting their width. (See Fig. No. 28).

Stensen's explanation was in sharp contrast to the prevailing contemporary explanations of muscular contraction, which assumed an increase in total muscular volume in order to explain the contraction proper. Only a decade later were Stensen's conclusions confirmed experimentally, when Glisson performed his plethysmometric test. Stimulated by Stensen's theory, Lower also came to a similar view in 1669, expressing the phenomenon of muscular contraction in the same geometrical terms. For a long time thereafter, these ideas prevailed in physiology, until a more accurate microscopical knowledge of the muscular fibres became available.

Among Stensen's other accomplishments was the recognition that the so-called testes mulierum or ovaries were the organs of egg formation. In 1667, he performed an experiment which still carries his name. He ligated the abdominal aorta of an animal with the resulting paralysis of the hind legs, an effect which could be reversed when the ligature was eliminated.

During his Italian sojourn, Stensen published another important work unrelated to physiology. Entitled *De solido intra solidum naturaliter contento dissertationis prodromus* (Florence, 1669), this book established the foundations for a scientific geology. Shortly thereafter, Stensen was named to the chair of anatomy at Copenhagen by a royal decree. In his last public dissection there, performed in 1673, Stensen had this to say: "Beautiful is everything we are able to see, and more beautiful those things we know, but by far the most beautiful of all is the unknown." He resigned his teaching post in 1675 in order to become a Catholic priest at Florence, where he concentrated his efforts on disseminating the faith. Later, he was named an apostolic vicar at Hanover, where Leibniz was already active. Stensen's asceticism and the burdens of his numerous missionary travels consumed his energy and he died as bishop of Schwerin on November 25, 1686, at the early age of forty-eight.[156]

In England, the example of authors such as Boyle, Lower, Hooke and others lingered on. At the end of the seventeenth century, another, although erroneous, attempt was made to determine the cardiac force solely from a measurement of the blood weight. Two authors, *James Keill* and the English physician *James Jurin,* both dealt with the same subject. Keill was one of the main representatives of the iatrophysical current in England, where he published his work *De vi cordis sanguinem per corpus pellente.* Jurin, on the other hand, carried out some remarkable optical experiments.

More important were the efforts of the English clergyman *Stephen Hales* (1677-1761) (see Fig. No. 29) who published his results in the celebrated work entitled *Statistical Essays: containing Haemastaticks* (London, 1733). Born at Kent, in 1677, Hales had shown great interest in physics but instead studied theology. In addition to his parsonage, he was occupied with anatomical, botanical and physical experiments, which made him famous. In 1717, he was elected to membership in the Royal Society of London.[157]

Fig. 29. *Stephen Hales* (1667-1761).

Not only did Hales carry out some important studies concerning the movement of sap in plants — *Vegetable Staticks* (London, 1727) — he also made some fundamental discoveries about the circulation of the blood. Of especial importance were Hales' pioneering efforts to measure the blood pressure. Using a horse, Hales inserted two long glass tubes, one into the jugular vein and the other into the carotid artery. At the venous tube, the blood ascended only one foot and a few inches (about 40 cm) while the blood column in the arterial cannula rose to nine feet, six inches (about 3m). With these measurements, Hales established the basis for more exact calculations of the total cardiac output, carried out later by the Dutch physician *Daniel Bernoulli* (1700-1782).

Less known, but equally important were Hales' studies regarding the vascular resistance offered by the capillaries. He postulated that the blood globules described by Leeuwenhoek were only slightly smaller than the diameter of the smallest vessels. Thus, the flow of these blood particles through the capillaries was accompanied with considerable friction and resistance, which actually slowed the circulation.

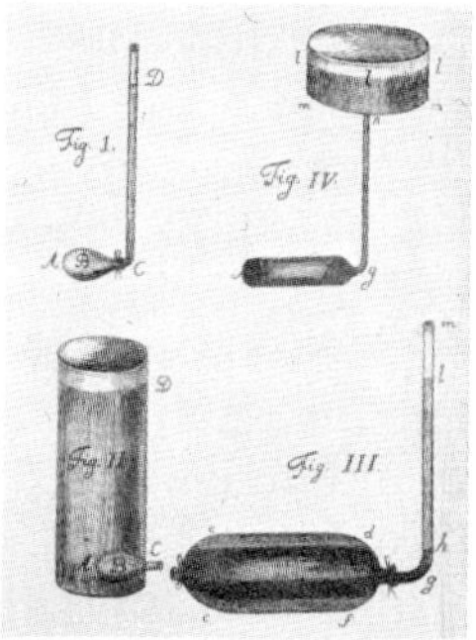

Fig. 30. *Passavant* — The work of the heart (from his book *De Vi Cordis:* 1748).
Fig. I: Bladder B is in equilibrium with with column of fluid C-D
Fig. II: Compressing bladder B in a cylindrical container full of water allows the outflow of its contents like the heart during systole.

From the preceding it is clear that Hales applied the new

knowledge of hydromechanics and hydraulics for the first time to the questions of blood circulation. An exact date for these capillary experiments has not been established, but it is probable that they were carried out several years prior to the publication of the *Haemastaticks.*[158]

The advances made in the field of hydromechanics also stimulated the mathematician Bernoulli, who in 1737 completed a new and much more correct calculation of the cardiac output by multiplying the stroke volume with the lifting force. His student, *Daniel Passavant* (1722-1799), computed the stroke volume at about 1½ ounces (about 45 gm), and using Hales' blood pressure values, arrived at figures of the left heart output which approximate the modern values (see Fig. No. 30).

Haller, who described these computations in detail, apparently did not understand their true physical character. He objected to the Bernoulli-Passavant calculations on the grounds that they had not taken into account the enormous resistances inherent in the circulation.

Thus, we conclude our journey through the physiology of the seventeenth century. Some important physiological questions such as the blood circulation, the movement of the lymph and the path of light in the eye had already been clearly solved with the aid of anatomy and the application of physical and chemical principles. Moreover, other topics, such as respiration and muscular contraction were considerably better understood.

6.
THE BEGINNINGS OF MICROSCOPICAL OBSERVATIONS AND THEIR IMPORTANCE FOR THE SOLUTION OF PHYSIOLOGICAL PROBLEMS (MARCELLO MALPIGHI — ANTON VAN LEEUWENHOEK)

In biology, form and function are inextricably related to each other, and any modern physiological interpretation has to take into consideration the morphology of the organic structures in which certain functions are being postulated. To be sure, the need for such a correlation dates back to antiquity, when Galen already questioned the "usefulness" of organs such as the liver, spleen, cardiac valves and kidneys. But as long as the anatomical knowledge was restricted to an awareness of the gross anatomy of larger organs, its importance for judging functions was negligible, thus allowing the unrestricted development of physiological speculations.

Moreover, even the considerable progress which anatomy experienced with the works of Vesalius, Fallopio and others, contributed only to the solution of relatively broad and general physiological problems, like the discovery of the blood circulation. Indeed, Harvey's evidence was almost exclusively based on gross anatomy, and Borelli's muscular mechanics were derived from phenomena observable with the naked eye. However, the explanation concerning the true process of muscular contraction already presented problems because of a lack of knowledge regarding the structure and disposition of the small muscular fibres. For the same reason, Willis accepted an erroneous vesicular structure for these fibres, and Stensen assumed a geometrical theory of muscular contraction.

In addition, it was nearly impossible to explain the secretory function of various glands from the few macroscopic details which could be recognized, such as tissues, vessels, nerves and ducts. Moreover, how could nervous transmission and the action of the sensory organs be interpreted as long as there was no further understanding about their microscopical composition? Thus, one can confidently state that physiology could not have advanced beyond its early level around the year 1650 without the assistance of magnifying glasses and microscopes, which opened new fields for physiological inquiry through the discovery of the microscopic structure of numerous organs.

Two divergent factors combined to facilitate the establishment of a truly microscopical investigation of organic tissues in the seventeenth century. In the first place, there was an improvement in the techniques related to the preparation of lenses, and secondly, the newly revived atomism stimulated more interest in the intimate structure of objects. Regarding the development of magnifying glasses, some eyeglasses with this property had already been in use in northern Italy between 1280 and 1300.[159] However, the discovery did not benefit anatomy, because of the inability to manufacture lenses with a sufficiently short focal length until the seventeenth century.

The first simple glasses were the "Perspicilia Pulicaria" (flea-glasses), used more for entertainment purposes than for studies and allowing only a nine to ten-fold magnification. Some skillful glassgrinders were producing small and very convex lenses with higher magnification, but they allowed much less light to pass through them. Among those devices left by the Dutch microscopist *Anton van Leeuwenhoek* (1632-1723), who ground his own lenses, were some which could apparently attain a 270th magnification.[160]

The application of magnifying glasses to the search of the minute components of the organism was aided decisively by the newly reawakened interest in the ultimate structure of matter. The Aristotelean-Galenic concept of elements, qualities and forms had become inadequate to explain the observations made by experimental chemistry at the beginning of the seventeenth century. The humoral doctrine had been rejected by Paracelsus, van Helmont and others, while the new anatomical discoveries discredited ancient knowledge and increased the lingering doubts about its degree of accuracy. Correlations between the macroscopic structures and their observed functions had already brought forth the idea that behind the visible organs, like the glands, there were some microscopic "machinulae" at work. In 1620, Francis Bacon urged in his *Novum Organum* the use of magnifying lenses in order to visualize the Democritean atoms. Galileo had already inverted his telescope in 1610, in order to observe insects and other structures invisible to the naked eye. By 1625, another Italian scholar, *Federigo Cesi* (1585-1630), presented some drawings of the microscopical anatomy of the bees.

Thus, a new *anatomia subtilis* or microscopical anatomy came to the fore, concerned with the analysis of the particles or atoms of living structures — *resolutio ad indivisibilia.* In 1645, the Italian anatomist *Marc A. Severino* (1580-1656) published a work entitled *Zoologia Democritaea*[161] and Galileo's disciple, Borelli, postulated in 1649 that microscopical particles were the causes of infections.

Sennert, Gassendi, Descartes and Boyle conceived matter to be formed of atoms, particles and corpuscles. According to Descartes, all the properties of natural bodies could be explained from the form, size, disposition and motion of small particles or elements. *Materia et motus,* matter and motion, became the key terms for the new understanding of physical and biological phenomena. The highly influential chemist Robert Boyle declared himself a follower of the doctrine of particles in 1661, a position which he confirmed later in his work *Introductio ad historiam qualitatum particularium* (1677).

The microscopists tried to visualize these hypothetical particles, and they encountered certain structures which seemed to fulfill their expectations. However, the contributions made by microscopy to anatomy in the first half of the seventeenth century were minor.[162] Among the physicians, the French Petrus Borellus or *Pierre Borel* (1620-1689), a personal physician of the "sun king" Louis XIV, made considerable use of the microscope.[163] However, his findings as published in his work *Observationum microscopicarum centuria* (1656) were rather incomplete and lacked systematization.

Only in the hands of the Italian physician *Marcello Malpighi* did the microscopical method of anatomical investigation finally flourish. Malpighi has the honor of having been the first to apply this approach systematically in order to elucidate the fine structure of the bodily organs.[164]

Malpighi (see Fig. No. 31) was born in Bologna, March 16, 1628. During his student years in his home city, he began to carry out microscopical observations under the direction of his teacher, Bartholomeo Massari, while other influential men and expert scholars rejected this method as being useless and confusing. After a short career as professor of theoretical medicine at the University of Bologna, Malpighi moved to the city of Pisa in 1656, where he established a close scientific and personal relationship with Borelli. Unquestionably, the two men stimulated each other greatly, and Malpighi was introduced to Borelli's circle and the "Philosophia Democritaea" — Democritus' atomism.

Thanks to his superior method of tissue preparation, Malpighi was able to discover in 1657, without the use of a microscope, the coiled direction of the cardiac muscular fibres. After many years of fruitful scientific work at Messina and Bologna, he was elected to membership in the Royal Society of London in 1669, and finally personal physician to Pope Innocence XII. The appointment, which he owed in part to his friendship with Cardinal Pignatelli, brought Malpighi to Rome where he died three years later, the victim of a stroke.

Fig. 31. *Marcello Malpighi* (1628-1694).

Malpighi's microscopical investigations were carried out in a variety of fields. Together with the English naturalist Nehemiah Grew (1641-1712), he was the founder of plant morphology with the work *Anatome Plantarum* (1675/79). He also examined many different animal organisms, thus advancing considerably the disciplines of comparative anatomy and physiology. Finally, Malpighi successfully completed his investigations already in progress regarding the development of the chick embryo in the egg.[165]

Malpighi's first publications, two letters addressed to Borelli concerning the structure of the lungs and entitled *De Pulmonibus Epistolae Duae ad Borellium* (Bologna, 1661), were of great importance for physiology. In them, Malpighi succeeded in demonstrating that the hitherto undifferentiated pulmonary parenchyma was actually composed of a series of vesicles. Moreover, Malpighi discovered the links between the arterial and venous systems while observing the capillary vessels of a frog's lung through the microscope in 1661. It may be recalled that this was the missing piece of evidence needed to confirm Harvey's circulatory scheme. (See Fig. No. 32.)

In addition, Malpighi also described the circulation of the blood in pulmonary and mesenteric vessels, making another important observation at the same time which was erroneously interpreted. He saw for the first time small red corpuscles floating in the circulating blood, and considered them to be merely fat droplets (*De Polypo Cordis,* 1665). Malpighi declared that: "the blood is not a mixture of four humors, but its substance consists of almost an infinite number of particles, white serum and red atoms."

Especially important were Malpighi's glandular investigations, published in the work *De Viscerum Structura* (1666). The anatomy of most glands and their mode of operation were still completely unknown in Malpighi's time. From the middle of the seventeenth century onwards, only the ducts of some major glands had been discovered and described in certain detail: ductus pancreaticus (J.G. Wirsun, 1642), ductus parotideus (N. Stensen, 1661) and ductus submaxillaris (T. Wharton, 1656).

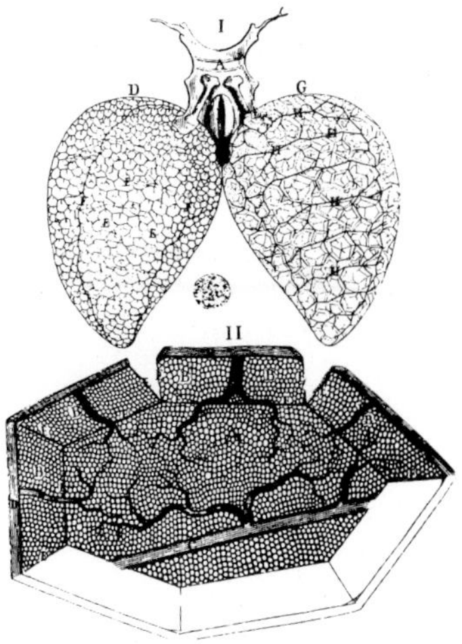
Fig. 32. Malpighi — *Capillaries of the lungs* observed in 1660 in the lungs of a frog.

Regarding liver functions, no advances had been achieved over the classical doctrine of blood, yellow and black bile formation in the liver tissue. The black bile was supposed to flow towards the spleen and from there to the stomach. The production of yellow bile occurred in the liver or the gallbladder. According to the English physician *Thomas Wharton* (1614-1673) glandular secretions were not primarily a product of the glandular parenchyma, but composed of "succus nerveus" — a nervous fluid — secreted through the nerves.

Malpighi recognized that the liver was a gland with secretory functions because of its lobular structure. He described how the bile was formed in the acini and subsequently flowed to the gallbladder through the biliary duct system. Malpighi investigated the spleen, which until then had also been vaguely considered to be a gland. He discovered the splenic capsule, the trabecules, the pulp as well as the networks of vessels now known as Malpighian bodies, and considered the entire spleen as a contractile vascular organ.

In his investigations of the kidneys, Malpighi went beyond the views previously held by Bellini. He described how the renal tubules ended in the renal calyces which in turn formed the renal pelvis. In the renal cortex, he observed the winding tubules and the small vascular structures — the glomeruli — which looked like apples hanging from a tree.

Malpighi was less successful with his microscopical examinations of the cerebral cortex, published in his work *De Cerebri Cortice* (1666), and the other centers of grey matter. Following Aristotle, he postulated that the entire grey matter was a glandular organ. By contrast, Malpighi was more fortunate in his studies of the tongue, where he discovered the innervation of the papillae, considering them *organs of touch.* Likewise, he observed somewhat similar papillae in the skin, ascribing to them an identical tactile function: *De Externo Tactus Organo* (1665).[166]

Thus, Malpighi was able to make many important discoveries, each one of them of great consequence for the subsequent development of physiology. He used mostly a microscope composed of two lenses, while Leeuwenhoek always worked with a single magnifying glass.

The microscopical method was also successfully employed by the Dutch naturalist *Jan Swammerdam* (1637-1680), a friend of Stensen and Leeuwenhoek. A very pious individual, Swammerdam compiled his many excellent observations, especially concerning the microscopic structure of insects, in his *Biblia Naturae,* a work published after the death of Herman Boerhaave in 1737.[167] The German edition of the work was entitled *Bibel der Natur* (Leipzig, 1752). Another of Swammerdam's notable contributions was his method of injecting liquid and colored wax into the vessels, a procedure frequently used thereafter for gross anatomical preparations.

More important for human physiology were the observations of *Anton van Leeuwenhoek.* As already mentioned, he had acquired an extraordinary skill in grinding magnifying lenses, and at the time of his death left no less than four hundred and ten of these instruments. However, Leeuwenhoek distrusted everybody and kept his skills and instruments a secret. Without a scientific background, he observed all kinds of objects indiscriminately, from the dust in the streets and stagnant water to butterfly wings, animal muscles and testes.

Beginning in 1673, Leeuwenhoek sent his observations in the form of about three hundred letters to the Royal Society of London.[168] Also, a great number of Leeuwenhoek's findings appeared in his work, entitled *Arcana Naturae* (Delft, 1695). He was the first to point out the striated nature of the muscular fibres, the netlike structure of the cardiac fibres and the fibrillar composition of the lens in the eye, hitherto considered to be a fluid called humor cristallinus.

Leeuwenhoek observed the layer of rods and cones in the retina, and identified the small globules in the blood — our erythrocytes — in 1673.[169] Moreover, he confirmed the presence of spermatocytes in various animal species, an observation already made by Jan Ham in 1677. Finally, Leeuwenhoek described the crystals remaining in dried solutions of certain salts, acids and several drugs. The various microscopical spheres, needles, pyramids, stars and quadrangles were to him "proof" of the peculiar particles composing matter. (See Fig. No. 33).

Wherever the human eye looked with the help of the microscope, a new world appeared before it. Without the help of this instrument the physiological explanations concerning the various organs would never have advanced beyond an aimless groping in the dark. The knowledge of the tissue

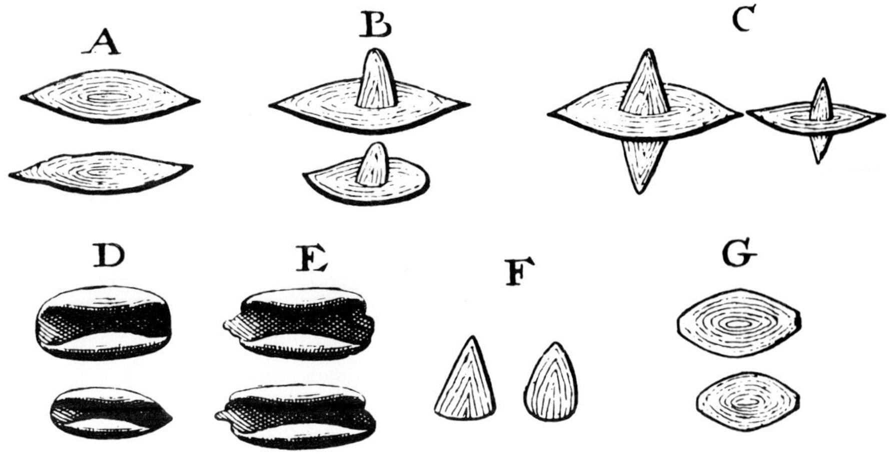

Fig. 33. *Leeuwenhoek:* Crystals from various solutions observed under the microscope.

structure especially advanced greatly through microscopical investigations. In the pre-microscopical era, only vessels, fibres, membranes and adipose tissue could be discerned within the parenchyma. Now, with the aid of the lenses, new possibilities emerged to differentiate these elements further, and clarify the intimate structure of the so-called "fibres."

In gross anatomy, the Dutch anatomist *Frederick Ruysch* (1638-1731) successfully applied the vascular injection techniques with dyed wax pioneered by Swammerdam. He discovered the existence of myriads of small vessels, a fact which led him to the dictum "Totum corpus ex vasculis." Ruysch was the first to describe the vasa vasorum, the vessels of the lamina chorioidea, the bronchial arteries and the arteria centralis retinae.

At the same time, the German anatomist *Johann N. Lieberkühn* (1711-1756) clarified the composition of the intestinal mucosa, demonstrating that the villi did not communicate with the lumen through free vascular apertures. Therefore, an entirely new situation was created regarding the mechanisms of food absorption.

In the ensuing decades the gradual advance of morphological knowledge created different viewpoints and opened new avenues of inquiry for physiology, which unfortunately cannot all be followed in detail here. Hence, we close the depiction of the seventeenth century, a period of classical investigations in the natural sciences, especially chemistry, physics, anatomy, histology and physiology.

REFERENCES

[1]Alfred R. Hall, *The Scientific Revolution 1500-1800,* the formation of the modern scientific attitude, 2nd ed., Boston, 1966.

[2]Herbert Butterfield, *The Origins of Modern Science 1300-1800,* rev. ed., New York, 1965.

[3]Charles D. O'Malley, *Andreas Vesalius of Brussels, 1514-1564,* Berkeley, 1964.

[4]G. Rath, "Pre-Vesalian anatomy in the light of modern research," *Bull. Hist. Med.* 35 (1961), pp. 142-148.

[5]K. Sudhoff, *Ein Beitrag zur Geschichte der Anatomie im Mittelalter, speziell der anat. Graphik nach Handschriften des 9. bis 15. Jh.s.,* Leipzig, 1908.

[6]C.D. O'Malley - E. Clarke, "The discovery of the auditory ossicles," *Bull. Hist. Med.* 35 (1961), pp. 419-441.

[7]See C.D. O'Malley's biography in *Dict. Scientif. Biogr.* 1 (1970), pp. 617-621.

[8]J. Berengario, *A Short Introduction to Anatomy,* transl. with intro. & notes by L.R. Lindand, Chicago, 1959.

[9]Harvey W. Cushing, *A Bio-bibliography of Andreas Vesalius,* 2nd ed., Hamden, Conn., 1962.

[10]C.E. Kellett, "Sylvius and the reform of anatomy," *Med. Hist.* 5 (1961), pp. 101-116.

[11]*The Illustrations from the Works of Andreas Vesalius of Brussels,* with annotations and transl., a discussion of the plates and their background, authorship and influence and biographical sketch of Vesalius, by C.D. O'Malley and J.B. de C.M. Saunders, Cleveland, 1950.

[12]The Preface of Andreas Vesalius to "De Fabrica Corporis Humani 1543", transl. by B. Farrington, *Proc. Roy. Soc. Med.* 25 (1931/32), pp. 1357-66.
Vesalius on the Human Brain (transl. of a section of his Fabrica of 1543), introd., transl. and notes by C. Singer, London, 1952.

[13]F. Guerra, "The identity of the artists involved in Vesalius' *Fabrica,* 1543", *Med. Hist.* 13 (1969), pp. 37-50.

[14]See C.D. O'Malley's biography in *Dict. Scientif. Biogr.* 4 (1971), pp. 519-521.

[15]His biography, also written by C.D. O'Malley appeared in *Ibid,* 4 (1971), pp. 487-488.

[16]Charles Singer, *The Evolution of Anatomy,* London, 1925.

[17]Ritter R. v. Tőply, *Studien zur Geschichte der Anatomie im Mittelalter,* Leipzig & Wein, 1898 & 1922.
E. Hintzsche, "Die Galensche Anatomie", *Ciba-Z.* 8 (1944), p. 3410.

[18]Michael Foster, *Lectures on the History of Physiology,* Cambridge, 1901.

[19]G. van Rijnberk, "Vésale comme physiologiste expérimentateur," *Arch. néerl. Physiol.* 1 (1918), p. 129.

[20]See biography by R. Granit in *Dict. Scientif. Biogr. 4 (1971), pp. 584-586.*

[21]*Charles S. Sherrington, Endeavors of Jean Fernel,* Cambridge, 1946.

[22]E.J. Holmyard, *Alchemy,* Harmondsworth, Middlesex, 1957.

[23]H. Schipperges, "Strukturen und Prozesse alchemistischer Überlieferungen," in *Alchimia, Ideologie und Technologie,* Munich, 1970, pp. 67-118.

[24]P. Walden, "Zur Problematik der Alchemie und ihrer Ausstrahlungen auf die modernen Naturwiss," *Naturwiss. 35* (1948), p. 225.

[25]G. Goldschmidt, "Der Ursprung der Alchemie," *Ciba-Z.* 5 (1938), p. 1950; and "Die mittelalterliche Alchemie," *Ciba-Z. 6* (1939), p. 2234.

[26]F.A. Yates, "The Hermetic tradition in Renaissance science," in *Art, Science and History in the Renaissance,* ed. by C.S. Singleton, Baltimore, 1968, pp. 255-274.

[27]Schipperges, "Strukturen," footnote 23. See also Mark Graubard, *Astrology and Alchemy: Two Fossil Sciences,* New York, 1953.

[28]P. Diepgen, "Studien zu Arnald von Villanova," *Arch. Gesch. Med. 3, 5, 6* (1910, 1912, 1913).

[29]W. Pagel, "Paracelsus and the neoplatonic and gnostic tradition," *Ambix* 8 (1960), pp. 125-166.

[30]Allen G. Debus, *The Chemical Dream of the Renaissance,* Cambridge, 1968.

[31]His complete works were published by: K. Sudhoff & W. Mathiessen, *Paracelsus (Theophrastus v. Hohenheim.) Sämtliche Werke.* 16 vols., Munich, 1922-1933.

[32]Walter Pagel, *Paracelsus. An Introduction to Philosophical Medicine in the Era of the Renaissance,* New York, 1958.

[33]Franz Strunz, *Theophrastus Paracelsus, sein Leben und seine Persönlichkeit,* Leipzig, 1903.

[34]*Four Treatises of Theophrastus von Hohenheim called Paracelsus,* trans. from the orig. German, with intro. essays, by C.L. Temkin, G. Rosen, G. Zilboorg and H.E. Sigerist, Baltimore, 1941.
Volumen Medicinae Paramirum, transl. from the German by K.F. Leidecker, Baltimore, 1949.

[35]Franz Strunz, *Theophrastus Paracelsus;* Idee und Problem seiner Weltanschauung, Salzburg & Leipzig, 1937.

[36]F. Taylor Sherwood, "The idea of the quintessence," *Science, Medicine and History,* ed. by E. Ashworth Underwood, 2 vols., London, 1953, I, pp. 247-265.
W. Pagel, "The prime matter of Paracelsus," *Ambix* 9 (1961), pp. 117-135.

[37]Allen G. Debus, *The English Paracelsians,* London, 1965.

[38]Harold Hartley, *Studies in the History of Chemistry,* London, 1971.

[39]See M. Foster, *Lectures,* ch. V, pp. 121-144: "Van Helmont and the rise of chemical physiology."

[40]Walter Pagel, *Johann Baptist van Helmont, Einführung in d. philos. Med. d. Barock,* Berlin, 1930.

[41]Franz Strunz, *Johann Baptist van Helmont. Ein Beitrag z. Gesch. d. Naturwiss,* Leipzig & Wien, 1907.
H. Redgrove & I.M.L. Redgrove, *Johannes Baptista van Helmont, Alchemyst, Physician and Philosopher,* London, 1921.

[42]W. Pagel, "The religious and philosophical aspects of van Helmont's science and medicine," *Suppl. Bull. Hist. Med.,* 2 (1944).

[43]R.P. Multhauf, "J.B. van Helmont's reformation of the Galenic doctrine of digestion," *Bull. Hist. Med.* 29 (1955), pp. 154-163.

[44]P. Rattansi, "The Helmontian-Galenist controversy in Restoration England," *Ambix* 12 (1964), pp. 1-23.

[45]W. Pagel, "The position of Harvey and van Helmont in the history of European thought," *J. Hist. Med.* 13 (1958), pp. 186-199.

[46]Foster, *Lectures,* pp. 145-173.

[47]A. Gubser, "The *Positiones Variae Medicae* of Franciscus Sylvius," *Bull. Hist. Med.* 40 (1966), pp. 72-80.

[48]See biography by K.E. Rothschuh in *Dict. Scientif. Biogr.* 2 (1970), pp. 237-238.

[49]See Geoffrey L. Keynes, *A Bibliography of the Writings of Dr. William Harvey,* Cambridge, 1953.

[50]Charles J. Singer, *The Discovery of the Circulation of the Blood,* London, 1922.

[51]Gweneth Whitteridge, *William Harvey and the Circulation of the Blood,* London, 1971.

[52]R. Kapferer, "Der Blutkreislauf, seine Darstellung in den hippokratischen Schriften," Hippokrates (1937), p. 697; (1938), p. 251; (1939), p. 1331; (1940), pp. 84 & 646.

[53]H. Diller, "Die Lehre vom Blutkreislauf, eine verschollene Entdeckung der Hippokratiker?", *Arch. Gesch. Med.* 31 (1938), p. 201.
L. Aschoff, "Über die Entdeckung des Blutkreislaufes," *Freiburger Forsch. z. Med.-Gesch. H.* 1, Freiburg, 1938.

[54]P. Diepgen, "Die Hippokratiker u. die Lehre vom Blutkreislauf," *Münch. med. Wschr.* (1939), p. 299.
P. Diepgen, "Nochmals zur Frage der Kenntnis des Blutkreislaufs in den hippokratischen Schriften," *Hippokrates* (1938), p. 385.
[55]E. Lesky, "Harvey and Aristoteles," *Sudhoffs Arch.* 41 (1957), pp. 289-316, 349-378.

[56]M. Meyerhof, "Ibn an-Nafis (XIIIth cent.) and his theory of the lesser circulation," *Isis* 22 (1935), pp. 100-120.

[57]E.E. Bittar, "A study of Ibn-Nafis," *Bull. Hist. Med.* 29 (1955), pp. 352-368 & pp. 429-447.

[58]L.G. Wilson, "The problem of the discovery of the pulmonary circulation," *J. Hist. Med.* 17 (1962), pp. 229-244.

[59]M. Herzfeld, *L. da Vinci, der Denker, Forscher und Poet,* 2. Aufl. Jena, 1906.
H. Hess, *Die Naturanschauung der Renaissance in Italien. Verh. d. kunstgesch. Seminars,* Marburg, 1924.

[60]H. Boruttau, "L. da Vincis Verhältnis zur Anatomie und Physiologie der Kreislauforgane," *Arch. Gesch. Med.* 6 (1913), p. 217.

[61]M. Holl, "L. da Vinci. Quaderni d'Anatomia," I-IV, *Arch. Anat.* [*u. Physiol*], *1911-1917.*

[62]*W. Pagel, "Vesalius and the pulmonary transit of venous blood," J. Hist. Med.* 19 (1964), pp. 327-341.

[63]Charles D. O'Malley, *Michael Servetus,* a transl. of geograph., med. and astrol. writings, Philadelphia, 1953.

[64]John F. Fulton, *Michael Servetus, Humanist and Martyr,* with a bibliog., New York, 1953.

[65]O. Temkin, "Was Servetus influenced by Ibn an-Nafis?", *Bull. Hist. Med.* 8 (1940), pp. 731-734.

[66]Walter Pagel, *William Harvey's Biological Ideas,* New York, 1967.

[67]J.J. Bylebyl in *Dict. Scientif. Biogr.* 3 (1971), pp. 354-357. See also the author's "Cardiovascular physiology in the 16th and early 17th centuries," (unpublished Ph.D. dissertation, Yale University, 1969).

[68]R.H. Bainton, "The smaller circulation: Servetus and Colombo," *Arch. Gesch. Med.* 24 (1931), p. 371.

[69]See biography by G. Muratori, in *Dict. Scientif. Biogr.* 3 (1971), pp. 40-41.
[70]K. J. Franklin, "Valves in veins: an historical survey," *Proc. Roy. Soc. Med.* 21 (1928), pp. 1-33.

[71]Hieronymous Fabricius, *De Venarum Ostiolis,* 1603, facs. ed. with intro., transl. and notes by K.J. Franklin, Springfield, Ill., 1933.

[72]W. Pagel, "The philosophy of circles—Cesalpino—Harvey," *J. Hist. Med.* 12 (1957), pp. 140-157.

[73]W. Pagel, "Die Stellung Caesalpins und Harveys in der Entdeckung und Ideologie des Blutkreislaufs," *Sudhoffs. Arch.* 37 (1953), pp. 319-328.

[74]J.P. Arcieri, *Considerations about Cesalpinus and Harvey's Works on the Blood Circulation Discovery,* New York, 1957.

[75]See English translation with annotations by Chauncey D. Leake, 3rd ed., Springfield, Ill., 1941.

[76]Kenneth J. Franklin, *William Harvey, Englishman, 1578-1657,* London, 1961.

[77]Geoffrey L. Keynes, *The Life of William Harvey,* Oxford, 1966.

[78]Kenneth D. Keele, *William Harvey: The Man, the Physician and the Scientist,* London, 1965.

[79]The quoted passages are from Harvey's *De Motu Cordis* in the English transl. of R. Willis, reprinted in *Great Books of the Western World,* Chicago, 1952, vol. 28, pp. 268-273.

[80]Alfred P. Fishman, Dickinson W. Richards, eds., *Circulation of the Blood: Men and Ideas,* New York, 1964.

[81]W. Pagel, "William Harvey revisited," *Hist. of Science* 8 (1969), pp. 1-31 and 9 (1970), pp. 1-41; C. Webster, "Harvey's *De Generatione:* its origins and relevance to the theory of circulation," *Brit. J. Hist. Sci.* 3 (1967), pp. 262-274.

[82]M.T. Ghiselin, "William Harvey's methodology in De Motu Cordis from the standpoint of comparative anatomy," *Bull. Hist. Med.* 40 (1966), pp. 314-327.

[83]G. Basalla, "William Harvey and the heart as a pump," *Bull. Hist. Med.* 30 (1962), pp. 467-470.

[84]A.G. Debus, "Robert Fludd and the circulation of the blood," *J. Hist. Med.* 16 (1961), pp. 374-393, and by the same author "Harvey and Fludd: the irrational factor in the rational science of the 17th century," *J. Hist. Biol.* 3 (1970), pp. 81-105.

[85]W. Pagel, "Giordano Bruno: the philosophy of circles and the circular movement of the blood," *J. Hist. Med.* 6 (1951), pp. 116-124.

[86]P.L. Entralgo, "Harvey in the history of scientific thought," *J. Hist. Med.* 12 (1957), pp. 222-231.

[87]W. Pagel—F.N.L. Poynter, "Harvey's doctrine in Italy: Argoli (1644) and Bonaccorsi (1647) on the circulation of the blood," *Bull. Hist. Med.* 34 (1960), pp. 419-429.

[88]K.E. Rothschuh, "Die Entwicklung der Kreislauflehre im Anschluss an William Harvey," *Klin. Wschr.* 35 (1957), pp. 605-612.

[89]See biography by L. Premuda in *Dict. Scientif. Biogr.* I (1970), pp. 315-316.

[90]A.E. Nielsen, "A translation of Olaf Rudbeck's 'Nova exercitatio anatomica...' 1653", *Bull. Hist. Med.* 2 (1942), p. 304.

[91]J.F. Fulton, "The early history of the lymphatics," *Bull. Hennepin County Med. Soc.* 9 (1938), p. 5.

[92]A.B. Davis, "Some implications of the circulation theory for disease theory and treatment in the 17th century," *J. Hist. Med.* 26 (1971), pp. 28-39.

[93]Carl J. Friedrich, *The Age of the Baroque 1610-1660,* New York, 1962.

[94]H. Schimank, *Epochen der Naturforschung. Leonardo, Kepler, Faraday,* Volksverband der Bűcherfreunde, Berlin, 1930.

[95]Edwin Arthur Burtt, *The Metaphysical Foundations of Modern Science,* Garden City, N.Y., 1954.

[96]Richard F. Jones, *Ancients and Moderns,* Berkeley, 1965.

[97]W. Houghton, "The English virtuoso in the 17th century," *J. Hist. Ideas* 3 (1942), pp. 51-73 and pp. 190-219.

[98]M. Ornstein, *The Role of Scientific Societies in the 17th Century,* Chicago, 1938.

[99]Alexander Koyré, *From the Closed World to the Infinite Universe,* New York, 1957.

[100]M. Boas, "The establishment of the mechanical philosophy," *Osiris* 10 (1952), pp. 412-541.

[101]Leonora C. Rosenfield, *From Beast-Machine to Man-Machine,* New York, 1940 and M.D. Grmek, "Réflexions sur des interprétations mécanistes de la vie dans la physiologie du XVII siècle," *Episteme* 1 (1967), pp. 17-30.

[102]E. Mendelsohn, "The changing nature of physiological explanation in the 17th century," in *Mélanges Alexandre Koyré,* Paris, 1964, Vol. I, pp. 367-386.

[103]E. Gotfredsen, "Nicolaus Cusanus", *Münch. med. Wschr.* (1937), p. 1821.

[104]H. Miessen, "Die Verdienste Santorii um die Einf. physikal. Methoden in die Heilkunde," (*Med. Diss.* Dusseldorf, 1940).

[105]M.D. Grmek, *L'introduction de l'expérience quantitative dans les sciences biologiques,* Université de Paris [Publ. du Palais de la Découverte], Paris, 1962.
M.D. Grmek, "L'énigme des relations entre Galilei et Santorio", [Atti del simposio su Galileo Galilei nella storia e nella filosofia della scienza] *Firenze-Pisa* 1964, pp. 155 ff.

[106]R.H. Major, "Santorio Santorio," *Ann. Med. Hist.* 10 (1938), pp. 369-381.

[107]Santorio Santorio, *Medicina Statica:* being the aphorisms of Sanctorius transl. into English with large explanations, by John Quincy, London, 1718.

[108]For a biography see A.C. Crombie in *Dict. Scientif. Biogr.* 4 (1971), pp. 51-55 and Jack R. Vrooman, *René Descartes; a Biography,* New York, 1970.

[109]J. Collins, *Descartes' Philosophy of Nature,* Oxford, 1971.

[110]K.E. Rothschuh, *R. Descartes. Über den Menschen (1632) sowie Beschreibung des menschlichen Körpers* (1648). Transl. 1st French ed. 1664, with intro. and notes, Heidelberg, 1969.

[111]T.S. Hall, "Descartes' physiological method: position, principles, examples," *J. Hist. Biol.* 3 (1970), pp. 53-79.

[112]E. Wallach, "Descartes und Harvey," *Arch. Gesch. Med.* 20 (1928), p. 301.

[113]K.E. Rothschuh, "Rene Descartes und die Theorie der Lebenserscheinungen," *Sudhoffs Archiv* 50 (1966), pp. 25-42.

[114]A.G. Berthier, "Le mécanisme cartésien et la physiologie au XVII siècle," *Isis* 2 (1914), pp. 37-89; (1920/21) pp. 21-58.
E. Gilson, *Etudes sur le rôle de la pensée dans la formation du système cartesien,* Paris, 1930. (Etudes de Philosophie Medievale 13).

[115]K.E. Rothschuh, "Technomorphes Lebensmodell contra Virtus-Modell," *Sudhoffs Archiv.* 54 (1970), pp. 1-35.

[116]See biography by T.B. Settle in *Dict. Scientif. Biogr.* 2 (1970), pp. 306-314.

[117]Borelli's work has been translated into German:
J.A. Borelli, *Die Bewegung der Tiere.* Übers u. mit Anm. vers. durch Max Mengeringhausen. (Ostwalds Klass. 221) Leipzig, 1927.

[118]Foster, *Lectures,* Ch. III: "Borelli and the influence of the new physics," pp. 55-83.

[119]See biography by H. Straub in *Dict. Scientif. Biogr.* 3 (1971), pp. 36-46.

[120]O. Spiess - F. Verzar, (eds.), "Bernoulli's De Vita," *Verh. Naturforsch. Ges. Basel* 52 (1941), p. 189.

[121]L.L. Langley, "A historical introduction to the physiology of anoxia," *Bull. Hist. Med.* 14 (1943), pp. 321-40.

[122]Kurt Lasswitz, *Geschichte der Atomistik vom Mittelalter bis Newton,* 2 vols., Hamburg and Leipzig, 1890.

[123]Theodore M. Brown, "The mechanical philosophy and the animal economy: a study in the development of English physiology in the 17th and early 18th century," (Ph.D. dissertation, Princeton Univ., 1967).

[124]See biography by M.D. Grmek, *Dict. Scientif. Biogr.* I (1970), pp. 391-392.

[125]M. Salomon, *Giorgio Baglivi und seine Zeit,* Berlin, 1889.

[126]Giorgio Baglivi, *The Practice of Physick,* Eng. transl., London, 1704, p. 135.

[127]McC. Brooks - P.F. Cranefield, (eds.), *The Historical Development of Physiological Thought,* New York, 1959.

[128]M. von Rohr, "Kepler und seine Erklärung d. Sehvorgangs," *Naturwiss.* 18 (1930), p. 941.

[129]F. Plehn, "Johannes Kepler. Paralipomena ad Vitellionem seu Astronomiae pars optica," *Arch. f. Optik* 1 (1908), pp. 75, 114, 156.

[130]H. Koelbing, *Renaissance der Augenheilkunde, 1540-1630,* Bern & Stuttgart, 1967.

[131]J. Kepler, "De modo visionis," a transl. from the Latin by A.C. Crombie of *Ad Vitellionem Paralipmena* and related passages on the formation of the retinal image, in *Mélanges Alexandre Koyré,* Paris, 1964, I, pp. 136-172.

[132]Arthur Koestler, *The Watershed: a Biography of Johannes Kepler,* New York, 1960.

[133]H.M. Koelbing, "Ocular physiology in the 17th century and its acceptance by the medical profession," in *Steno and Brain Research in the 17th Century,* Oxford, 1968.

[134]S.M. Straker, "Kepler's optics: a study in the development of 17th century natural philosophy," (doctoral dissertation, Yale University, 1971).

[135]O. Temkin, "The classical roots of Glisson's doctrine of irritation," *Bull. Hist. Med.* 38 (1964), pp. 297-328.

[136]See Croone's biography by L.M. Payne in *Dict. Scientif. Biogr.* 3 (1971), pp. 482-483.

[137]L.G. Wilson, "William Croone's theory of muscular contraction," *Notes and Records Roy. Soc. London* 16 (1961), pp. 158-178.

[138]Hansruedi Isler, *Thomas Willis (1621-1675) Doctor and Scientist,* New York, 1968.

[139]E. Clarke, "The doctrine of the hollow nerve in the 17th and 18th centuries," in *Medicine, Science and Culture,* ed. by L. Stevenson & R. Multhauf, Baltimore, 1968, pp. 123-141.

[140]A. Meyer, R. Hierons, "On Thomas Willis' concepts of neurophysiology," *Med. Hist.* 9 (1965), pp. 1-15 & 142-155.

[141]R. Hierons - A. Meyer, "Willis' place in the history of muscle physiology," *Proc. Roy. Soc. Med.* 57 (1964), pp. 687-692.

[142]G. Canguilhem, *La formation du concept de réflexe aux XVIIe et XVIIIe siecles,* Paris, 1955.

[143]E.C. and Ph. M. Hoff, "The life and times of Richard Lower, physiologist and physician, 1631-1691," *Bull. Hist. Med.* 4 (1936), pp. 517-535.

[144]Richard Lower, A Facsimile Edition of *Tractatus de Corde,* pref., introd. & transl. by K.J. Franklin, Oxford, 1932.

[145]Margaret Espinasse, *Robert Hooke,* Berkeley, 1956.

[146]Robert Hooke, *The Posthumous Works of Robert Hooke,* New York, 1969.

[147]J.R. Partington, "The life and work of John Mayow (1641-1679)", *Isis* 47 (1956), pp. 217-230 & pp. 405-417.

[148]W. Bőhm, "John Mayow und die Geschichte des Verbrennungsexperiments," *Centaurus* 11 (1965), pp. 241-258.

[149]L.G. Wilson, "The transformation of ancient concepts of respiration in the 17th century," *Isis* 51 (1960), pp. 161-172.

[150]W. Bőhm, "John Mayow and his contemporaries," *Ambix* 11 (1963), pp. 105-120. See also: John F. Fulton, *A Bibliography of Two Oxford Physiologists: Richard Lower, 1631-1691; John Mayow, 1643-1679,* Oxford, 1935.

[151]See biography by M. Boas Hall in *Dict. Scientif. Biogr.* 2 (1970), pp. 377-382.

[152]Robert Boyle, *Robert Boyle on Natural Philosophy,* an essay, with select. writings by M. Boas Hall, Bloomington, Ind., 1965.

[153]D.E. Hall, "Mayow to Haller; a history of respiratory physiology in early 18th century," (unpublished doct. dissertation, Yale Univ., 1966), and M. Speter, "Lavoisier und seine Vorläufer," *Samml. chem. techn. Vorträge* 15 (1910), p. 109.

[154]J. Studtmann, *Nicolaus Steno, der grősste Naturforscher seiner Zeit, ein Apostel der norddeutschen Diaspora,* Hildesheim, 1934.

[155]Gustav Scherz, *Ein Pionier der Wissenschaft. Niels Stensens Schriften in Auswahl* Kopenhagen, 1963, and Gustav Scherz, (ed.), *Steno and Brain Research in the 17th Century,* Oxford, 1968.

[156]Raffaello Cioni, *Niels Stensen, Scientist-Bishop,* New York, 1962.

[157]Archibald E. Clark-Kennedy, *Stephen Hales, D.D., F.R.S.; an 18th Century Biography,* Cambridge, 1929.

[158]G.E. Burget, "Stephen Hales (1677-1761)," *Ann. Med. Hist.* 7 (1925), pp. 109-116.

[159]R. Greeff, *Kurze Geschichte des Brillen-und Optikerhandwerks.* Hrsg. v. Reichsinnungsverband d. Optiker — u. Feinmechanikerhandwerks. Weimar, 1938.

[160]A. Schierbeek, *Measuring the Invisible World. The Life and Works of Antoni Van Leeuwenhoek,* New York, 1960.

[161]L. Belloni, "Zur Geschichte der tierforschenden Mikroskopie," *Nova Acta Leopoldina N.F.* 30 (1965), Nr. 173, pp. 443-458 and "Auf dem Wege zur Elementardrüse als Sekretionsmachine," in: Med. Geschichte im Spektrum, *Sudhoffs Archiv.* Beiheft 7, Wiesbaden 1966, pp. 11-29.

[162]F.W. Bayer, "Die Anfänge mikroskopischer Forschung in der Medizin," *Klin. Wschr.* 23 (1944), pp. 31-33 and
G. Miller, "Leeuwenhoek's observations on the blood and capillary vessels," in *Medicine, Science and Culture,* Baltimore, 1968, pp. 114-122.

[163]See article by E. Farber in *Dict. Scientif. Biogr.* 2 (1970), pp. 305-306.

[164]J.S. Wilkie, "Marcello Malpighi (1628-1694)," in *Late 17th Century Scientists,* ed. by D.W. Hutchings, Oxford, 1969 and
M. Cardini, *Malpighi,* Rome, 1927.

[165]Howard B. Adelmann, *Marcello Malpighi and the Evolution of Embryology,* 5 vols., Ithaca, N.Y., 1966.

[166]E. Clarke, "The brain 'glands' of Malpighi elucidated by practical history," *J. Hist. Med.* 23 (1968), pp. 309-330.

[167]F.J. Cole, "Jan Swammerdam," *Nature* 139 (1937), p. 218.

[168]Clifford Dobell, *Antony van Leeuwenhoek and his "Little Animals",* New York, 1958.

[169] K. Heinemann, "Zur Geschichte der Entdeckung der roten Blutkörperchen," *Janus* 43 (1939), pp. 1-42.

CHAPTER IV

The Physiology of the Enlightenment

1.
HERMAN BOERHAAVE, FRIEDRICH HOFFMANN, GEORG E. STAHL

A profound change in Western thought occurred at the beginning of the eighteenth century. Called the "Enlightenment," this new era in the development of the human spirit emerged in the culture of Europe.[1] Appropriately, the French author Pierre Costa translated into his native language the work of the English philosopher *John Locke* (1632-1704) entitled *Essay Concerning Human Understanding,* and published it in 1700. This event led to a rapid diffusion of the Lockean philosophy on the Continent, making Leibniz comment: "Finis saeculi novam rerum faciem aperuit."

The roots of the Enlightenment can be traced far back into the Renaissance, Reformation and Counter-Reformation. Its successful progress had begun in England where both philosophers Thomas Hobbes and John Locke proclaimed that all knowledge had to be based on experience. They felt that even when experience was rather limited, it still provided answers about our way of life — a goal which was preferred over the attempt to solve metaphysical problems.

The French Enlightenment, led by the philosophers *Jean le Rond D'Alembert* (1717-1783) and *Denis Diderot* (1713-1784) continued the earlier efforts to displace the "era of theology and faith," replacing these disciplines with "science." The term addressed itself to the natural sciences which had reached a temporary culmination in Newton's impressive work entitled *Philosophiae naturalis principia mathematica* (1687), which was based on the empirical method. Therefore, to collect facts, interpret them critically and rationally, and then test the conclusions in the concrete world of the phenomena, became the generally accepted method which promised success in the sciences. In the religious realm, a new rational kind of religion emerged which opposed the older and revealed one.[2]

The spirit of the Enlightenment developed somewhat slower on German soil than in England and France. At the turning point of the 18th century lived a great universal thinker, the German philosopher *Gottfried W. Leibniz* (1646-1716). Mathematician, logician, jurist, technician, politician and historiographer, Leibniz attempted to reconcile existing and opposing viewpoints. He tried to eliminate the contrasts between the general and particular, matter and spirit, mechanism and teleology, experience and innate knowledge as well as freedom versus necessity, religion versus philosophy.[3]

The Enlightenment developed fully in Germany only towards the middle of the eighteenth century, culminating with the works of the philosopher *Immanuel Kant* (1724-1804). In 1784, Kant characterized the new intellectual movement as follows: "Enlightenment is man's leaving his self-caused immaturity. Immaturity is the inability to use one's intelligence without the guidance of another. Such immaturity is self-caused if it is not caused by lack of intelligence but by lack of determination and courage to use one's intelligence without being guided by another. Sapere Aude! Have the courage to use your own intelligence! is therefore the motto of the Enlightenment."

Side by side with this strong wave of rationalism, a second current stressing more feeling and religion existed throughout the eighteenth century. The movement was reflected in the rise of the German pietism and the ideas expressed by the French philosopher *Jean-Jacques Rousseau* (1712-1778), and eventually was instrumental in overcoming the Enlightenment altogether.

During the so-called "enlightened" period, man was highly receptive to the various sciences. Indeed, the scientist achieved a high standing within society, above poets and writers, and he was in favor with the public. The salons concerned themselves with the current activities of the scientific community, and the conviction existed that a systematic cultivation of the sciences was advantageous for the development of society and would benefit human life conditions. Thus, Leibniz demanded in 1700, at the foundation of the Prussian Academy of Sciences, that this organization would concern itself not only with *curiosa,* but also with *utilia.*

The practical usefulness of the scientific inquiries appeared in all fields of endeavor. The eighteenth century developed and improved clocks, thermometers, air pumps, low and high compression steam-engines, air balloons, mechanical looms, jennies and chlorine bleaching. In addition, porcelain and glass manufacturing, weaving mills, ironworks and many other industries were developed.

The exact natural sciences made a great deal of progress in this period. In physics, the study of gases, electricity and heat — all important subjects for the evolution of physiology — advanced considerably. Equally important was the progress in chemistry. Oxygen, nitrogen and hydrogen were discovered, and many new substances isolated. Gradually the chemical affinities and combinations, as well as the reactions of acids and salts, were better understood. The process of combustion — oxidation — became more clear towards the end of the century, thanks to the work of the French chemist Antoine Lavoisier who destroyed the "phlogiston" theory.

Therefore, physics and chemistry furnished a great number of newer methods and ideas towards the solution of physiological problems. However, the first half of the eighteenth century exhibited only a relatively small number of significant discoveries. Rather, efforts were made during that time to order and systematize the already available knowledge, incorporating the useful parts into medicine's theoretical and practical aspects. Thus, only the second half of the century witnessed the appearance of new viewpoints for the interpretation of physiological questions, such as the attempts by Haller, Lavoisier and Galvani to solve some individual problems of physiology.

Among the most notable eighteenth-century systematists of physiological knowledge were, before 1750 and in chronological order, *Georg E. Stahl* (1659-1734), *Friedrich Hoffmann* and *Herman Boerhaave* (1668-1738). Their works remained and continued to be discussed until the end of the century. Both, Friedrich Hoffmann and Herman Boerhaave, had very similar ideas and careers, although Hoffmann began about ten years earlier than his Dutch counterpart. They were both great teachers, theoreticians and clinicians, as well as popular, influential and wealthy individuals.

Friedrich Hoffmann left an enormous volume of writings which in contrast with other authors have been only sparsely studied. Like Boerhaave, he tried to incorporate into a medical system the results obtained by the new physics, chemistry, anatomy and microscopy based on the work of Descartes, Leeuwenhoek, Boyle, Willis, Borelli, Malpighi and Baglivi. His first two publications, *Medicinae Mechanicae Ideae Universalis* and *Fundamenta Medicinae* (Halle, 1695), already contained the bases of a system erected on the physico-chemical concepts of Descartes and Boyle.[4] Hoffmann's medical system was published in its final form under the title *Medicina Rationalis Systematica* (1718/40).

According to Hoffmann, motion and matter determined all natural phenomena (Descartes), and there were no "formae substantiales." The bodies acted according to divine laws and mechanical necessity and therefore, all vital actions had to be explained mechanically. Immaterial substances such as the soul had no motive forces of their own, although Hoffmann granted that the soul could influence the direction of the bodily movements and the onset of nervous action.[5]

For Hoffmann, motion was necessary for the preservation of life. The primary motive impulse was provided by the "subtilissimae aethereae elasticae particulae" which were present in the arterial blood at the level of the lungs. These particles formed the most subtle and mobile component of the blood, and caused also the "motus intestinus" (Willis). These fine parts flowed through the cardiac nerves and vessels into the muscle fibres of the heart, causing their systolic swelling and contraction. In diastole, Hoffmann postulated that the fibres returned to their initial position of rest, following their inherent elasticity. Moreover, all other fibres in the solid parts of the body had their own elasticity or tonus, which was maintained by the ethereal particles of the blood and nerves.

The blood consisted of liquid and solid particles (Boyle), and was in constant inner motion, thus providing vitality and animal heat. Hoffmann's "subtle particles" which in his system played the role of prime movers of the human machine, originated in the air, where they existed in the form of vesicular and elastic particles. They arrived in the body with the foodstuffs and through the body's pores, not by breathing air. Once in the blood, these particles contributed to the motion, mixture and separation of the blood's components, a task which at the level of the lungs gave rise to the vital effects of respiration.

In addition to the blood's "motus intestinus," there was a second source of movement: the action of the heart which projected the blood to the periphery and into the venous system. According to Hoffmann, glandular action was possible because the glands allowed the excretion of certain particles according to the size and form of their pores.

In the brain, Hoffmann thought that the "globuli aetherei" and a nourishing nervous fluid were forced from the blood into the nerves thanks to the systole and diastole of the dura mater. The distribution of this ethereal compound occurred through the medulla oblongata, the spinal cord and the peripheral nerves to the muscles. Muscular contractions were therefore produced by the nervous fluid, which dilated the fibres and shortened them in the process (Descartes). On the other hand, sensation was possible because of a centripetal oscillation of the nervous fluid, while the soul could initiate and direct the muscular movements through a "commercium" with the ether circulating in the nerves.

As already mentioned, Hoffmann was strongly influenced by Descartes and not by Leibniz. Most of his physiology was Cartesian in character, with some elements taken from Thomas Willis, Robert Boyle and other adherents of the corpuscular and mechanical philosophy such as Malpighi, Bohn and Borelli.

Herman Boerhaave spent almost his entire life at Leyden, a small city not far from Amsterdam.[6] At the University of Leyden, founded in 1575, famous men such as Franciscus Sylvius, Jan de Waal, Frelincourt and Nuck had already been active during the seventeenth century. But only in the eighteenth century did Leyden attain its highest international fame with a galaxy of famous teachers which included not only Boerhaave, but the physician *Frederick Dekkers* (1648-1720), the anatomist *Bernard Albinus* (1653-1721) and his son *Bernard S. Albinus* (1697-1770), the naturalist *Gottfried Bidloo* (1649-1713) and finally the medical systematist *Jerome D. Gaub* (1705-1780). A stream of young people from the various European countries came to study at Leyden, attracted mainly by the prestige which the name Boerhaave came to represent in Western medicine[7] (see Fig. No. 34).

Boerhaave's significance for the development of clinical medicine will remain only a peripheral concern here.[8] To understand his importance and

Fig. 34. Lecture room of Boerhaave in Leyden.

significance for medical theory it should be pointed out that there was no coherence among the numerous and bold hypotheses formulated by the late seventeenth-century iatrochemists and iatromechanists regarding the genesis of diseases. Nevertheless, the investigation conducted in physics, chemistry, anatomy and physiology, had uncovered a great number of new and valuable insights, still awaiting inclusion into practical medicine.

This application of the recently acquired knowledge was the great task which Boerhaave tried to accomplish. Thanks to his wide knowledge in mathematics, botany, physics and the familiarity with authors such as Boyle, Sennert, Baglivi, Swammerdam, Huygens, Leeuwenhoek, Borelli as well as Bohn, Eustachius, Sydenham, Ruysch, Stensen and Stahl, Boerhaave set out to accomplish his goal. Moreover, he was acquainted with the writings of philosophers such as Descartes, Leibniz and Gassendi. Under those circumstances, Boerhaave wanted to provide medicine with a firm foundation, based on the new and factual knowledge obtained in the natural sciences.[9]

To be sure, Boerhaave made no great new discoveries of his own, but was content to sift critically the available knowledge and integrate it into a series of well-received works. Boerhaave, (see Fig. No. 35) was born January 1,

1668 in Voorhoot, near Leyden, during an era of great Dutch prosperity and flowering. After a hard period of theological instruction, followed by his medical studies, Boerhaave was named in 1701 a lecturer in medical theory — the so-called "Institutes of Medicine" (approximately equivalent to physiology). Eight years later he was appointed to the chair of botany, followed by that of practical medicine in 1714 and chemistry in 1718. For many years he carried this extremely heavy teaching load, and generations of students received their medical instruction from him. Some of the most famous disciples were the physicians Gaub, the French *Julien Offroy de La Mettrie* (1709-1751), *Andrew St. Clair* (1669-1760), *John Pringle* (1707-1782) and *Robert Whytt* of Britain, as well as *Johann T. Eller* (1689-1760), *Georg G. Richter* (1694-1773), *Lorenz Heister* (1683-1758), *Johann N. Lieberkühn* of Germany, and especially the Swiss *Albrecht von Haller* (1708-1777) and the Dutch *Gerhard van Swieten* (1700-1772). Boerhaave died, highly honored, in 1738.[10]

Fig. 35. Portrait Herman Boerhaave.

Like Hoffmann, Boerhaave was greatly influenced by the Cartesian philosophy and the atomism of Gassendi, and therefore mechanical considerations and interpretations played an important role in his medical system. He also possessed a great chemical knowledge, as reflected in his work entitled *Institutiones et experimenta chemiae*. The mechanical tendency is reflected in his *Oratio academica de usu mechanici in medicina* (1709) where he wrote: "Mechanices in Medicina usum esse summum, necessitatem maximum." Boerhaave further emphasized that the human body was truly a machine whose solid parts were primarily composed of vessels. These vascular structures contained, directed, modified, divided and secreted the body's fluids, while the other solid structures acted like mechanical instruments which could support or determine certain motions according to their own form and composition.

The fluids were supposed to have similar mechanical qualities of coherence and elasticity as well as possess a relative amount of the various blood particles. For Boerhaave, the motion, obstruction and stagnation of these fluids was a matter of life or death, health or disease.

Regarding the workings of the bodily organs, which Boerhaave explained in his physiological treatise *Institutiones Medicae* (Leyden, 1708) (see Fig. No. 36), he adopted predominately mechanical and some chemical principles. What follows are some examples of the Boerhaavian physiological interpretations. The nerve-fibres were considered to be hollow allowing the flow of a subtle liquid or spirit from the brain, which Boerhaave, like Malpighi, thought to be a gland. The particles composing this nerve-liquor were viewed as the most dense, fine, simple and mobile elements available in the body. In turn, the finest muscular fibres were hollow extensions of the nerves, surrounded by a lubricating fluid which was supposedly secreted by the muscle's arterioles. The muscular contraction occurred when the nervous fluid was impelled into the muscular fibres themselves.[11]

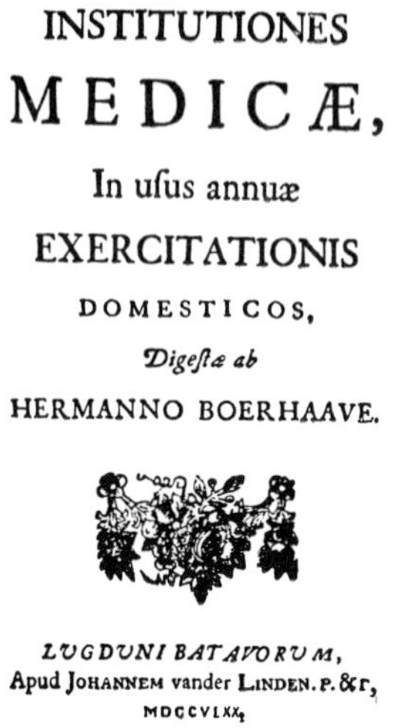
INSTITUTIONES
MEDICÆ,
In ufus annuæ
EXERCITATIONIS
DOMESTICOS,
Digestæ ab
HERMANNO BOERHAAVE.
LUGDUNI BATAVORUM,
Apud JOHANNEM vander LINDEN. P. &F,
MDCCVIXX.

Fig. 36. Title page of *Boerhaave's Institutiones Medicae.*

The cardiac pulsations were due to the adequate supply of blood from the coronary vessels and the arrival of the nerve-liquor. During systole, the blood was forced into the major arteries, dilating them, while the supply of nervous fluid was promptly interrupted. Thus, diastole represented the natural relaxation of the contracted fibres, while the body heat was the result of friction between the blood particles and the walls of the vessels.

The usefulness of breathing air was not clearly recognized in the Boerhaavian system of physiology. According to Boerhaave, respiration was not designed to cool the blood nor was it prepared to furnish the body with nitrous particles. Rather, the lungs played a role in the preparation of the blood. The venous blood, rich in chyle and composed of many porous, tenacious and sluggish particles, was subjected during its pulmonary transit to a series of pressures. These mechanical influences produced the fragmentation and division of the blood's particles, making them finer and more globular in form.

The digestive functions were viewed by Boerhaave in strict physicochemical terms. His concepts were not original, but taken mainly from the various doctrines espoused by the iatrochemists.

Of interest is Boerhaave's physiology of the sense organs. Regarding the eye's accommodation to see objects located near the face, he postulated that the vitreous humor actually displaced the crystalline lens towards the cornea.

Nearsightedness occurred when the diameter of the eyeball was increased or when the cornea exhibited an excessive convexity. By contrast, farsightedness was the result of an approximation between the cornea and the retina.

In regard to the hearing process, Boerhaave came close to the modern views on the subject. He wrote: "For by this Contrivance it is possible for an infinite Number of small tremulous Chords to be assigned, of an equal Degree of Tensity in this Lamella, in a Direction from its Basis terminating in its Apex; so that among these almost innumerable Chords there will be always some, which being in harmonical Consent, will tremble with every Sound, so as to be capable of representing the same, and of conveying it as such to the common Sensory; so that the nice Distinction of different Sounds is performed by means of the Membrane belonging to the Fenestra ovalis, while the Office of the Membrane of the Fenestra rotunda is only to give notice of a Noise or indistinct Sound."[12]

Boerhaave's views can probably be traced back to the ideas of the French anatomist *Joseph G. Duverney* (1684-1730) who had postulated that the tympanic membrane prepared the perception of the individual sounds, which were subsequently transmitted to the brain thanks to the vibrations of the stapes and the cochlea.[13]

The relationship between mind and body was, for Boerhaave, an insoluble problem. He wrote: "...there is such a reciprocal Connection and Consent between the particular Thoughts and Affections of the Mind and the Body, that a Change in one always produces a Change in the other, and the reverse, also."[14] In this respect, Boerhaave almost followed the ideas expressed by Leibniz.

If one compares the physiologies of Friedrich Hoffmann and Herman Boerhaave, they both seem to agree in their fundamental assumptions regarding the mechanical character of the human organism. However, Hoffmann, in developing the principles of matter and motion, dealt with many physiological problems in a more systematic and detailed fashion. By and large, Boerhaave's physiology did not provide any new facts. Yet, his attempt to use physical and chemical explanations in order to interpret organic functions, and the systematization of the available empirical knowledge for theoretical and therapeutical purposes, prompted other physicians to carry out similar efforts.[15]

One of Boerhaave's students, Haller, became interested in some of these questions of organic function, and was instrumental in preparing the way for further physiological progress in the second half of the eighteenth century. Likewise, the ideas of the influential German physician *Johann T. Eller* at the Charité in Berlin, departed very little from the basic Boerhaavian framework. Eller's work entitled *Physiologia et pathologia medica seu philosophia corporis humani* was translated into German by *Johann G. Zimmermann,* in 1748.

From 1708 onward, Boerhaave's *Institutiones Medicae* appeared in successive editions and was translated into French, English and German. As already mentioned, the impact of the title carried by this publication prompted a similar designation to the academic chair both in British and American medical schools. Under the name of "Institutes of Medicine," physiology became a recognized subject both for teaching and studying in the true Boerhaavian sense. In Edinburgh, a chair with this denomination was inaugurated there in 1726, and a student of Boerhaave, Andrew St. Clair, selected to occupy it. The term "Institutes of Medicine" was finally replaced at Edinburgh for the more familiar name "Physiology" in 1874.[16]

The third prominent medical systematist of the earlier eighteenth century was *Georg E. Stahl.* (See Fig. No. 37.) Unfortunately, his ideas were not useful for the acquisition of further isolated physiological facts. Rather, Stahl developed the concept of organism, an "animistic" view of vitality which was diametrically opposed to the mechanical current prevailing in contemporary physiological thought.[17] Thus, Stahl's animism played a dominant role in the formation of several vitalistic currents during the latter part of the century.[18]

Fig. 37. *Georg E. Stahl* (1659-1734).

Stahl was born October 22, 1659, in Ansbach, and he studied medicine at Jena together with Friedrich Hoffmann, graduating in 1684. After teaching a few years at his alma mater, Stahl became the personal physician to the Duke of Weimar between 1687 and 1694. Hoffmann called him to the newly founded University of Halle in 1694 where he was appointed professor of medicine. Stahl carried out his teaching at the university until 1715, when he was summoned to Berlin in order to become the personal physician of Frederick William I of Prussia. He died there May 14, 1734.

Stahl was an excellent physician and a first rate chemist, a very religious but also harsh person. The historical verdict regarding his person and work remains undecided.[19] He was an antimechanist, both by reasoning and feeling, and his fundamental theoretical ideas had already coalesced by 1684. Stahl's concepts were systematically expressed in the work *Theoria Medica Vera* (Halle, 1708), and further elaborated in a polemical publication entitled *Negotium Otiosum seu Skiamachia* (1720). In the latter, Stahl defended himself against thirty-three "Animadversiones" or reproaches, issued by the philosopher Gottfried W. Leibniz against the *Theoria Medica Vera* in 1708.[20]

The fundamental ideas expressed by Stahl in his theory of the organism can be summarized as follows.[21] God had created for man's salvation a body and an "anima" or soul. The body was merely the temporary companion for the immortal soul, and tended towards corruption because of its very composition. However, as long as the organism was alive, this dissolution did not take place thanks to the influence exerted by the "anima," which controlled the body's activities and directed its vital motions. Moreover, the soul had an aversion towards death and feared for the fragile bodily frame, constantly making the necessary adjustments for the preservation of the body with great wisdom and foresight.

The bodily machine itself had no activity nor will. All its parts were passive and every movement had to be communicated to them by the soul. Stahl viewed the "anima" as the "agens" while the body was only the "patiens," both, however, united and constituting the organism. Under these circumstances, Stahl granted that the body was a mechanism, in the sense that its various parts could be used for restricted mechanical movements. Moreover, the soul was capable of promoting and controlling these motions, thanks to its own intrinsic and incessant movement.[22]

Since the "anima's" own motive power was something incorporeal for Stahl, the mobility of the entire organism, as an effect of this soul, had likewise the same characteristics. Therefore, the incorporeal "motus" was constantly active within the body. Stahl's "anima" was also responsible for all local changes occurring in the organism. It could think and act without the need of any previous memory, experience or imagination, and many of the soul's actions were indeed unconscious. However, and while operating directly on the various organs without any intermediaries such as spirits, the soul still depended on the mobility and receptivity of the body, two qualities which followed the laws of matter.

One cannot expect within such a theoretical framework a great deal of interest for special anatomical and physiological questions. However, Stahl's animism or psychomorphic model of the organism exposed the inherent weaknesses of the mechanistic viewpoint, and its feeble explanations regarding some specifically living functions. In addition, the underlying Cartesian philosophy was viewed with a great deal of suspicion in Lutheran circles, as a probable atheistic movement since it opposed the Aristotelean tradition. Thus, there were several contemporary currents which favored the spread of the Stahlian ideas, and the ensuing opposition to the mechanical philosophy took a number of different forms in the remaining eighteenth — as well as early nineteenth — centuries.[23]

Stahl's animism influenced the ideas of the French physician *François*

Boissier de la Croix des Sauvages (1706-1767), professor at the University of Montpellier since 1732. Based on the concepts of Sauvages, and the thought of the French philosopher *Etienne de Condillac* (1715-1780), another vitalistic French physician, *Théophile de Bordeu* (1722-1776), established a living and spiritual principle which he called "la nature." His student, the physician *Paul J. Barthez* (1734-1806)[24] of Montpellier, and, somewhat later, the anatomist and pathologist *Marie François X. Bichat* (1771-1802), also held vitalistic ideas.

In Germany, the physician *Johann F. Blumenbach* (1752-1840)[25] promoted his own vitalistic views at Göttingen, where he was active as anatomist, physiologist and anthropologist. Impressed by the remarkable regeneration of the polyps, a fact discovered by the Swiss naturalist *Abraham Trembley* (1710-1784), Blumenbach postulated the existence of a "nisus formativus" or "Bildungstrieb" — a molding power — which had plastic and regenerative properties.

Blumenbach, representing the epigenetic point of view, contributed substantially to the knowledge of embryology. His originality and lively personality, as well as his philosophical position, captivated generations of students who came from all parts of Europe to attend his lectures. Finally, it should also be mentioned here, that the notable British surgeon *John Hunter* (1728-1793) was likewise strongly influenced by Stahlian ideas.

After this short digression, we should return once more briefly to Stahl and mention his most important contribution in the field of chemistry. Stimulated by the German chemist *Johann J. Becher* (1635-1682), Stahl developed the so-called "phlogiston" theory in the work entitled *Zymotechnia fundamentalis sive fermentationis theoria generalis* (Halle, 1697). This doctrine included all the phenomena later connected with the process of oxidation. Stahl assumed that during the combustion or calcination of metals, as well as in the process of respiration and fermentation, a combustible component escaped. Initially called by Becher "brennbare Erde" — combustible earth — the element was designated "phlogiston" by Stahl.

To be sure, the "phlogiston" theory was incorrect, since we know today that the process of oxidation is not characterized by the removal of an element, but rather by the incorporation of oxygen. However, Stahl's theory was important for the development of inorganic, organic and even physiological chemistry. For the first time, the hypothetical processes of "phlogistication" (incorporation of phlogiston) and "dephlogistication" (removal of phlogiston) postulated the reversibility of chemical actions which became a general principle. Moreover, these phenomena indicated that similar processes could occur in both the living and inert realms of matter.

The phlogiston theory was adopted by the most prominent chemists of the eighteenth century, such as *Joseph Priestley* (1733-1804) and *Henry Cavendish* (1731-1810) of England, *Carl W. Scheele* (1742-1786) of Sweden, and the Germans *Carl F. Wenzel* (1740-1793) and *Jeremias B. Richter* (1762-1807). The theory, as indicated above, was especially attractive because it promoted the reversibility of combustion. Stahl did not overlook the fact that during the "dephlogistication" (oxidation) the weight of the calcinated metals actually increased. However, he was more impressed by the common property of all combustible elements to generate heat and light. Thus, the apparent conversion of energy was the most important factor, and Stahl considered "phlogiston" to be the cause for the entire process.[26]

The repercussions to Stahl's chemical theory can only be summarily sketched here. At first, Cavendish considered his newly discovered hydrogen to be the pure "phlogiston." He also isolated the so-called "mephitic air" — nitrogen — from the atmospheric air, although Daniel Rutherford preceded him in communicating the discovery. Cavendish was the first to determine correctly the composition of the air and the water, thereby opening new possibilities for physiological research on respiration and metabolism.

The same was true with the discovery of oxygen, called the "fiery air" by Scheele (1771/2) and "dephlogisticated air" by Priestley (1774).[27] Priestley had already observed that this gas conferred to the blood in the lungs its peculiar red color. Hence, Stahl's chemical theories continued to influence scholars well into the last third of the eighteenth century, leading them to acquire new insights and make further discoveries. These findings culminated in the complete overthrow of the "phlogiston" theory by the French chemist Antoine Lavoisier, and the establishment of the modern theory of oxidation.[28]

2.
ALBRECHT VON HALLER AND THE STATE OF PHYSIOLOGY IN THE MIDDLE OF THE EIGHTEENTH CENTURY

Albrecht von Haller (see Fig. No. 38) was one of Boerhaave's most notable and critical students, whose writings typically reflect the advancements made by physiology over several centuries. Born October 16, 1708 at Bern, Switzerland, as the son of pious Calvinist parents, Haller began his literary and scientific endeavors early. A willful and sentimental person, he matriculated when only 15 years old, at the University of Tübingen. In 1725, Haller went to Leyden, where he studied for two years under the direction of Boerhaave, Albinus and Gaub.[29]

Haller was greatly stimulated by the scientific atmosphere at Leyden which became his new spiritual home. After his graduation there in 1727, he travelled extensively throughout Europe, visiting the British physician *James Douglas* (1675-1742) and *William Cheselden* (1688-1752) in London, the Danish anatomist *Jacques B. Winslow* (1669-1760) in Paris, and finally the

Swiss mathematician Bernoulli in Basel. At the age of twenty-one, Haller settled down in Bern and began to practice medicine. However, his keen interest in scientific subjects could not be satisfied by medical practice alone, and he began to carry out anatomical and botanical studies which brought him in touch with the contemporary scientific investigations.

In 1734, Haller was nominated to a post in the small "Anatomical Theater" of Bern, which had been built especially for him. In order to extend his activites even further, he applied for a chair at the newly founded University of Göttingen. Haller succeeded with his application and was named professor of botany, anatomy and medicine at the German university in 1736, barely twenty-eight years old.

Fig. 38. *Albrecht von Haller* (1708-1777).

For the next seventeen years, Haller carried out his multiple academic duties successfully and untiringly. It was said that he virtually devoted all his time to the work at hand, barely getting four hours of sleep daily between his activities. He founded a botanical and an anatomical institute at the university, performing in the latter no less than 360 dissections. Thanks to his successful technique of vascular injections, Haller was able to contribute valuable new knowledge in the anatomy of vessels, depicting his findings in superb illustrations: *Icones Anatomicae* (1743/56). Beginning in 1739, Haller also republished Boerhaave's lectures, which were accompanied by his own extensive commentaries. Moreover, in 1747 he assumed the editorship of the local journal entitled *Commentarii Societatis Regiae Scientiarum Gottingensis* or "Göttinger Gelehrtenanzeigen," contributing many book reviews to the various issues.

The fruit of Haller's physiological work was the book *Primae Lineae Physiologiae* (1747) and the monograph *De partibus corporis humani sensilibus et irritabilibus,* the latter being presented first in 1752 at the "Göttingischen Gesellschaft der Wissenschaften" — the Academy of Sciences of Göttingen. Haller attempted to distinguish in this monograph between the sensible and irritable (contractile) parts of the human and animal organisms.[30] Frictions at Göttingen, and an unstilled nostalgia for his native mountains caused Haller to resign his post in 1756. He returned to Switzerland, where he occupied a series of official posts, including the direction of a salt mine. However, his new activities allowed him to continue his restless scientific pursuits, resulting in the publication of various works, among them the first comprehensive eight volume treatise on physiology: *Elementa physiologiae corporis humani.*

In addition, Haller was concerned with embryological, botanical and anatomical problems. Highly respected throughout Europe, he died at Bern on September 12, 1777.[31] Haller's multiple accomplishments have been recently examined by a series of authors.[32,33]

In his work and intellectual position, Haller superbly represented the early Enlightenment in Europe.[34] In obvious contrast with the French philosopher Voltaire, he could still combine the Christian belief in creation and redemption with all the other rational demands of his time. Haller's faith allowed him to consider the existence of an all-powerful, wise and benevolent Deity in the contemplation of natural events. Moreover, he could question not only the causes but also the goals and meaning of those vital manifestations which occurred in the organism. Therefore, there was *no schism* between Haller the Christian believer, and Haller the scientist.[35]

For Haller, man had a perishable body subjected to natural laws, while his soul was a free moral entity endowed with immortality. Thus equipped with his physical and mental qualities, man could relate to nature and acquire some knowledge concerning natural events. Haller thought that this information was gathered by the senses which disclosed the existence of a real world independent from the observer. However, only thought was capable of transferring these perceptions into well-grounded experiences.

It follows that Haller viewed physiology as an empirical science. The organism was a bodily machine composed of passive and extended matter kept in motion by a series of incorporeal forces which could — following the Newtonian model — only be known by their effects. Thus, nature was determined by matter and motion, a viewpoint which closely follows the Cartesian philosophy and the ideas of Boerhaave and Hoffmann. On the other hand, Haller's empiricism can be traced back to Francis Bacon, John Locke and the examples of the great British scientists of the seventeenth century.[36]

Hence, as mentioned before, there was no hiatus between the worldview of the poet and scientist Haller, and his counterpart, the believer in revealed religion. Therefore, he was challenged and engaged in lively debates with Voltaire and the materialist physician La Mettrie. Haller's view of nature and his philosophy are reflected in the line of a poem which he wrote in 1731: "Ins Innere der Natur dringt kein erschaffner Geist, zu glücklich, wenn sie noch die äussere Schale weist" (from Haller's poem "Falschheit menschlicher Tugenden).

Like few men before him, Haller succeeded in solving some of the mysterious puzzles of vitality. His anatomical descriptions and botanical accomplishments cannot be considered here. Instead, Haller's physiological works will be analyzed, beginning with the *Primae Lineae Physiologiae* which offers a typical cross-section of the knowledge available in physiology around the middle of the eighteenth century. Essentially a short version of the later monumental work, the *Elementa,* Haller's *First Lines of Physiology* became a highly popular book, going through several editions.[37]

According to Haller, the solid parts of the body were composed of fibres formed by earthy particles which were held together by a gelatinous substance. There were three fundamental types of fibres: one group constituted the cellular tissue as a layer of fibres connecting the various organs of the body. Its fibres not only formed membranes, vessels and tunicae or organ coats, but also provided support for the vessels and nerves, shaping the various organs, glands and the inner layers of the bowels.

In addition to these connective fibres, Haller distinguished the muscular and nervous fibres, attributing to each of these types a specific structure and function. Haller's pioneering achievement lay precisely in this recognition of different functions which were related to certain characteristic cellular structures.[38]

Based on innumerable experiments with animals, Haller published in 1752 his monograph entitled *On the irritable and sensible parts of the body,* an epoch-making work in the history of physiology.[39] Haller concluded that only certain specific parts of the body reacted to pain, disclosing the presence of *sensibility,* while other structures without any obvious sensibility reacted with contractions when stimulated. The latter portions of the organism, which responded to mechanical, chemical and electrical agents, were thought to be endowed with *irritability.* Among the sensible organs, Haller placed the skin, the bulk of the muscular tissue, tongue, peritoneum, heart and nerves, while the cellular tissue, muscular tendons, articular ligaments, periosteum and dura mater did not possess any sensitivity.

Haller considered the heart, skeletal muscles, as well as the smooth muscle layers of bladder, uterus and intestines, to have irritability (contractibility). By contrast, the skin, lungs, liver, spleen, cellular tissue and nerves had no capacity for contraction. Thus, in Haller's view, the muscular fibres exhibited irritability while the nervous fibres possessed sensibility. Both were specifically vital properties, which Haller conceived as forces inherent in the living fibres themselves, rather than ascribing them to an immaterial principle such as Stahl's anima. Hence, Haller's position was closer to Boerhaave's mechanism than to Stahl's animism, although Haller's views regarding these specific properties for the living tissues stimulated the subsequent development of vitalistic thought during the remainder of the eighteenth century.

According to Haller, the heart was the most irritable organ of the body, composed of various layers of muscular fibres. These fibres, he claimed, were stimulated by the inflowing blood and responded with a systolic contraction. The cardiac nerves were not involved in the cardiac cycle, since their elimination did not stop the heart-beat, provided the organ continued to receive blood and air. Even isolated portions of the cardiac muscle maintained their rhythmic contractions, leading Haller to conclude that the cardiac cycle had no relationship to the pressure of the blood upon the cardiac nerves, a theory postulated earlier by Boerhaave.

Haller's assumption was confirmed by the fact that differently structured hearts of various animal species also exhibited a sequence of systole-diastole. In this context, Haller also demonstrated that the coronary arteries were not empty during systole, a fact evidenced by the stream of blood which issued forth from the vessels when they were punctured in vivisection experiments. Haller believed that the cardiac force had to be necessarily great in order to overcome the vascular friction, and should be measured in pounds, but he did not understand the calculations of Bernoulli.

Following Harvey, Haller postulated that the blood flowed from the heart through the arteries and returned to the central organ in the veins. The connection between the arterial and venous systems was possible because of the existence of small and tenuous vessels which could even be demonstrated with the aid of injection techniques. The blood flowed continuously, even in the face of a pulsating impact made by the new masses of the vital fluid. Haller declared that the smaller peripheral vessels eliminated the differences in the speed of the blood-flow which existed during systole and diastole.

The pulse originated when a wave of blood was ejected from the heart and went on to distend the arterial vessels. It was for Haller a local phenomenon, dependent upon a sudden increase in the amount of circulating fluid, which had to be accommodated. In his view, the arterial filling acted in turn as a stimulus, causing the arterial wall to contract and thus propel the blood further towards the periphery. Under those circumstances, there was a great deal of friction in the vessels: on the walls, at the points of ramification and among the various blood components themselves. This phenomenon not only imposed a great strain on the heart, but was also responsible for the warm temperature of the blood.

Regarding the pulmonary circulation, Haller believed that the inspiration of air facilitated the inflow of blood to the lungs, while the expiration accomplished the opposite results. These actions were mainly explained in terms of changes occurring in the pulmonary vessels as a result of vascular friction. Haller calculated that a stroke volume of 1½ ounces circulated 23½ times throughout the body in one hour.

The blood was composed of water, salt, earth, iron and coagulable elements, being neither alkaline nor acid in character. According to Haller, it also contained red corpuscles which possessed iron and acquired their spherical shape as a result of the vascular friction. Haller estimated that their total amount in circulation was about twenty-eight pounds, postulating that the blood's warm temperature resided in these red particles rather than being the result of mysterious agents such as the vital force of "Lebenskraft."

Haller thought that there was basically no difference between the arterial and venous blood, and rejected the idea that the blood absorbed some air at the level of the lungs. Moreover, he considered Mayow's hypothesis of an absorption of nitre particles to be incorrect. Respiration occurred through a widening of the thorax, and the lungs played only a passive role in the process, allowing the inflow of outside air which was attracted by the relatively rarefied atmosphere prevalent within the lungs themselves. The elevation and expansion of the ribs was caused by both the internal and external intercostal muscles. Haller's experiments demonstrated that the lungs were situated in close proximity to the thoracic wall. The pleural space between them was free of air, and when opened, caused the collapse of the lungs and their withdrawal from the chest cage.

Regarding glandular functions, Haller believed that the secretions actually transpired from fine pores existing in the vessels of the gland. His purely mechanical theory envisioned various secretions, formed by specific, small particles selectively filtered through the vascular walls.

The brain was abundantly irrigated with blood, containing about a sixth part of the total amount circulating in the body. Arguing against the ideas of Baglivi and Pacchioni, Haller denied that the cerebral membranes had any motions of their own. He felt that these structures merely pulsated with respiration and possessed no sensitivity. The gray matter was also rich in vessels, but Haller rejected Malpighi's notion that it was glandular in nature. In his experiments, he discovered that pressures exerted on certain areas of the brain and spinal cord resulted in the interruption of nervous action. Likewise, a nerve subjected to compression lost its ability to transmit impulses, a fact which caused Haller to oppose the idea of an electrical nervous transmission.

For Haller, the seat of the soul was not in the pineal gland nor in the corpus callosum. In opposition to Willis' theory, he also did not ascribe any special role to the cerebellum in the operation of the internal organs. Haller also argued against Borelli's assertion that the nerves were under tension and thus susceptible to mechanical oscillations during the nervous transmission. Instead, he asserted that the nervous fibres were hollow, and conducted a non-electrical fluid. This liquid circulated to the periphery in order to generate motions and centripetally in order to communicate sensations to the brain.

Haller's muscles were composed of solid fibres which were not arranged in a geometrical fashion as postulated by Stensen. These fibres possessed irritability — the capacity to contract after being stimulated. The normal stimuli were received from the nerves, since the arterial blood merely contributed the necessary nourishment and did not participate in the process of contraction. The muscles of the heart and the intestines experienced their contractions independently from the nervous system, through the action of special stimuli.

The phenomenon of hearing occurred through the atmospheric transmission of vibrations coming from the objects, the tone depending on the oscillatory frequency. In Haller's view, the sound travelled at the same speed in each tonal range. Then, the oscillations of the tympanic membrane were transmitted across the tympanic cavity by the three auditory ossicles. The vestibular or oval window was the point of departure for the cochlea and semicircular canals. For Haller the spiral structure of the cochlea suggested the presence of strings of various lengths, with those situated lowest being the longest. However, the precise operation of the semicircular canals and the cochlea in hearing still remained shrouded.

According to Haller, the cornea of the eye collected all the incoming rays through a process of refraction. In turn, the lens, which was not a humor, and the vitreous body, both completed this refraction, allowing the projection of an inverted and sharp image onto the retina. Both optic discs were blind spots (E. Mariotte, 1668). Haller realized that in order to always achieve a clear picture of distant or near objects, a sharp image had to be conveyed to the retina at all times. In the past (see Boerhaave), it had been postulated that the lens experienced a certain displacement in order to achieve the necessary refraction, but Haller could not discover any structure in the eye which was capable of promoting this movement. Thus, he believed that perhaps a pupillary contraction accomplished near vision. On a related topic, Haller felt that the single image perceived by the viewer was not dependent on the anatomical fusion or crossing of the optical nerves. Instead, the single vision occurred when the images of the perceived objects were projected on similar areas of the two retinas.

Haller declared that the saliva accomplished a digestive function in the mouth with the aid of a fast-acting force. Moreover, in his view, the gastric juice was neither acid nor alkaline. The hunger sensation was a result of friction between the gastric mucosal folds when the stomach was actually empty. The pancreatic juice supported, as the saliva, the breakdown of foodstuffs, while the liver allowed the passage of fluid from the portal system to the biliary ducts. Once in the gut, the bile overcame the acidity of the food, dissolving the curdled milk and the oily substances, stimulating also intestinal motility. Thereafter, the intestinal juices continued the breakdown of the nutritious substances, leading to their absorption by the mesenteric vessels and lymphatic channels.

Finally, the kidneys separated water and salts from the blood by forcing these elements out during their passage through small renal vessels, while thicker elements, such as the cells and fibrin, remained within the blood.

We complete this summary of Hallerian physiological views with a final assessment of Haller's position in the history of physiology. In the first place, Haller was an anatomist, "ut vix quidquam nos in physiologicis scire persuadetur, nisi quae per anatomen didicimus." Haller judged the function of organs in the light of their anatomical structure. His experiments were almost always anatomical in character, such as the frequent ligation, preparation and injection of vessels. In that sense, Haller's physiology was mostly an "anatomia animata" — animated anatomy. He favored the use of vivisections, which he considered justified from the scientific viewpoint. However, single, not repeated or confirmed experiments were conducive to error: "Nullum umquam experimentum administratio nulla. Plurima sunt aliena, quae se in experimenta inmiscent."

Haller stressed the importance of chemistry and physics (hydraulics) for physiology, although he placed less value on physical interpretations for organic functions than did his predecessors. Regarding his bibliographical efforts he deserves high praise and admiration for the comprehensive and critical presentation of the hitherto available medical literature. Not only was Haller a polyhistorian and encyclopedist but also a fair critic who judged the contemporary accomplishments, and enriched many areas of knowledge with his own inquiries.

Haller's most original physiological contributions were the establishment of both concepts of sensibility and irritability, recognition of the cardiac automatism and confirmation of an airless pleural space. In addition, Haller rejected the concept of an active systole occurring in the cerebral membranes, and the previous attempts to localize the seat of the soul. Moreover, he clarified the relationships between the cerebellum and the motion of the heart as well as other vegetative functions.

Haller's efforts to solve hemodynamic problems has been the subject of recent studies.[40] According to these views, Haller correctly interpreted the origin of the pulse wave and assessed well the speed of the blood-flow. As stated before, he described the laminar flow, was aware of changing vascular diameters and the hydrostatic weight of the blood column. However, many errors and vague interpretations still remained for some time, such as the idea that the cardiac contraction was triggered by a stimulus from the blood. Also, Haller's concept that animal heat was the result of a friction between blood-particles persisted, together with the denial that nervous functions could be related to electrical processes and that air particles were important for respiration.[41]

Haller's prestige and authority stimulated a series of investigations in most fields of physiological endeavor while stifling a few others. His concept of sensibility and irritability as two specific vital forces was rapidly adopted by the contemporary medical systematists and incorporated into their theories of disease. The French vitalists in particular, used and expounded on Haller's views. Bordeu declared that even the glandular secretion was a specifically vital action.

At Göttingen, Blumenbach considered several true vital forces: 1) contractility of the cellular tissue; 2) irritability; 3) sensibility or nervous force; 4) specific vital forces and 5) a moulding power.[42] In his famous speech, the physician and biologist *Karl F. Kielmeyer* (1765-1844) distinguished five different forces in 1793: sensibility, irritability, a reproductive power, and a secretory as well as propulsive force.[43] These properties dominated alternatively within the body according to the evolutionary position achieved by a given living organism, a theory which provided an important point of departure for the subsequent interpretations of the romantic philosophers of nature.

On the other hand, the Scottish physician *William Cullen* (1710-1790) established his medical theories on alterations of one supreme nervous principle.[44] His student, the physician *John Brown* (1735-1788), followed these ideas to a great extent in fashioning his own system of medicine. In 1780, Brown published his work entitled *Elementa Medicinae,* in which he declared that life was merely a passive state which had to be continuously maintained through the action of stimuli.[45] Health was achieved if the stimulations were adequate and resulted in a normal amount of excitability in the body. However, insufficient or excessive stimuli caused a decrease or increase in excitability, and thus disease ensued. Therefore, Brown distinguished two fundamental forms of disease, one of excessive excitability called "sthenia," and another of inadequate excitability known as "asthenia."[46] Brown's medical system was widely accepted throughout Europe, especially in Germany[47] and Italy, generating a flood of favorable and critical publications.[48] Brownianism was another basis for the establishment of the so-called romantic ideas in medicine.

The medical concepts of the German physician *Christoph W. Hufeland* (1762-1836) were based on the idea that the patient had a certain capacity of reaction. Even the law of specific sensory capacity of the famous German physiologist *Johannes P. Müller* (1801-1858) was founded on the notion that the organism was able to react to stimuli in accordance with its irritability and organization.

Thus, Haller's original discovery was interpreted in various ways, producing a series of unexpected developments of which the Swiss physician would not always have approved. Haller's physiology was predominantly qualitative in character, since it corresponded to the idea of an "anatomia animata." Quantitative measurements still remained in the background, even though authors such as Borelli, Kepler, Stensen, Hales and Boyle had already used them. To be sure, it is difficult to harmonize Haller's qualitative attitude with the contemporary state of physical and chemical measuring techniques. The Hallerian approach did not favor the introduction of measurement and numerical expressions for the elucidation of the vital functions. Hence, Haller's physiology remained close to the knowledge of structure, and became both the source of his fame as well as his limitation.

3.
PHYSIOLOGY AT THE END OF THE EIGHTEENTH CENTURY

During the second half of the eighteenth century, a series of major physiological discoveries were made, especially regarding the digestive system, the brain, spinal cord and nerves, as well as respiration and the exchange of gases, animal heat and electricity.

In 1752, an article by the French naturalist *René A.F. de Réaumur* (1683-1757) relating to digestion, appeared in the "Mémoires" of the Parisian Academy of Sciences. Réaumur (see Fig. No. 39) concerned himself with many questions of physics, chemistry, technology, mathematics, mining and biology. Among other accomplishments, he improved the thermometer, establishing the well-known Réaumur scale. Especially interesting for physiology were his experiments on the digestive process. In the article already mentioned, entitled "Sur la digestion des oiseaux," Réaumur described how he fed small tubes filled with meat and other substances to tamed kites. Then, the naturalist examined the contents of these same tubes after they had been excreted again by the animals, and noted that the meat had been digested to a great extent in the stomach, more so than the starches. Moreover, he obtained almost pure gastric juice with the use of small sponges, which the birds first swallowed and then vomited again.

Fig. 39. *René A.F. de Réaumur* (1683-1757).

Réaumur's observations were supplemented by those of the Italian naturalist *Lazzaro Spallanzani* (1729-1799), who for a long time held the chair of natural history at the University of Pavia (see Fig. No. 40). One of his inquiries concerned the question whether infusoria originated through spontaneous generation or from some seeds. Long before Pasteur, Spallanzani concluded that no microscopic organisms could appear in heated and hermetically sealed glass containers until they were opened and exposed to the atmospheric air.[49]

Fig. 40. *Lazzaro Spallanzani* (1729-1799).

In his work entitled *Expériences sur la génération* (1787), Spallanzani followed in Réaumur's footsteps and tried to determine whether the male frog issued forth a fertilizing substance during copulation.[50] For this purpose he designed some covers made of waxed cloth for the males and prevented fertilization of the females. In 1780, Spallanzani carried out successfully, for the first time, an artificial insemination in a female frog. His studies on gas exchange, regeneration of the nervous system and circulation have already been examined in detail.[51]

Spallanzani's experiments regarding the action of digestive juices went considerably beyond the tests conducted by Réaumur. With remarkable fortitude, he himself swallowed small linen bags containing meat or bread, perforated wooden tubules with food, and some sponges, all of which he vomited again after certain intervals. The courageous naturalist observed the partial dissolution of these food particles, concluding that the gastric function — stressed by Boerhaave and Haller — played only a very minor role in the digestive process.

Spallanzani perceived that the gastric juice acted similarly in vitro on the meat and bread, a process of decomposition which was accelerated by heat. He concluded that the gastric digestion was not a genuine fermentation since no gas was produced. Indeed, the gastric juice actually prevented the putrefaction of the foodstuffs, although the Italian physiologist could not demonstrate chemically the acid nature of this secretion. Thus, Spallanzani demonstrated that once the food reached the stomach, a peculiar chemical process of disintegration ensued. This phenomenon was not entirely dependent on heat and mechanical friction, nor could it be interpreted as a putrefaction or fermentation.

Similar experiments were carried out by the British surgeon and anatomist *John Hunter* at London[52] (see Fig. No. 41). We are indebted to him for a series of anatomical discoveries, such as the description of the branches belonging to the N. olfactorius, the arterial supply of the uterus and the lacrimal apparatus in man. Moreover, Hunter carried out a series of careful comparative anatomical and physiological examinations in various species of animals: *Essays and Observations on Natural History, Anatomy, Physiology, Psychology and Geology* (London, 1861).[53]

Fig. 41. *John Hunter* (1728-1793).

In 1786, Hunter severely and unjustly criticized Spallanzani, accusing him of gross errors. Hunter, strongly influenced by vitalists ideas, had been misled to assume that the process of digestion was not a chemical but a vital occurrence, in which the foodstuffs were transformed into animal tissue. He considered all living actions to be the effects of a vital principle which was the immediate cause for every bodily phenomenon.

According to Hunter, the living principle resided primarily in the blood and thus exerted its vital action throughout the organism. Hence, he believed that the blood, on its own volition, could expand or contract in the vascular system. In turn, the blood-clotting and clot-retraction were manifestations of the supreme vital action in the dying blood, in some way comparable to muscular contractions.[54]

Further and essential contributions to the physiology of the blood were made by the British physician *William Hewson* (1739-1774). As already mentioned, blood coagulation had been viewed as the result of a cooling process or due to a loss of vitality within the blood itself. In his work *Experimental Enquiry into the Properties of the Blood* (1771-72), Hewson demonstrated that the cold could only delay blood-clotting but not avoid it. Instead, the student of both brothers William and John Hunter discovered that the blood actually possessed a coagulable lymph (today's fibrinogen) which was absolutely necessary for the clotting to take place.[55] The so-called "crusta phlogistica" or clot which followed any puncture for blood-letting was, in Hewson's view, not a substance, but merely coagulated lymph. Also, for the first time, he reliably demonstrated the presence of colorless and round particles in the blood's serum, the leucocytes. By contrast, the red blood cells were flat, small bodies, a question of cellular morphology only completely elucidated in the nineteenth century by Johannes Müller in Germany.[56]

Decisive progress was made in physiology towards the end of the eighteenth century regarding the nature and combination of the atmospheric gases. As mentioned, Scheele and Cavendish had discovered the oxygen, and Cavendish had isolated hydrogen and oxygen. However, in the biological metabolic processes, another gas was of the utmost importance, carbonic acid. Van Helmont had already mentioned its existence, calling it "gas sylvestre," but the Scottish chemist *Joseph Black* (1728-1799) studied in detail the appearance of this gas in both chemical and biological reactions.

Black showed that this so-called "fixed air" which could be precipitated in lime water, also originated during fermentations and the heating of carbon, and was exhaled during respiration. Moreover, the excessive accumulation of this gas was lethal for living beings and responsible for the extinction of the flames.

On August 17, 1771, Priestley brought a series of living plants into a room filled with "fixed air," in which a flame had just been extinguished because of the prevailing carbonic acid concentration. Ten days later, after opening the tightly closed chamber for the first time, Priestley was able to again burn a flame in the same air. The conclusion drawn from the experiment was that plants had the capacity of "cleaning" the air. In spite of these and other subsequent valuable insights — he obtained "dephlogisticated air" by heating mercuric oxide with a burning glass and sunlight in 1774 — Priestley was unable to clearly recognize the real nature of the chemical actions which occurred during metabolism, respiration and oxidation. These accomplishments were made through the experiments of the French chemist *Antoine L. Lavoisier* (1743-1794) (see Fig. No. 42).[57]

Fig. 42. *Antoine L. Lavoisier* (1743-1794).

Lavoisier, a Parisian by birth, received a careful education at the College Mazarin, and then, following his own inclinations, pursued various topics concerning physics and chemistry. In 1768, he became a member of the Parisian Academy of Sciences, and in the same year, received a position as tax collector which furnished him with the financial means to carry out his scientific research. Somewhat later, in

1776, he became inspector of the French powder and saltpetre works. From 1777 on, Lavoisier published a series of pioneering works which provided the actual basis for chemistry as well as the chemical foundations for respiration, metabolism and animal heat.[58]

On May 3, 1777, Lavoisier presented to the "Académie des Sciences" in Paris his views regarding combustion, opposing Priestley's conception that phlogiston was given up into the air during that process. According to Lavoisier, the real chemical occurrences proceeded in an opposite fashion. When mercury was burned, it gained as much weight as the air lost. Moreover, if the compound mercuric oxide was, in turn, heated, it yielded the pure metal and combustible air, the metal actually losing weight in the process.

Another of Lavoisier's observations was concerned with the components of atmospheric air. He realized that the air inhaled by an animal until suffocation contained only one-sixth of the so-called "fixed air." After absorbing this component in lime water, another gas remained, useless for respiration and incapable of sustaining a flame. This gas, which seemed unnecessary for the maintenance of life was called "azot" (nitrogen).[59]

Lavoisier recognized that the portion of air needed for respiration was transformed into gaseous carbonic acid in the body, a process which also occurred during chemical combustions. Thus, both phenomena of respiration and combustion were linked to a single process in which the so-called "burnable air" — Lavoisier termed it "oxygéne" in 1777 — was utilized. In a subsequent work, carried out with the French mathematician *Pierre S. de Laplace* (1749-1827), Lavoisier interpreted respiration as a process of slow oxidation or combustion of carbon in the body: *Mémoire sur la chaleur* (1783). Two years later, he reported on his first "respiration experiments" in animals. During these investigations, Lavoisier discovered that the volume of expired carbonic acid was smaller than the amount of assimilated oxygen. From this fact he deduced that part of the oxygen combined in the organism with hydrogen in order to produce water.[60]

Working together with the French chemist *Armand Séguin* (1767-1835), Lavoisier improved his experimental methods of studying respiration by replacing the consumed oxygen, *Mémoire sur la respiration des animaux* (1789). His investigations finally led to the conclusion that hydrogen combined with oxygen, gives off heat, at the level of the lungs. Thus, these organs were conceived to be the producers of animal heat. During physical exertion and digestion, the consumption of oxygen increased in close correspondence with the intensity of the organic combustion, with a resulting increase in animal heat (see Fig. No. 43). It should be stressed that these studies exposed for the first time the basic facts about the theory of energy transformation.

In 1791, Lavoisier's conclusions were criticized by another French chemist, *Jean H. Hassenfratz* (1755-1827), a student of the mathematician *Joseph L. Lagrange* (1736-1813). Hassenfratz pointed out that the processes of oxidation and combustion could not exclusively occur in the lungs, because in such a case these organs had to possess a greater temperature than the rest of the body. The alternative was to conceive that the postulated chemical reactions took place also in the blood and the entire organism. Thus, the blood first absorbed the oxygen, then allowed the combination of its own carbon with the oxygen to yield carbonic acid while the union with hydrogen gave water.

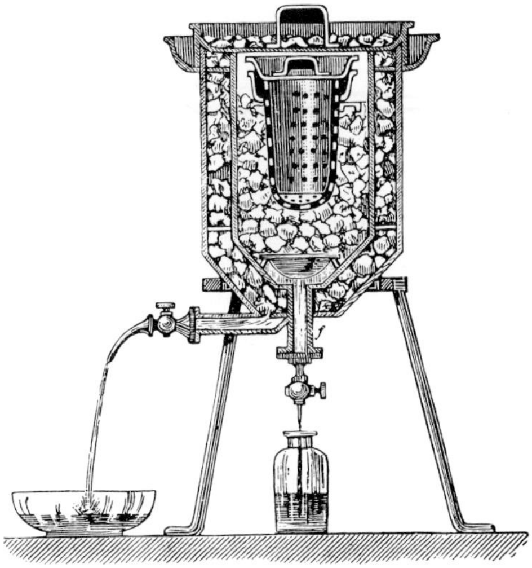

Fig. 43. The calorimeter of Lavoisier and Laplace for the measurement of combustion heat (1780).

Hence, the theories of respiration and metabolism enunciated towards the end of the eighteenth century approximated our modern views on the subject to a great extent. Although the oxidation process was thought to occur mainly in the blood, Spallanzani's last experiments pointed more in the direction of the bodily tissues. Using in his investigations some lower animals without lungs, the Italian physiologist discovered that these organisms also required oxygen in order to keep alive. He therefore concluded that a direct fixation of oxygen had to take place in the bodily fibres themselves, expressing for the first time the idea that both respiration and combustion occurred at the tissue level, *Mémoires sur la respiration* (Geneva, 1803).[61]

In the meantime, Lavoisier became one of the victims of the French Revolution. Sentenced to death by the guillotine, he lost his life on May 8, 1794 after the Revolutionary Tribunal had expressed the now famous phrase, "Nous n'avons plus besoin des savants." On the other hand, Laplace, talking about Lavoisier's execution, was quoted to have lamented that it took only one instant to cut the head of a man that could probably not be replaced within a whole century.[62]

During the discussion of Haller's physiology, it became obvious that the functions of specific parts of the nervous system still remained obscure. However, during Haller's lifetime and thereafter, several scholars began to study the problem, and systematic experiments stimulating or extirpating selective portions of the nervous system were attempted. Haller had already observed that dogs without a cerebellum continued to live for some time following the removal of this organ. Moreover, other investigations seemed to demonstrate that the meninges and cerebral cortex had no sensibility. In turn, Haller's stimulations of the cerebral surface did not trigger any motions, and responses were only obtained when deeper cortical areas were touched.

Experiments conducted by the German anatomist *Johann G. Zinn* (1727-1759) demonstrated that the medulla oblongata played an especially important role in the maintenance of vitality: *Experimenta circa corpus callosum, cerebellum* (Göttingen, 1749). In 1760, another French physician, *Anne C. Lorry* (1726-1783) discovered that the damage of a circumscribed portion of the medulla produced the instant death of the affected animal.

In turn, the French surgeon *Nicolas Saucerotte* (1741-1814) was able to cause paralysis of the front and hind legs through selective trephining of a dog's skull in 1768. These functional losses occurred always on the opposite side of the cerebral lesion, a fact already known to man since antiquity. Regarding the same topic, the French physiologist *François Pourfour du Petit* (1664-1741) had furnished a tentative explanation of the contralateral paralysis by describing the medulla's pyramidal decussation. In 1795, the British surgeon and chemist *William C. Cruikshank* (1745-1800) pointed out the importance of the medulla for respiration on the basis of his experiments with dogs. Therefore, as the eighteenth century came to an end, a series of observations and assumptions regarding certain localized nervous functions had gradually emerged.[63]

Reflex movements in frogs were observed early, but no proper understanding of the phenomenon involved had been achieved. Descartes explained certain unconscious reactions with the help of a very hypothetical mechanical process occurring in the brain, while Willis postulated a cerebral function analogous to the reflection of the light (1670).

In 1739, the English physician *Alexander Stuart* (1673-1742) observed that when he lightly touched the cut surface of the upper spinal cord in a decapitated frog, a strong contraction of the animal's legs took place (see Fig. No. 44). He explained this result by postulating that the touch actually pressed the nervous fluid from the cord further into the nerves and muscles. The Scottish physician *Robert Whytt* (1714-1766) also made some remarkable observations, described in his work *An Essay on the Vital and other Involuntary Motions of Animals* (Edinburgh, 1751). Among Whytt's discoveries was the abolition of the pupillary reflex to light after the destruction of the corpora quadrigemina (1768). However, Whytt declared that muscular motion was mediated by the nervous system but actually originated through the actions of a sensitive soul.

In 1751, the Swiss physician and disciple of Haller, *Johann G. Zimmermann* (1728-1795) demonstrated that decapitated frogs were still capable of

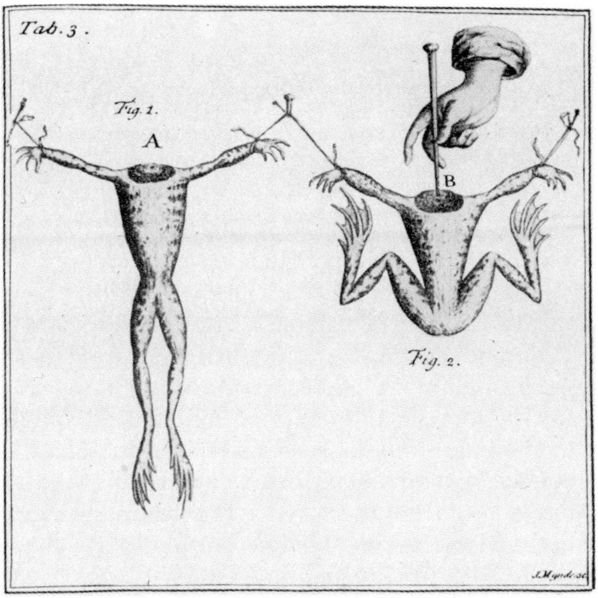

Fig. 44. Alexander Stuart and his experiments on frogs concerning spinal reflexes.

carrying out coordinated jumping movements and other defensive actions *(De irritabilitate,* Göttingen, 1751). This observation later led to the establishment of the concept of a "spinal cord soul."

At the turn of the eighteenth century, the Czechoslovakian physiologist *Georg Prochaska* (1749-1820) (see Fig. No. 45) also carried out experiments on the spinal cord of frogs. Prochaska, a well-known ophthalmologist and discoverer of the oliva[64] concluded that provoked motions occurred in the frog as a result of unconscious sensations. These sensations travelled in centripetal directions thanks to a nervous force until they reached the "sensorium commune" — the common sensory center — located in the central nervous system. In turn, emanating from this same focal point, came the

Fig. 45. *Georg Prochaska* (1749-1820).

response in the form of a centrifugally flowing order for the adequate motor reaction.[65]

Published in a work entitled *De functionibus systematis nervosi in observationes anatomicae-pathologicae* (Prag, 1784), Prochaska's scheme clearly outlined the basic ingredients of the reflex mechanism. However, the author spoke of motor and sensory nerve pathways which had not, as yet, been properly demonstrated. Moreover, he literally used the term reflection in order to explain how incoming sensory impressions gave rise to motor impulses within the common center.

The "sensorium commune" had its seat in the medulla, the crura cerebri and cerebelli, but especially in the spinal cord itself. Prochaska considered some of these reflected reactions useful, such as sneezing, coughing, the closing of the eye lids, all of which occurred unconsciously. Moreover, the reactions were also automatic and could not as a rule be influenced by the will. One can only admire Prochaska's theoretical interpretations of these reflective phenomena, which he described so accurately without the knowledge of the proper anatomical pathways.[66] The necessary structural knowledge only became available during the nineteenth century, as a result of the work carried out by the British anatomist *Charles Bell* (1774-1842), the French physiologist *François Magendie* (1783-1855) and the German Johannes Müller.

Finally, something remains to be said about the great advances made in the field of electricity which were applied to eighteenth-century physiology. The electrical machine, derived from the knowledge that friction creates an electrical charge, was used as a stimulating agent in various biological experiments. Likewise, the great speed of electrical conduction was early compared to the swiftness exhibited by the nervous processes in general. Thus, from 1744 the idea that the nervous action was perhaps electrical in nature gradually came to the fore and was expressed by authors such as Sauvages, the French chemist *Charles F. d. C. Du Fay* (1698-1739), and the Italian authors *Tommasseo Laghi* (1660-1734) and *Felice Fontana* (1720-1805).

Since 1745, the Leyden jar or precursor of the condenser became widely available, leading the Italian anatomist *Leopoldo M.A. Caldani* (1725-1813) to use it frequently for his experiments. Caldani, a professor of theoretical medicine and anatomy at the University of Padua, employed the electrical discharges of the jar as stimulants for his studies of animal tissues. Even though he was unsuccessful in his attempts to stimulate the heart, Caldani obtained a series of strong contractions in skeletal muscles (1757) with the sparks.

In this connection it should be mentioned that many people had already remarked about the strange phenomena — crackling noises and sparks — which occurred during the brushing and combing of hair. Rightly, these events were interpreted as being electrical in nature. Although the observations were erratic and appeared to be strange, they furthered the belief that the electric fluid was also related in one form or another to man and the animals.[67] What remained unclear, however, was whether the electricity played a significant role in physiological processes.

In 1766, Haller had denied that the nervous action was related in any way to the passage of an electrical fluid. Only after the rather surprising observations made by the Italian physician *Aloisius Galvani* (1737-1798) was the entire question brought anew to the limelight of physiological research and speculation.[68]

Galvani, a native of Bologna, had been a professor at the local university since 1762. In his diary, dated November 6, 1780, he made an entry recording the first observation of a contraction achieved in the legs of a frog stimulated by an electrical current. For the next ten years, Galvani continued systematically to observe these phenomena with the help of a Leyden jar, an electrical machine, atmospheric electricity and a metallic arch. Then, in 1791, he published his results in a work entitled *De viribus electricitatis in motu musculari commentarius*.[69]

Galvani's experiments were immediately reproduced and confirmed by others, although not all authors agreed with the physiological significance of the phenomena described by the Italian physician. Unfortunately, in spite of his international reputation, Galvani was stripped of all his academic posts for refusing to take an oath of allegiance to the new Cisalpine Republic. This event took place shortly before his death on December 4, 1798.

Basically, Galvani had observed that each time a spark from his electrical machine touched the frog's leg preparation, a strong muscular contraction ensued, provided the crucial nerve was in contact with a metallic conductor. He wrote: "Hereupon, I became extremely enthusiastic and eager to repeat the experiment so as to clarify the obscure phenomenon and make it known."[70] The observation was confirmed under a series of varying conditions, both in warm and coldblooded animals (see Fig. No. 46). Instead of using the sparks which emanated from an electrical conductor, lightning was observed to produce the same effect if the preparation was placed on a metallic hook or wire outdoors.

Subsequently, Galvani discovered that the muscles also contracted without the need of stimulating sparks of lightning, when a brass hook which had pierced the spinal column of the frog touched the iron plate on which the animal preparation was laying. This last effect gave rise to the speculation that the animal possessed both positive and negative electricity like a Leyden jar. Further experiments led to the conclusion that the muscle was the seat of both types of electricity, while the nerve acted merely as a conductor, transporting the nervous fluid in its center while the peripheral layers were formed by an isolating substance.

Galvani postulated that the source of this electricity was probably located

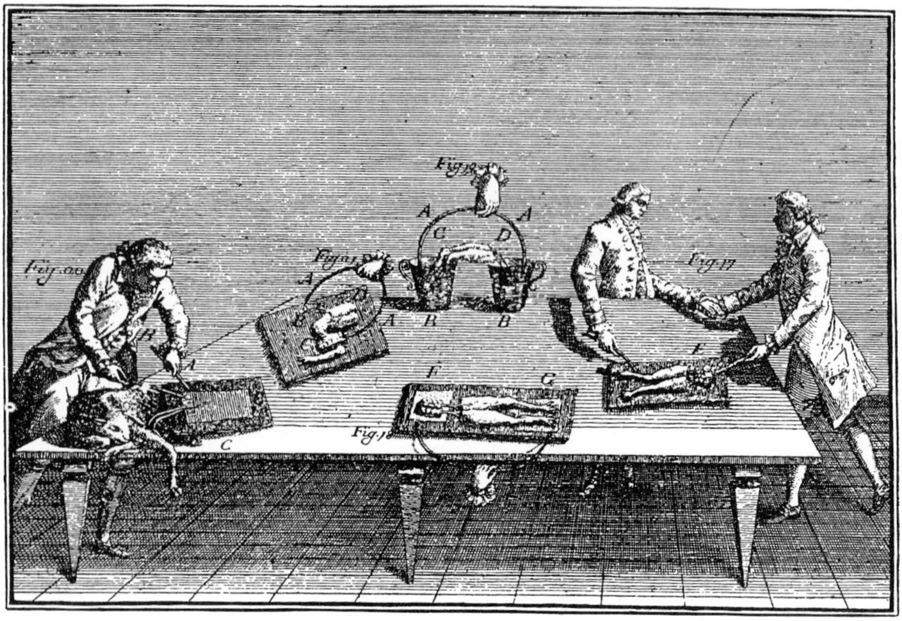

Fig. 46. *Galvani's experiments* about the origin and transmission of "animal electricity."

in the brain, whereas the muscles represented the true containers of electrical fluid in the image of the Leyden jar: their surface was charged negatively while the inside had a positive charge. Thus, during muscular contraction, the electricity arrived through the nerve and produced a temporary exchange and mutual cancellation of electrical charges between the surface and inside of the muscle. This phenomenon was responsible for the contraction of the muscular fibres.

Soon, Galvani's experiments were subjected to a rigorous analysis by the Italian physicist *Alessandro Volta* (1745-1827) at Pavia.[71] Under those circumstances, it became clear that almost all of the phenomena described by Galvani could be explained as the results of an electrical current which occurred following the contact between two different metals. Therefore, in Volta's view, the assumption of a so-called "animal electricity" was both unnecessary and totally unwarranted. In the ensuing controversy between the two scientists, Galvani and Volta, the former finally emerged victorious since he was able to demonstrate that the muscular contractions could also be produced without the contact of metals.

The results of these crucial investigations were published anonymously in 1794. In one of them, for example, Galvani placed the nerve of a muscle-nerve preparation on the fresh section of another muscle, eliciting contractions when the nerve and muscle established contact. For him, this was the decisive proof of the existence of an "animal electricity." Galvani's final experiments were confirmed in 1797 by the German naturalist *Alexander von Humboldt* (1769-1859), who published his findings in a two volume work entitled *Versuche über die gereizten Muskel — und Nervenfasern.*[72]

The Italian physician *Eusebio Valli* (1755-1816) carried Galvani's theory one step further in his book *Experiments on Animal Electricity, with their Applications to Physiology and some Pathological and Medical Observations* (London, 1793). Valli assumed that the muscles had a positive charge on their surface and a negative one inside during rest. Then, during contraction, this electrical equilibrium was disturbed, leaving the muscle without electricity, virtually unloaded, for a short period of time. Throughout this interval, the muscle could not be stimulated into contraction under any circumstances. Only after the electrical charge had been reestablished, did the muscle regain its excitability.

Valli's electrical interpretation of the latent period had been provoked by the controversy between Whytt, Haller and Fontana regarding the contraction and relaxation of the cardiac muscle.[73] As already indicated, Haller was of the opinion that the cause for a cardiac systole was furnished by the incoming blood, while diastole occurred only after all the stimulating blood had been expelled. However, Whytt had already established as early as 1751, that the heart could continue to contract even after the entire bloodflow had been stopped through the use of ligatures around the major vessels. Moreover, despite the presence of trapped blood in the ventricles, these cardiac chambers experienced diastolic relaxation.

Thus, Whytt's experiment seemed to demolish Haller's theory of the stimulating action of the blood. On this basis, Fontana, and Caldani conducted further studies regarding the problem. Fontana, in particular, conducted a series of carefully designed experiments regarding the relationship between a stimulus and the ensuing reaction of the heart. In the course of his investigations, published in 1760, 1767 and 1775, Fontana observed that a small needle prick was enough to trigger a cardiac contraction. Indeed, all that was necessary was the stimulation of one single fibre in order to obtain a response from the entire organ.

However, during the cardiac contraction and for a short period thereafter, the heart muscle remained totally refractory to any stimulations, a situation which lasted until the later phases of the diastolic pause. Thus, Fontana was undoubtedly the first to clearly observe and describe the physiological state in the heart later known as the refractory period. His analogies between the spark gunpowder and stimulus-contraction relationships anticipated the "all or none law."

These various observations related to the so-called "animal electricity" or Galvanism at the turn of the eighteenth century profoundly influenced the medical and physiological thought of the period. Among those who continued to study the problem were the Swiss physician *Christoph Girtanner* (1760-1800) and the German physician *Christoph H. Pfaff* (1773-1852). Some authors even expressed the hope that the Galvanic process was going to unlock the mystery of organic life.

In the meantime, further studies related to the physical manifestations of the electrical phenomena were carried out. Volta constructed his so-called "Voltaic pile" (see Fig. No. 47) which was later used extensively for physiological experiments.[74] Above all, Volta's discovery was a source for obtaining continuous electric currents and higher electrical voltage, which the German physicist *Johann W. Ritter* (1776-1810) used in order to decompose electrically conductive solutions: the birth of electrochemistry. Moreover, Ritter fascinated his circle of romantic friends in Jena with his work *Beweis, dass ein beständiger Galvanismus den Lebensprozess in dem Tierreiche begleite* (Weimar, 1798), in which he equated organic life to a Galvanic process.

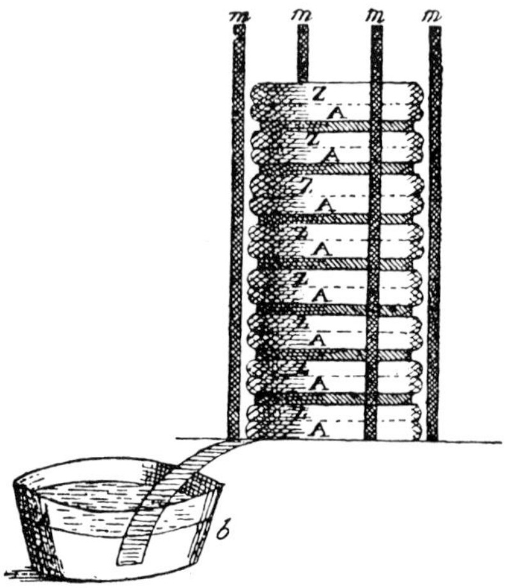

Fig. 47. Voltaic pile.

Under those circumstances, the German poet *Johann W. von Goethe* (1749-1832) also became interested in Galvanism. Goethe had already been concerned with various scientific topics, but we can only sketch here his own theory of colors (1810).[75] One can understand Goethe's viewpoint on the theory of colors only in the light of his methodological approach to other natural sciences. Thus, Goethe reintroduced, to a large extent, the method of dealing by analogy with subjects such as plant and animal morphology. He recognized the variations exhibited by the anatomical structures in different animal species as transformations of one basic and general prototype. Starting from this premise, the poet discovered the existence of the intermaxillary bone independently from another German biologist, *Lorenz Oken* (1779-1851). Moreover, Goethe was convinced that the skull was composed of a series of modified vertebral bones.

Goethe's comparative procedure led him to postulate the existence of an elementary plant or "Urpflanze" in botany, the prototype which underlies all the manifold forms and parts of plants: leaf, sprouts, thorn, calyx and corolla. The method of discerning elementary pattern was extremely fruitful in morphology, where it can be still applied today as a procedure of general induction.[76]

Hence, the diversity which existed in the phenomenal world could be sifted and ordered through comparisons and with the help of similarities. Goethe believed that this sensory experience and contemplation were appropriate to grasp immediately the order inherent in nature. He declared that only the reality which we could spontaneously experience with our senses, represented the extent of the knowable world. Goethe therefore opposed the use of microscopes and telescopes since they distorted the typical human standpoint and range of perception, widening the world of experience into areas which were otherwise impenetrable due to the limitation of our senses.

In Goethe's view, the experimental sciences were not in a position to consider the world of sense perception as the complete and correct image of reality. To the contrary, he felt that there was much more to the world than the small section which could be perceived by the human senses. Moreover, Goethe declared that the simple empirical observations were especially inadequate to acquaint us with those invisible forces of nature only known through their effects. More important, the empirical approach did not allow man to recognize the general and abstract lawfulness of nature behind the concrete reality grasped by his senses.[77]

To determine such natural laws, Goethe only accepted the use of the causal, analytical and experimental method in order to arrive at the correct induction. However, he was not willing to dissolve all sensual reality into a schematic system of forces, causes and effects. Thus, Goethe strongly opposed Newton's concept of the composite nature of white light, since it implied that an apparently simple sensory perception was actually complex, thereby demeaning the importance of the experiential world.

Instead, Goethe, as a phenomenalist, described and interpreted a series of subjective experiences with colors without considering the objective and physical aspects which underlied these occurrences. In the end, a physiological theory of color could only be satisfactory for him if it first outlined the possible physical phenomena which acted as stimulants. Secondly, the subjective events of color perception had to be all grouped together, and thirdly, all data had to be integrated and explained from the standpoint of a physico-biological event.[78]

Unfortunately, Goethe's concept was incomplete and onesided. He approached the subject from a general biological point of view which accepted the identity of external and internal phenomena. Thus, the external presence of a color was considered only in relation to the seeing eye. He wrote in the introduction to his work: "If the eyes were not related to the sun, how could we see the light? We could not enjoy Divine works unless we ourselves are pervaded by God's forces."

Goethe gathered a great number of observations in the area of color perception and tried to bring order into his experiences. His research later stimulated further investigations, carried out by Johannes Müller and the Bohemian physiologist *Johann E. Purkinje* (1787-1869). Goethe believed in the existence of three basic colors: red, yellow and blue. From the combination of these colors he explained the entire spectrum: yellow-red, yellow, green, blue, blue-red, red.[79] He was aware of the problem of color blindness and described a series of visual effects like colored images and shadows.

Goethe's distaste for the experimental and analytical approach in the natural sciences emerged repeatedly in his scientific writings, in a manner which is sometimes difficult to understand. To be sure, even Johannes Müller displayed in his early years a somewhat analogous attitude. Still attractive to the contemporary physiologist is the reading of Goethe's small treatise *Der Versuch als Vermittler von Objekt und Subjekt* (1792).

In retrospect, the eighteenth century was a period in which some important progress was made in a variety of physiological subjects. Gradually, the discipline had begun to evolve into an independent field of study, although its research and instruction were still connected with anatomy and practical medicine. Moreover, decisive impulses for further physiological development came more frequently from other sciences in progress such as physics and chemistry.

The single most important insight into the workings of the human organism during the eighteenth century was the further elucidation of combustion, metabolism and respiration. Likewise, the demonstration of animal electricity was of great importance. Haller's efforts regarding the distinction of irritability and sensibility contributed towards a better understanding of muscular contraction as well as the relationship between stimuli and the responses of the irritable organs (Fontana).

In addition, the source of animal heat was no longer attributed to a physical process of friction, but the whole phenomenon was now seen as a chemical reaction. The physiology of the blood was advanced through the discovery of the "clotting lymph" and the white blood cells. Conversely, the theories of vision and hearing were placed on more accurate anatomical foundations and also related to fundamental physical phenomena.

However, the general biological concepts were still strongly influenced by vitalistic points of view. Therefore, many problems were shunted into fruitless channels of inquiry or apparently solved with the help of metaphysical principles. The situation was only resolved during the nineteenth century, first in France, then in Germany where the development of a truly scientific and experimental physiology was delayed for some decades because of the so-called "romantic interlude."

REFERENCES

[1]Peter Gay, *The Enlightenment: an Interpretation,* 2 vols., New York, 1966/69.

[2]Paul Hazard, *European Thought in the Eighteenth Century,* transl. from the French by J. Lewis May, New York, 1965.

[3]R. Callinger, G. Risse, J. Marchand, *The Influence of Early Enlightenment Thought upon German Classical Science and Letters,* New York, 1972.

[4]Friedrich Hoffmann, *Fundamenta Medicinae,* transl. into English by L.S. King, New York, 1971.

[5]L.S. King, "Medicine in 1695: Friedrich Hoffmann's Fundamenta Medicinae," *Bull. Hist. Med.* 43 (1969), pp. 17-29.

[6]Gerrit A. Lindeboom, *Herman Boerhaave. The Man and his Work,* London, 1968. A short biographical sketch by the same author is contained in *Dict. Scientif. Biogr.* 2 (1970), pp. 224-228.

[7]L.S. King, *The Background of Herman Boerhaave's Doctrines,* Leiden, 1965, pp. 3-20.

[8]P. Diepgen, "Herman Boerhaave u. die Medizin seiner Zeit," *Hippokrates* (1939), pp. 289, 345.

[9]G.A. Lindeboom, "Boerhaave's concept of the basic structure of the body," *Clio Medica* 5 (1970), pp. 203-208.

[10]G.A. Lindeboom, (ed.), *Boerhaave and His Time,* Papers read at the Internat. Symposium in Leiden, 15-16 Nov. 1968. Leiden, 1970.

[11]See Lester S. King, *The Growth of Medical Thought,* Chicago, Ill., 1963, pp. 177-185.

[12]Herman Boerhaave, *Dr. Boerhaave's Academical Lectures on the Theory of Physic being a Genuine Translation of His Institutes,* 6 vols., London, 1745, IV, pp. 212-213.

[13]A. Kreidel, 'Zur Geschichte der Hörtheorien," *Arch. néerl. Physiol.* 7 (1922), p. 502.

[14]Boerhaave, *Academical Lectures,* I, p. 65.

[15]L.S. King, "Rationalism in early 18th century medicine," *J. Hist. Med.* 18 (1963), pp. 257-271.

[16]J.F. Fulton, "The influence of Boerhaave's Institutiones Medicae on modern physiology," *Nederl. Tijdschr. Geneesk* 82 (1938), p. 4860.

[17]L.S. King, "Stahl and Hoffmann: a study in 18th century animism," *J. Hist. Med.* 19 (1969), pp. 118-130.

[18]J. Roger, *Les sciences de la vie dans la pensée française du XVIII siècle* Poitiers, 1963.
R. Koch, "War Georg Ernst Stahl ein selbständiger Denker?" *Sudhoffs Archiv.* 18 (1926), pp. 20-50.

[19]B.W. Gottlieb, "Bedeutung und Auswirkungen des Hallischen Professors und königlich-preussischen Leibarztes Georg Ernst Stahl auf den Vitalismus des 18. Jahrhunderts, insbesondere auf die Schule von Montpellier," *Nova Acta Leopoldina N. F.* 12 (1943), pp. 425-502.

[20]L.J. Rather, J.B. Frerichs, "The Leibniz-Stahl controversy," *Clio Medica* 3 (1968), pp. 21-40 & 5 (1970), pp. 53-67.

[21]See also K.E. Rothschuh, "Leibniz, die prästabilierte Harmonie und die Ärzte seiner Zeit," *Akten des Internationalen Leibniz-Kongresses Hannover.*, Bd. II. Wiesbaden, 1969, pp. 231-254.

[22]L.J. Rather, "G.E. Stahl's psychological physiology," *Bull. Hist. Med.* 35 (1961), pp. 37-49.

[23]L.S. King, "Some problems of causality in 18th century medicine," *Bull. Hist. Med.* 37 (1963), pp. 15-24.

[24]See biography by R.S. Cowen, in *Dict. Scientif. Biogr.* 1 (1970), pp. 478-479.

[25]A biography by W. Baron appeared in *Dict. Scientif. Biogr.* 2 (1970), pp. 203-205.

[26]P. Walden, *Drei Jahrtausende Chemie.,* Berlin, 1944.

[27]S.M. Jörgensen, *Die Entdeckung des Sauerstoffs,* Stuttgart, 1909.

[28]J.R. Partington, *A History of Chemistry,* 2 vols., London, 1961.

[29]For a short biography see Henry E. Sigerist, *The Great Doctors,* transl. by E. and C. Paul, New York, 1958, pp. 173-186.

[30]G. Imhof, "Albrecht von Haller als Physiologe," *Arch. Gesch. Med.* 6 (1912), p. 52.

[31]E. Hintzsche, "Anatomia animata," *Ciba-Z.* 10 (1948), p. 4042.

[32]E. Hintzsche, "Einige kritische Bemerkungen zur Bio- und Ergographie Albrecht von Hallers," *Gesnerus* 16 (1959), pp. 1-15.

[33]Rita Schaer, *Albrecht von Hallers neue anatomisch-physiologische Befunde und ihre heutige Gültigkeit,* Bern, 1958.

[34]Stephen d'Irsay, *Albrecht von Haller,* Leipzig, 1930.

[35]R. Toellner, "Anima et Irritabilitas. Hallers Abwehr von Animismus und Materialismus," *Sudhoffs Archiv* 51 (1967), pp. 130-144.

[36]See introduction by L.S. King to the reprint of Haller's *First Lines of Physiology,* New York, 1966.

[37]H. Buess, "Albrecht von Haller and his Elementa Physiologiae as the beginning of pathological physiology," *Med. Hist.* 3 (1959), pp. 123-131.

[38]A. Berg, "Die Lehre von der Faser als Form- und Funktionselement des Organismus," *Virchows Arch.* 309 (1942), p. 33.

[39]Albrecht von Haller, *A Dissertation on the Sensible and Irritable Parts of Animals,* Engl. transl. with an introd. by O. Temkin, Baltimore, 1936.

[40]P. Wobmann, "Albrecht von Haller, der Begründer der modernen Hämodynamik," *Arch. Kreislaufforsch.* 52 (1967), pp. 96-128.

[41]For a total assessment see Richard Toellner, *Albrecht von Haller, . . .* Wiesbaden, 1971.

[42]Johann F. Blumenbach, *An Essay on Generation,* transl. by A. Crichton, London, 1792.

[43]Karl F. Kielmeyer, "Uber die Verhältnisse der organischen Kräfte," a speech reprinted in *Sudhoffs Arch.* 23 (1930), pp. 247-267.

[44]William Cullen, "Physiology," contained in *The Works of William Cullen, M.D.,* ed. by J. Thomson, Edinburgh and London, 1827.

[45]John Brown, *The Elements of Medicine,* transl. from Latin, London, 1788.

[46]G.B. Risse, "The Brownian system of medicine: its theoretical and practical implications," *Clio Medica* 5 (1970), pp. 45-51.

[47]Guenter B. Risse, "The history of John Brown's medical system in Germany during the years 1790-1806," (unpublished Ph.D. dissertation, University of Chicago, 1971).

[48]G.B. Risse, "The quest for certainty in medicine: John Brown's system of medicine in France," *Bull. Hist Med.* 45 (1971), pp. 1-12.

[49]P. Capparoni, *Lazzaro Spallanzani,* Torino, 1941.

[50]Lazzaro Spallanzani, *Expériences pour servir a l'histoire de la génération des animaux et des plantes,* Paris, 1787.

[51]F. Bottazzi, *Lazzaro Spallanzani nel secondo centenario della sua nascita,* Riv. Fisio. Neoscolastica, 1930.

[52]John Kobler, *The Reluctant Surgeon,* Garden City, N.Y., 1960.

[53]J. Dobson, "John Hunter's anatomy," *Ann. Roy. Coll. Surgeons* 41 (1967). pp. 493-501.

[54]K.E. Rothschuh, "Zur Geschichte der Physiologie des Blutes in der ersten Hälfte des 19. Jh.s." *Klin. Wschr.* (1941), p. 621.

[55]G.H. Bailey, "William Hewson (1739-1774). An account of his life and work," *Ann. Med. Hist.* 5 (1923), pp. 208-224.

[56]William Hewson, *The Works,* London, 1846.

[57]Henry Guerlac, *Lavoisier — the Crucial Year: The Background and Origin of His First Experiments on Combustion in 1772,* Ithaca, 1961.

[58]J.C. Hemmeter, "Lavoisier and the history of the physiology of respiration," *Bull. Hopkins Hosp.* 29 (1918), p. 254.

[59]R. Foregger, "Respiration experiments of Lavoisier," *Arch. Int. Hist. Sci.* 13 (1960), pp. 103-106.

[60]Kenneth Davis, *The Cautionary Scientists: Priestley, Lavoisier and the Founding of Modern Chemistry,* New York, 1966.

[61]Everett Mendelsohn, *Heat and Life,* Cambridge, Mass., 1964.

[62]Sidney J. French, *Torch and Crucible; The Life and Death of Antoine Lavoisier,* Princeton, 1941.

[63]Max Neuburger, *Die hist. Entwicklung der experimentellen Gehirn — und Rückenmarksphysiologie vor Flourens,* Stuttgart, 1897, and
S.W. Jackson, "Force and kindred notions in 18th century neurophysiology and medical psychology," *Bull. Hist. Med.* 44 (1970), pp. 397-410 & 539-554.

[64]M. Neuburger, "Neurol. Forschung in Wien während des 18. Jahrhunderts," *Wien med. Wschr.,* (1909), p. 2132.

[65]V. Kruta, "Prochaskas and J.E. Purkyne's contributions to neurophysiology" in Rothschuh: *Von Boerhaave bis Berger,* Stuttgart, 1964, pp. 134-156.

[66]V. Kruta, "The physiologist George Prochaska (1749-1820) and the reflex theory," *Epilepsia* 3 (1962), pp. 446-456. See George Prochaska, *A Dissertation on the Functions of the Nervous System,* transl. by T. Laycock, London, 1851.

[67]K.E. Rothschuh, "Von der Idee bis zum Nachweis der tierischen Elektrizität," *Sudhoffs Archiv* 44 (1960), pp. 25-44.

[68]H.E. Hoff, "Galvani and the pre-Galvanian electrophysiologists," *Ann. of Science* 1 (1936), pp. 157-172 and
M. Sirol, "Galvani et le Galvanisme. L'Electricité animale," (Thése Méd.), Toulouse & Paris, 1939.

[69]Luigi Galvani, *Commentary on the Effects of Electricity on Muscular Motion,* transl. by M.G. Foley with introd. by I.B. Cohen, Norwalk, Conn., 1953.

[70]*Ibid.,* p. 47.

[71]K.E. Rothschuh, "Die neurophysiologischen Beiträge von Galvani und Volta," in *Essays on the History of Italian Neurology,* ed. by Belloni, Milano, 1963, pp. 117-130.

[72]K.E. Rothschuh, "Alexander von Humboldt und die Physiologie seiner Zeit," *Sudhoffs Archiv* 43 (1959), pp. 97-113.

[73]H.E. Hoff, "The history of the refractory period," *Yale J. Biol & Med.* 14 (1942), pp. 635-672.

[74]A. Mauro, "The role of the Voltaic pile in the Galvani-Volta controversy concerning animal vs. metallic electricity," *J. Hist. Med.* 24 (1969), pp. 140-150.

[75]A. von Tschermak-Seysenegg, "Goethes Farbenlehre in ihrer Bedeutung für die physiol. Optik der Gegenwart," *Forsch. u. Fortschr.* 8 (1932), p. 68.

[76]O. Schuepp, "Goethe als Botaniker," *Gesnerus* 6 (1949), pp. 144-158.

[77]Ernst Grünthal, Fritz Strauss, *Abhandlungen für Goethes Naturissenschaft,* Bern, 1949.

[78]R. Matthaei, *Versuche zu Goethes Farbenlehre mit einfachen Mitteln. Ein Aufriss der Farbenlehre,* Jena, 1939.

[79]R. Matthaei, "Goethes Spektren und sein Farbenkreis," *Erg. Physiol.* 34 (1932), p. 191.

CHAPTER V

Nineteenth Century Physiology: The Beginnings

1. GENERAL LINES OF DEVELOPMENT IN NINETEENTH CENTURY PHYSIOLOGY

The nineteenth century is not only chronologically, but also in terms of its goals, a direct continuation of the eighteenth. The spirit of the Enlightenment, which aimed at an analysis of the laws of nature, history and religion with the help of reason, remained a strong impulse. Moreover, the detachment from worldly and spiritual authorities progressed even further, leading to the development of liberalism, individualism and democracy on the social level. Members of the urban middle class became the true cultural pillars of society, while the religious principalities, cloisters and convents vanished during this process of secularization.

The development of newer scientific methods and modes of thought led to unexpected and unprecedented triumphs, making science the idol of the entire century, not only in fashionable salons, but also in the streets, schools and factories. In 1859, the book entitled *The Origin of Species,* written by the English naturalist *Charles Darwin* (1809-1882) was published. Life was now viewed as the result of a large evolution, and the illusion of man's and earth's special place within the creation vanished.

Everywhere, painstaking research discovered reliable laws in nature and disclosed the inescapable determinism inherent in natural phenomena. The physical sciences seemed to dissolve in mechanics, while a newly popular materialism dominated the thought of large intellectual circles. However, the belief in progress was stronger than the feelings of insecurity which derived from a newly exposed existence without goals or obligations.

The development of medicine was greatly accelerated, especially towards the end of the nineteenth century, while the natural sciences, such as physics and chemistry, marched from discovery to discovery and established a series of general laws. Thus, the nineteenth century truly laid the foundations of knowledge on which most twentieth century technological accomplishments were based. If one contemplates the tempo of this nineteenth-century progress in the natural sciences and attempts to express it mathematically, the speed of the process would have to be represented in a logarithmic rather than arithmetical fashion.

The same situation also holds true for the development of nineteenth century physiology.[1] The first decades were still greatly under the influence of vitalistic concepts. However, as long as the course and laws of physiological phenomena were attributed to a vital principle or a series of living forces, there was no need for seeking further explanations. Hence, under those circumstances, no questions requiring experimental confirmation or physico-chemical analysis emerged. This explains the slow progress of physiology in Germany, Austria, France and Italy during the first decades of the century.[2]

German physiology, in particular, became associated from the turn of the eighteenth century with the idealistic philosophy of nature, giving rise to a so-called "romantic" movement in medicine and biology. This philosophical influence redirected science from an interest in particular problems to the consideration of general concerns regarding life and disease, and from experimental methods to merely speculative reasoning. Under the circumstances, German physiology became stagnant, especially during the period 1810-1815, when the so-called "romantic physiology" achieved its highest expression.

Thereafter, a new generation of men with novel ideas came gradually to the fore, and the experimental approach regained its lost position. As a reaction to the hitherto prevailing era of the "Lebenskraft" or vital force, and the romantic interpretation of nature, the newer scientific advocates were strongly empirical and favored sober, "unbiased" observations and experiments.

An empirical physiology was first launched in France, especially through the work of the physician and physiologist *François Magendie.* In fact, Magendie assured continued French leadership in this field until the 1840's, when German physiology under direction of Johannes Müller gradually closed the gap. The French approach was largely based on vivisections, and the experiments which were conducted concentrated on the resection of organs, interruption of nerve pathways, injection of certain drugs and observation of the inner organs in the living animal.

Thus, France cultivated a physiology strongly based on organic morphology and concerned with experiments affecting the entire living organism. In addition, some researchers preferred to be their own subjects in problems connected with the physiology of sense perceptions, as in the case of Müller, Purkinje and others. Only isolated attempts were made to interpret the vital phenomena in strict physico-chemical terms, such as the so-called "wave theory" elaborated by *Eduard F. Weber* (1806-1871) and his brother *Ernst H. Weber* (1795-1878). The reason for this attitude was that the basic knowledge necessary for such a physico-chemical approach was still largely undeveloped at the time.

In spite of the above outlined situation, the fundamental disciplines connected with physiology, such as histology, chemistry and physics, developed rather rapidly in the first decades of the nineteenth century. However, physiology followed slowly in their wake, and the so-called "physico-chemical period" in physiology only came into being around the middle of the century. It was inaugurated by a generation of physiologists who were disciples of Johannes Műller, such as *Hermann von Helmholtz* (1821-1894), *Emil H. Du Bois-Reymond* (1818-1896), *Ernst W. von Brűcke* (1819-1892) and especially *Carl F. W. Ludwig* (1816-1895). The actual outline for such an approach was expressed in Ludwig's *Lehrbuch der Physiologie* published in 1852.[3]

Thereafter, a conscious separation from structure and morphology occurred in physiological thought. Now, the problems of function were considered in relation to physical as well as chemical principles. The first culmination of this exacting, quantitative approach bolstered by a precise physico-chemical methodology occurred around the middle of the nineteenth century. At that time, with the exception of the French physiologist *Claude Bernard* (1813-1878) and his students, Germany became the leading country in physiological research. Fifty years earlier, everybody had gone to France in order to learn the methods, techniques and experience necessary to carry out physiological studies. Now, after 1850, Germany became the host for those men who wanted to study physiology both from Europe and the entire Western world.

Indeed, physiology emerged from these German physiological laboratories as an independent field of knowledge after the middle of the nineteenth century. Before that time, it had been connected everywhere with general, comparative and pathological anatomy as well as clinical medicine.[4] Although physiological lectures had already been given earlier, physiological laboratories in which experiments could be performed were only established in Germany in 1821. This first institute was founded at the University of Freiburg, and placed under the direction of the anatomist *Carl A.S. Schultze,* according to the records of the university.[5] Moreover, the first German course in experimental physiology was also given at Freiburg.[6] Somewhat later, in 1824, Purkinje's laboratory followed at Breslau.

In 1828, there were only seven professors (Dozenten) at six German universities who taught physiology as an experimental science.[7] As already mentioned, Purkinje was only the second professor who inaugurated a physiological laboratory. Other such laboratories were not furnished at other institutions of higher learning until many years later. When they finally became realities, most of them were housed at first in one or several small rooms of the anatomy building, remaining connected and often in a dependant role to the anatomical science. Only later came the realization that the physiological institutes needed independent facilities for their research, as reflected in Purkinje's new laboratories in 1839 at Breslau, and later in 1851 at Prague.

Even the physiologist Johannes Müller had to lecture on anatomy, physiology, ophthalmology, surgery and other subjects during his active teaching career at Bonn. Still later, he taught both anatomy and physiology together until his death in 1858. Only thereafter were his successors able to separate the two disciplines, and thus Du Bois-Reymond took over the chair of physiology while Carl B. Reichert began to teach only anatomy. Likewise, at Jena, the final separation between physiology, zootomy, embryology, comparative and systematic anatomy, occurred in 1859. Karl Gegenbaur kept the anatomical chair while the teaching of physiology was entrusted to the physiologist *Albert von Bezold* (1836-1868).[8]

At Würzburg, *Rudolf A. von Kölliker* (1817-1905) kept both chairs, anatomy and physiology, until 1864, and the separation between them was only effected at Giessen in 1891, according to recent and reliable information about the establishment of independent physiological chairs at the various German universities. "Separation" and "independence" in terms of physical facilities and budgetary matters were not always necessarily synonymous. In any event, the detachment of physiology from anatomy had been completed almost everywhere during the last quarter of the ninteenth century. A symbolic reflection of this division was the separation which took place in 1877 in Müller's journal entitled *Archiv für Anatomie und Physiologie,* then under the joint editorship of Du Bois-Reymond and Reichert. Henceforth, the publication was to be issued in two independent sections, one dealing with anatomy under the direction of Reichert, and the other purely physiological, with Du Bois-Reymond as editor.

New physical facilities devoted exclusively to physiology appeared in chronological order at the following universities: 1868 at Tübingen, 1869 at Leipzig, followed by Munich in 1874, Heidelberg, 1875, Berlin, 1877 and Bonn, 1878. Moreover, Kiel opened its laboratories in 1879, Halle, 1881, Strassburg, 1884, Würzburg, 1888, Greifswald and Marburg, 1889, and finally Freiburg in 1890.

The formal association of people with similar interests and creation of societies has always been a sign of the advanced development of any given subject. Therefore, the circumstances surrounding the foundation of the *Deutsche Physiologische Gesellschaft* should be briefly noted. In 1822, following an initiative of the biologist *Lorenz Oken* (1779-1851), the *Gesellschaft Deutscher Naturforscher und Ärzte* came into being, holding its first meeting at Leipzig. Six years later, in response to a suggestion made by Alexander von Humboldt, this society was divided into seven sections, one of them comprising anatomy, physiology and zoology. Many years later, in 1877, physiology became at last an independent section during the fiftieth session of this organization held in Munich, and the physiologist *Karl von Voit* (1831-1908) was named its chairman. However, the anatomists, physiologists and zoologists often continued to meet together in subsequent gatherings of the national association. Finally, during the 75th session, held at Kassel in 1903, a decision was made to create the *Deutsche Physiologische Gesellschaft,* and the new organization met the following year independently at Breslau, but in conjunction with the *Versammlung Deutscher Naturforscher und Ärzte.*

The year 1904 can be truly viewed as the founding year for the German Physiological Society. In the following year, the group met at Marburg, this time completely separated from the parent organization both in time and place. Compared with other countries, the German physiologists formally organized themselves relatively late. However, they had already created a forum for the exchange of ideas in 1875, when they established the *Berliner Physiologische Gesellschaft,* whose history has already been written.[9]

Germany had begun early to publish a journal dealing with physiological subjects. On July 1, 1795, the physician *Johann C. Reil* (1759-1813) published the *Archiv für Physiologie,* the first journal of its kind anywhere in the world. After 1801, this publication was coedited both by Reil and the well-known physician *Johann H. F. von Autenrieth* (1772-1835), and in 1805 it became the *Deutsches Archiv Für Physiologie* under the direction of the anatomist *Johann F. Meckel* (1781-1833). Somewhat later, in 1826, the journal again changed its name, becoming Meckel's *Archiv für Anatomie und Physiologie,* and in 1834 Johannes Müller took over as editor of the publication, now called *Archiv für Anatomie, Physiologie und wissenschaftliche Medizin,* until his death in 1858.[10]

Without taking into account the *Comptes Rendus,* France initiated the publication of its first physiological journal in 1821, under the direction of Magendie. Its title was *Journal de physiologie expérimentale et pathologique.* The first British journal in the field appeared in 1896 when the distinguished physiologist *Michael Foster* (1836-1907) published the *Journal of Physiology.* Finally, in 1898, in the United States, *William Townsend Porter* (1862-1949) and others began editing the *American Journal of Physiology.*

Thus, it is evident from this rather sketchy review, that physiology was predominantly cultivated in both Germany and France during the first half of the nineteenth century. Only in the last decades of that period did British physiology experience a rapid ascendancy which set the stage for the accomplishments of physiologists such as Michael Foster, *Charles S. Sherrington* (1857-1952), *Ernest H. Starling* (1866-1927), *William M. Bayliss* (1860-1924), *John N. Langley* (1852-1952) and others.

Other countries also participated in the development of physiology, such as Holland, Italy, the Scandinavian countries and the United States. Their accomplishments will be depicted in some detail later. After this general panoramic view of the nineteenth century we must now turn to the beginnings of that period, when the Napoleonic armies marched across the German lands, and a new intellectual movement, called romanticism, began to influence physiology.

2.
THE SO-CALLED "ROMANTIC" INTERLUDE IN PHYSIOLOGY

One of the most bizarre episodes in the history of physiology occurred during the period of the so-called "romantic" medicine.[11] It all began with the arrival of the idealist philosopher *Friedrich W.J. von Schelling* (1775-1854) in Jena during the year 1798. The new approach culminated in the years 1810-1815, and slowly declined thereafter.[12] The entire period was characterized by a tendency to derive medical theory and practice from philosophical considerations about nature rather than from experience. The same attitude was also displayed in all the natural sciences,[13] history, literature and jurisprudence. It was an outgrowth of the romantic movement which dominated Germany's contemporary intellectual life.

The concepts which shaped the "romantic" physiology can be described briefly. These ideas were based primarily on Schelling's "Naturphilosophie," although there were several schools and currents of opinion. Therefore, "romantic" physiology was, to a large extent, synonymous with a current of physiological thought based on the postulates of the "Naturphilosophie."

It should be stressed that such a "romantic" physiology was only *one expression* of contemporary physiology. Although many scholars were greatly impressed by the philosophical constructions and adhered to their basic postulates, another so-called *"empirical current"* existed in Germany. This non-philosophical physiology was based on observations and experiments, and vigorously opposed the "romantic" movement.

The superb bibliography of Ernst Hirschfeld[14] contains all the important medical publications which were written in the philosophical spirit. Among the most prominent leaders of the emerging "Naturphilosophie", the philosopher Schelling ranked in first place. Next to him, in close connection with the idealistic philosophy of nature, came the poet *Friedrich von Hardenberg* (1772-1801) better known by his pen name *Novalis,*[15] with his literary fragments or *Fragmente.* Novalis conceived the natural appearances as symbols of a hidden and transcendental reality.[16]

Among the other non-medical authors of the period, the poet *Joseph von Görres* (1776-1848) held a special position with his work *Exposition der Physiologie* (1805). Others in the group were *Johann J. Wagner* (1775-1841), *Stephan A. Winkelmann* (1780-1806) and Schelling's own brother, the physician *Carl E. Schelling* (1783-1854). In turn, the group of prominent physicians and biologists which developed the "romantic" physiology was formed by *Lorenz Oken,* the young *Karl F. Burdach* (1776-1847), *Ignaz P.V. Troxler* (1780-1860), *Ignaz Döllinger* (1770-1841, *Philipp F. von Walther* (1782-1849) and especially Giessen's physiologist, *Johann B. Wilbrand* (1779-1846).[17]

Next to the above mentioned authors, there were others who, although concerned with the philosophy of nature, were also capable of empirical research. Their allegiance to the dominating philosophical climate was reflected in the inordinate concern with basic physiological concepts and questions. Among these authors were the physician *Friedrich L. Augustin* (1776-1854), the biologist *Gottfried R. Treviranus* (1776-1837), the physicians *Johann J. Dömling* (1771-1803) and *Carl G. Carus* (1779-1868), as well as the young Alexander von Humboldt and even Johann C. Reil.

By contrast, the empirical current in physiology which clearly rejected Schelling's "Naturphilosophie," was represented by the physicians *Johann H.F. von Autenrieth, Georg Prochaska* (1749-1820), *Carl A. Rudolphi* (1771-1832) and *Johann F. Meckel.*

An analysis of the personal relationships of those physicians and biologists who adhered to the "Naturphilosophie" reveals that most of them were directly or indirectly influenced by their contact with Schelling himself. Thus, certain clusters of philosophically inclined authors appeared in centers such as Jena, Bamberg and Würzburg.[18]

The decisive impulse for the establishment of a romantic current both in the arts and in science, came in the last years of the eighteenth century. The movement originated in a circle of young men, composed of the German idealist philosopher *Johann G. Fichte* (1762-1814), *August W. von Schlegel* (1767-1845) and his brother *Friedrich von Schlegel* (1772-1829), both poets, Novalis and especially the twenty-year old Schelling. By virtue of his idealistic synthesis which purported to amalgamate the ideas and new scientific observations of the late eighteenth century, Schelling became the teacher for the so-called older romantic school and the leading theoretician for the "romantic" natural sciences.[19] His grandiose philosophy of nature or "Naturphilosophie" soon constituted the "Leitmotiv" and program for a series of writings composed by partisan physicians and scientists.[20]

Schelling had originally expressed his philosophical ideas in a work entitled *Ideen zu einer Philosophie der Natur* (1797), which was followed by two more explicit publications: *Von der Weltseele* (1798) and *Erster Entwurf eines Systems der Naturphilosophie* (1799). The most comprehensive summary of Schelling's philosophy was later furnished by the author himself, in a work called *Vorlesungen über die Methode des akademischen Studiums* (1803).[21] Since I have already described in more detail the roots and fundamental ideas of Schelling's "Naturphilosophie" elsewhere,[22] only the most important ideas will be summarized here.

Most of Schelling's particular philosophical ideas were not original. Their roots can be found in the eighteenth century, but their general formulation was new and important. Among the components included in the novel philosophy of nature were a series of recent discoveries which had profoundly stirred the contemporaries towards the end of the century: animal electricity (Galvani-Volta), Kleist's bottle and the effects of Galvanism (J.W. von Ritter), and the physical as well as animal magnetism (F.A. Mesmer). Almost simultaneously, chemistry had opened up new avenues of research with the already mentioned discoveries of oxygen, hydrogen and other elements. Finally, Haller had furnished two specific properties, called irritability and sensibility, in order to explain many of the vital properties inherent in the living organisms.[23]

The latter discovery also deeply influenced medical theory. The Scottish physician *John Brown,* a student of Cullen at Edinburgh, made his "excitability" the basis for a new system of medicine which became very popular in Germany at the turn of the eighteenth century. In addition, Brown's ideas decisively helped to shape Schelling's conception of vitality.

One can speculate that Schelling was also influenced by Goethe's ideas which found expression in his *Metamorphose der Pflanzen,* since the Jena group was closely connected with the circles of Weimar. Also apparent is the influence exerted on Schelling by the philosopher *Johann G. Herder* (1744-1803) with his work *Ideen zur Philosophie der Geschichte der Menschheit.* Similarly, Schelling received some inspiration from the biologist *Karl F. Kielmeyer.* Kielmeyer delivered a famous lecture at Stuttgart on February 11, 1793, entitled "On the relationships among organic forces in the various organic organizations and the laws and consequences of these connections," which proved to be of great importance for the development of "Naturphilosophie." The author analyzed the effective inner forces of nature and concluded that the most important ones were: 1) a reproductive force of "Bildungskraft;" 2) irritability or "Reizbarkeit," and 3) sensibility or "Empfindlichkeit," a triad of powers present in all living beings.

Schelling's point of departure in his philosophy of nature was the apparent contrast between *nature* and human *spirit.* His solution for this dichotomy was as follows: nature and spirit are actually identical, since both form part of the Absolute. Their identity actually suffered only a separation in the world of experience, appearing there as nature and spirit, or a real and ideal world. Yet, argued Schelling, both aspects really formed part of an indivisible whole, nature being the external manifestation while the spirit with its ideas constituted the inner core of the Absolute.

In this sense, Schelling felt that all natural appearances were expressions of ideas elaborated by the human mind. He declared that "nature was visible spirit while the mind was invisible nature." Therefore, in Schelling's view, nature was the manifestation, the object of ideas, which constituted the true essence of nature. Hence, to penetrate into the deepest and innermost aspects of nature, Schelling urged that we attempt to grasp the leading ideas which had shaped nature behind the external facade of phenomena.

As a result of its origin, the cosmos was considered to be one unified entity shaped like an organism. Thus, in Schelling's scheme, the various natural realms were essentially like organs of the cosmos, intimately related to each other and mutually influential. Schelling declared that living nature could only be distinguished from the so-called inert nature by the degree of perfection which the leading ideas of nature had attained in these living creatures. Both realms, the organic and the inorganic one, were organized according to similar and basic laws and principles.

Schelling concluded that since nature was simply the object of certain ideas, the proper task for science merely consisted of grasping the leading concepts in action behind the external phenomena. Those who concentrated on particular and individual occurrences were, in the philosopher's view, not practicing true science, since this approach failed to penetrate into the intimate structure of nature. To find these fundamental laws of the natural world, Schelling offered a useful tool, analogy — the "magic formula of romantic medicine." Since the ideas and laws valid for one step of nature's development were thus supposed to reappear at a higher level in a more perfected form, the task at hand was very clear. Schelling urged scientists to look for the similarities among the phenomena occurring at different stages of the natural evolution in order to discover the underlying general pattern and therefore find the guiding ideas and laws which supported the external world. (See Fig. No. 48)

Schelling's second major principle was the existence of "polarities." According to him, nature had arranged everything in opposing pairs and could only function as a result of the tensions which developed between the opposites. Hence, it was the duty of the scientist to start his search from the premise that the whole cosmos was established on the basis of polarities. Examples of these dualities were the physical phenomena of attraction and repulsion, the existence of a positive and negative magnetic point, the organic properties of sensibility and irritability, and the evidence of two opposite sexual principles: a positive male, and a negative female one.

According to these principles, Schelling interpreted all the appearances of nature. The essential forces of the inorganic realm were chemical principles, electricity and magnetism. At a higher level of natural development, the organic and living nature, analogous forces were operative: a reproductive power, irritability and sensibility. Schelling's view of the proper distribution of the above mentioned forces is of interest. Plants were supposed to be dominated by the *reproductive power* — "Bildungstrieb" — while the worms reflected the opposition between this molding force and *irritability*.

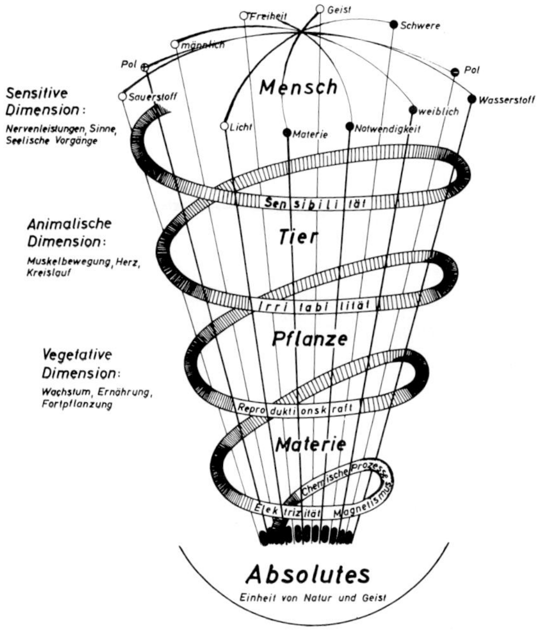

Fig. 48. Scheme of the fundamental ideas of the romantic philosophy of nature.

In turn, some lively insects were thought to reflect the dominant position of the irritable power. Schelling interpreted the amphibious creatures as evolutionary forms which mirrored a struggle between irritability and *sensibility*. The latter force was dominant in mammals and in man, situated at the top of the creation.

Since the individual organism was actually a replica of the entire cosmic organism for Schelling, the three fundamental forces which shaped the natural world were also present in each living *individual*. The philosopher believed that growth and the performance of the body's vegetative functions were the result of the postulated reproductive force, while muscular and nervous functions were expressions of irritability. Finally, the highest sensory and intellectual functions in man were derived from sensibility.

The effect of Schelling's philosophy of nature was remarkable. Within a short period of time, the new "Naturphilosophie" spread through the German universities and medical institutions, and it became fashionable to express oneself in Schelling's abstract language.[24] Among the interested physicians who adhered to Schelling's views were *Andreas Röschlaub* (1768-1835), professor of medicine at Bamberg and later Landshut, *Adalbert F. Marcus* (1753-1816), the director of the Bamberger Hospital, *Ignaz Döllinger,* professor of medicine at Würzburg, and *Philipp F. von Walther,* professor of surgery at Landshut.[25]

The physiologists were likewise affected by the new philosophical ideas. The so-called "romantic" physiology stressed its departure from investigations dealing with particular subjects, thus dismissing the observation and analysis of individual phenomena. Instead, the new approach concerned itself only with general physiological issues. For example, rather than studying specifically the circulation of the blood, the "romantic" physiologists debated the general idea of circulation in nature, since for them physiology was the knowledge of the major ideas operative in the living cosmos.

The result of such a shift in emphasis was the neglect of experiments and careful evaluation of individual experiences. Moreover, the principle of analogy was carried to extremes difficult to understand today. If two appearances in the macro — as well as microcosm showed any potential similarities, a concentrated effort was launched to achieve complete correspondences between all other properties related to the analogy. Some examples will illustrate this point. To the question whether the blood itself or the nervous system was the primary force responsible for circulation, one of the "romantic" authors, the physician *Johann H. Osterreicher* (1805-1843) responded: "Regarding this problem, analogy can be of help".[26] Oesterreicher then proceeded to declare that since planetary motion occurred around the sun which was the determining but resting principle involved, the nervous system assumed in the bodily circulation the role of a directing sun while the blood, like the planets, was merely the circulating element.

Another example, taken from an account of the medical historian Paul Diepgen,[27] concerned an analogy between the rotation of the earth and the motions of the blood cells. Morphologically, both the earth and the red blood cells were round and had somewhat flattened poles. In addition, both seemed to have a firm nucleus or core, a sort of vaporous outer envelope and they rotated around a given axis. Therefore, wrote the physician *Franz X. Horn,* due to the striking similarities between these two bodies, we are allowed to conclude that all the properties inherent in the blood cell can also be detected on the earth and vice versa.

The seriousness with which these analogies were considered and established, was reflected in the considerations of Johann J. Wagner's work entitled *Uber die Natur der Dinge* (1803). Wrote Wagner: "the atmosphere is the arterial system of the earth, the rivers are the veins, the oceans are the heart. Likewise, all creatures are the organs of an earthly body, plants being the portal vein, animals representing the lungs, light assuming the role of nerves while the sun is the brain. Let it be known that if I establish these comparisons, I am not revelling or dreaming but simply recognizing nature, trying to find in the small units what is already known in the larger ones."[28]

But enough about the misuse of analogy in "romantic" medicine. The movement was truly an infatuation with similarities, and their use to solve the greatest natural problems without bothering to consult nature herself has been herewith documented. A somewhat identical abuse was committed by the philosophical physicians with the concept of polarity. Everywhere, throughout the writings of the speculative physiologists, Schelling's ideas of polarity or the basic tensions and oppositions which eventually are reconciled, can be readily found.

In his *Einleitung in die dynamische Physiologie* (Göttingen, 1803), the German author *August Winkelmann* (1780-1806) wrote: "Nature is a competition of forces, the conflict between a positive and a negative energy. Life is positive, death negative, nitrogen and hydrogen are positive gases while oxygen is negative and their combination leads to an indifferent compound: water. Likewise, the nervous system should be viewed in the same context as being positive and the blood negative. Their polarity is dissolved in the lymphatic system. Finally, the magnetic attraction and repulsive force is positive, in contrast with another negative magnetic action and force of attraction".[29] There were several books written according to Schelling's "Naturphilosophie" which contained numerous pairs of opposites, and in many instances it is difficult today to understand and interpret properly the ideas which prompted the establishment of such dualisms.

Unfortunately, the use of polarities was not only restricted to the general philosophical issues of nature. In some instances, it was also employed in order to solve some specific medical problems. Such is the case of the physiologist *Johann B. Wilbrand,* one of the most prominent and interesting figures of the so-called "romantic" physiology.[30] (See Fig. No. 49) According to Wilbrand, polarity was the cause of the circulation of the blood. He wrote: "The movement of blood to the heart and from it to all other parts of the body, can be viewed as two motions which truly opposed each other. They act like two manifestations of magnetism or similar to the positive and negative aspects of electricity. Without this internal polarity in the motions of the blood, the circulation would actually be inconceivable."[31]

The sexual male-female duality was also used frequently in the theoretical conceptions of this philosophical medicine. In addition, the principle of *metamorphosis,* already used by Goethe and Schelling, played a significant role in numerous explana-

Fig. 49. Johann B. Wilbrand (1779-1846).

tions. In the contemporary physiology based on "Naturphilosophie," the idea of gradual changes in form was expressed typically by Wilbrand himself. He declared: "I deny the existence of a complete circulation of the blood." For Wilbrand, the blood actually flowing to the periphery was transformed into living tissue, which in turn liquified one portion of its structure as venous blood for the return to the heart. Thus, circulation was a continuous metamorphosis, since there were no capillaries in which the blood circulated directly from the arterial to the venous side without suffering changes.

In a similar fashion, respiration was, in Wilbrand's view, also a process of metamorphosis in which the blood was transformed into pulmonary tissue and then again into blood. Under those circumstances the absorption of oxygen and discharge of carbonic acid were only secondary phenomena, subordinated to the structural changes occurring at the level of the lungs. From the above descriptions, it is obvious that the philosophical speculations discarded in many instances some prior notions already proven by experimental physiology.[32]

Finally, Schelling's distinction of three fundamental vital forces, reproduction, irritability and sensibility, was largely imitated in other physiological and pathological theories. As a rule, the textbooks dealing with these subjects divided the vital functions into these three categories. Moreover, diseases were also grouped into reproductive, irritable and sensible categories. The popularity of this method of classification was attested by the German physician and psychologist *Carl G. Carus* (see Fig. No. 50) who stated: "We were tortured with irritability, sensibility and reproductive forces, and faced the classification of diseases based on these criteria long before we knew the particulars about cell formation, nervous irritability and fibrilar contractibility."

Fig. 50. Carl G. Carus (1789-1869).

The poet and physician *Joseph Görres* (1776-1848) was strongly influenced by Schelling's "Naturphilosophie," but in many ways also a highly original thinker. He expressed his physiological views in the works entitled *Prinzipien einer neuen Begründung des Lebens durch Dualismus und Polarität* (1802), *Aphorismen über die Organonomie* (1803) and especially *Exposition der Phy-*

siologie (1803). Both principles, polarity and the identity between the macrocosm and microcosm, can also be found in Görres' "physiology."[33] The following table depicts his duality of the natural and spiritual realms. (See Fig. No. 51)

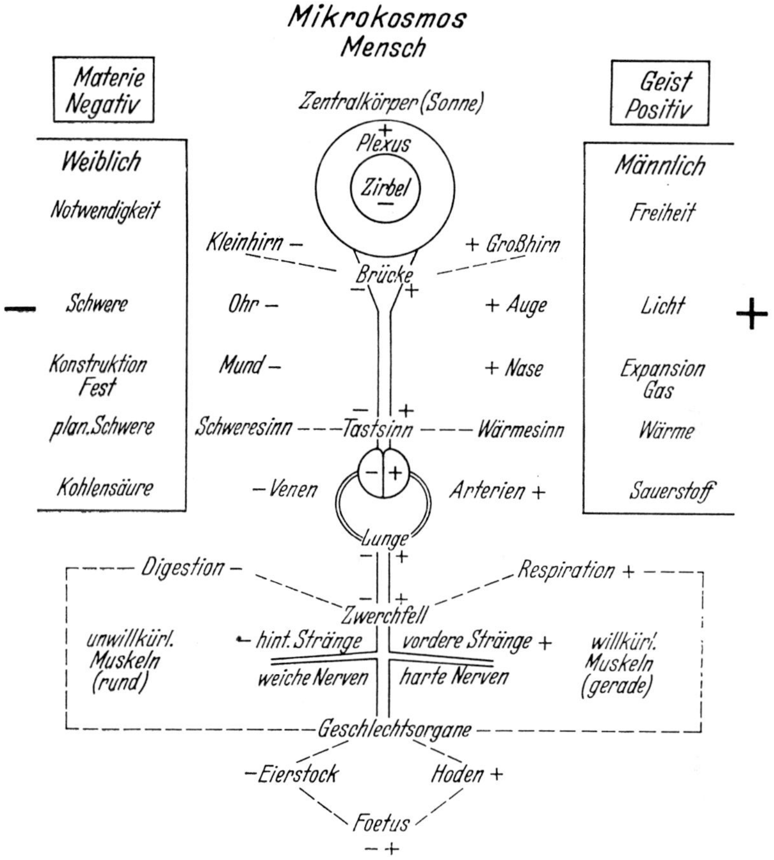

Fig. 51. Cosmic polarity of forces acting on man, the microcosm. Interpretation of Joseph Görres.

	INTELLIGENCE	
Masculinity		*Femininity*
First step	spiritual domain	First step
REASON	notion, idea	UNDERSTANDING
Positive-negative		Positive-negative
Second step	emotional domain	Second step
IMAGINATION	sentiment	TEMPER
Positive-negative		Positive-negative
Third step	organic domain	Third step
IRRITABILITY	life	EXCITABILITY
Positive-negative		Positive-negative

	EXTERNAL NATURE	
Masculinity		*Femininity*
First step	cosmic domain	First step
LIGHT	heat	GRAVITY
red-violet		highest-lowest
Second step	physical domain	Second step
ELECTRICITY	Galvanism	MAGNETISM
positive-negative		positive-negative
Third step	chemical domain	Third step
"COMBURIRTER STOFF"	"Comburirter Stoff"	COMBUSTIBLE ELEMENT
Oxygen-nitrogen		Carbon-hydrogen

According to Görres, the principles and oppositions existing in the macrocosm also recurred in each organic system, determining the laws by which these microcosmic entities functioned. In his search for the central organic structure, Görres assumed that the brain was the solar representative for the entire organism, although each organ had, in turn, its own directing sun. Accordingly, the heart was viewed as the sun for the circulatory system, while the pineal gland fulfilled an analogous role in the brain.

In Görres' view, the cerebrum was the organ of freedom and the cerebellum represented necessity, although both were united in the totality of nature. He wrote: "Like the sun changes into planets, so does the nervous string become a muscular fibre." The anterior motor roots of the spinal nerves belonged to a mobile, masculine principle in contrast with the posterior sensory one which was feminine. Görres believed that at the level of the spinal cord the tension between cerebrum and cerebellum was bridged by the union of the anterior nerve roots and the spinal ganglion.

The heart was likewise involved in a sexual kind of polarity and was, in fact, hermaphroditic in nature. For Görres, the left chambers were masculine, together with arteries and the expansible oxygen, while the right side, veins and heavier carbonic acid were feminine. Both principles were, however, inseparably united within the circulatory system.

Hence, we would like to conclude with the description of the so-called "romantic" physiology, a short intermission or interlude in the history of science whose grandiose expectations were not fulfilled. Those authors born after 1800 did not follow as a rule the siren-song of "Naturphilosophie," although some of them succumbed to these ideas for a limited time only — see Johannes Müller. Therefore, the philosophical physiology can be seen as an isolated phenomenon restricted to only one generation. Even scholars of high intellectual rank, such as Carus and Reil became adherents of Schelling's philosophy of nature. All of them expected to achieve a true understanding of the inherent natural order through philosophical principles rather than by gathering innumerable and individual empirical facts. The question foremost in their minds was to find the reason for the existence of certain natural appearances, their relative rank within the total order of nature rather than the investigation of the immediate causal factors.

The philosophical approach, which stressed a deeper understanding of events instead of merely their control, was diametrically opposed to the tendencies and aims which characterized the subsequent nineteenth century and prevail in our own times. Thus, the philosophical view of nature remained a useless pursuit, since it failed to influence, measure, manipulate and dominate the vital appearances. Nevertheless, it is fair to say that "Naturphilosophie" left a legacy of ideas which proved to be powerful impulses for the further exploration of general and comparative physiology, embryology, animal electricity, physiology of the senses and psychology. The real danger of the approach lay primarily in its disdain for empirical findings.

Therefore, the naturalist Alexander von Humboldt expressed a common hope when he wrote to the philosopher Schelling in 1805: "The 'Naturphilosophie' cannot be harmful to the development of empirical science. To the contrary, this philosophy reduces the discovered facts to principles, which in turn may lead to new findings. If in the process a new generation of persons has emerged which believes that it is more comfortable to carry out chemical investigations with the brain rather than getting its hands dirty, this is neither your fault nor that of the philosophy of nature."

3.
EMPIRICAL PHYSIOLOGY IN FRANCE AND GERMANY 1800-1850 (BICHAT — MAGENDIE — PURKINJE — WAGNER — WEBER)

After 1815 the so-called "romantic" physiology gradually declined and most of its representatives died in the decade 1830-1840. When Carus published his work *System der Physiologie* in three volumes between 1838 and 1840, the spirit of "Naturphilosophie" still pervaded large parts of the book. However, the new physiological accomplishments of men such as the German physiologist *Johannes Müller,* the French investigator *François Magendie* and the Czech author *Johann E. Purkinje* were also included in Carus' publication.

The new empirical current in physiology displaced the philosophical approach gradually. It returned to Haller's so-called "animated anatomy," a tradition which was still alive, and the new empiricists also borrowed from French sensualism and vitalism prevalent in the eighteenth century. Even though France experienced a certain romantic intellectual current, the movement had little effect on scientific matters. Instead, vitalism dominated the scene and was widely recognized in biology, anatomy and physiology, especially at the University of Montpellier.

Therefore, it is not surprising to find that the French anatomist *François X. Bichat,*[34] a student from Montpellier, expressed vitalistic concepts. In his work entitled *Anatomie Génerale,* Bichat established a new theory regarding the structure of tissues, almost without the help of the microscope. He viewed all essential biological actions in the organism as mere effects of vital forces. Organic sensibility and contractibility acted in the lower and relatively simple structured organisms, and both were also responsible for circulation, absorption, nutrition. digestion, secretion and excretion. Moreover, Bichat argued that the sensibility was also involved in the sensory phenomena, while the contractibility was associated with all true motions. Thus, Bichat traced all physiological and pathological appearances back to the action of these two vital properties already studied by Haller. After becoming a physician at the Hôtel Dieu in Paris, he died in 1802 at the early age of thirty one.[35]

Bichat greatly influenced French medicine and physiology, especially with his tendency to observe phenomena and refrain from a search for ultimate causes. His empirical physiology did not go beyond the facts, stressing primarily observations and descriptions, an approach which was to be superseded by a measuring, causal-analytical physiology in the mid-nineteenth century. Hence, the French physiology pioneered by Bichat concentrated on the observation of the phenomena, carrying out simple *experiments* such as ligatures, sections and ablations. These attempts were qualitative, not quantitative in nature. Thanks to the efforts of Bichat, as well as the physiologists *Julien J.C. Legallois* (1770-1840), *Marie-Jean P. Flourens* (1794-1867) and *François Magendie,* France took the lead in empirical physiology early in the nineteenth century.[36]

By contrast, the new approach rose to prominence in Germany in only the two decades between 1830 and 1850. To some extent, the empirical movement was mirrored in art, where a sober, descriptive and observing style began to permeate the literature and graphic arts of the period. Known as the "Biedermeier" style, this new attitude was even carried over into the political philosophies which prevailed before the year 1848.[37] Among the most prominent German physiologists of the period were the already mentioned Johannes Müller, Purkinje, as well as *Karl F. Burdach, Friedrich Tiedemann* (1781-1861) and *Karl A. Rudolphi.*

In England, the empirical physiology likewise came into a more prominent position, thanks to the works of the physiologists *Alexander Philip Wilson* (1772-ca.1851), *Charles Bell, Benjamin C. Brodie* (1783-1862) and *Marshall Hall* (1790-1857), the latter especially stressing in his works "observation" and "experiment."

As already mentioned, Bichat's distinguished contemporaries were Legallois and Flourens, who greatly advanced the understanding of the nervous system. Legallois, who had graduated in 1801 and was a physician at the Bicêtre Hospital, was one of the most skilled experimental physiologists of his day. His vivisection experiments regarding the nervous system and the vestibule of the ear were admirable for his time, although they now appear to have been somewhat cruel from our modern viewpoint. Legallois stressed the importance of the medulla oblongata in the processes of respiration, circulation and thermoregulation.

Flourens (see Fig. No. 52), on the other hand, will always be associated with the discovery of the "noeud vital": the respiratory center. In a series of skillful experiments, he also demonstrated the importance of the cerebellum and the semicircular canals for the maintenance of equilibrium during walking.[38]

Fig. 52. Marie Jean Pierre Flourens (1794-1867).

The true founder of modern ex-

Fig. 53. François Magendie (1783-1855).

perimental physiology and medicine was *François Magendie* (see Fig. No. 53). Born October 6, 1783, at Bordeau, the son of a surgeon, Magendie studied medicine at Paris, where he was greatly influenced by the works of Bichat.[39] However, Magendie was a sceptic, and his critical spirit was opposed to the tendency of the comtemporary vitalists to explain all unknown living phenomena as effects of likewise unknown vital forces. From this time onwards, Magendie exhibited a great aversion for all theoretical considerations, stressing that physiology was still excessively speculative in character.

Magendie declared that all living forces which had been hitherto accepted to explain the animal functions, were nothing more than arbitrary assumptions. These explanations had been used for many centuries in order to hide the ignorance concerning the origins of life, which in his view would remain forever a secret. Magendie concluded that such vitalistic explanations had hindered the development of physiology, which as a science was therefore still in its infancy.

Magendie consciously attempted merely to observe the phenomena occurring during his experiments.[40] He was even reluctant to draw any general conclusions from his investigations, but nevertheless enriched physiology through a great number of new and important facts. In the process, great numbers of animals, especially dogs, lost their lives at the hands of Magendie in order to furnish new insights for a series of physiological and pathological problems. These vivisections, carried out on living, not anesthetized animals, enriched almost all areas of experimental physiology. Magendie viewed himself — he expressed this to his student, Claude Bernard — as a "scavenger," who with rake in hand collected whatever he could possibly find.

Magendie's early work, entitled *Précis élémentaire de physiologie* (1816-17), set the tone for his subsequent publications, in which he reported the results of his innumerable experiments.[41] In 1836, he became professor at the "Collège de France," dying in 1855 after having clarified many clinical and pharmacological questions, and contributed to most aspects of physiology.[42]

Among Magendie's numerous discoveries I shall only mention his confirmation of the fact that the posterior roots in the dog's spinal cord were sensitive, the investigations regarding the motions of the chylus and the experiments which pointed out the routes of absorption from the gut. Moreover, the attempts to feed animals with pure foods, the experiments concerning the elasticity of the vascular system, the origin of cardiac sounds as well as the roles of both cardiac diastole and systole were other important contributions made by Magendie. He discovered the cerebro-spinal fluid and its various communications within the ventricular system — foramen Magendie. A series of specific punctures and sections of the various parts of the brain and cerebral nerves helped to clarify some of their functions.

Finally, Magendie's lectures concerning vital phenomena and functions as well as diseases of the nervous system were mines of important information: *Leçons sur les phénomènes physiques de la vie* (Paris, 1839) and *Leçons sur les fonctions et les maladies du système nerveux* (Paris, 1839). Most of Magendie's works were translated into German and contributed significantly there to the eventual supremacy of the empirical, experimental approach in physiology.

Next to the philosophically inclined physicians in Germany, there were a fair number of empirically oriented and competent scholars. Although some of them partially adhered to the contemporary speculative tendencies, these physicians on the whole pursued a physiology based on critical observations and experiments. One of them was the already mentioned *Johann H.F. von Autenrieth,* professor at the University of Tübingen since 1797 and the author of a work entitled *Handbuch der empirischen menschlichen Physiologie* (1801-02). Autenrieth attempted to explain disease from a physiopathological standpoint. He was a contemporary of other well-known authors such as Rudolphi, Treviranus, Burdach and Döllinger.

Karl A. Rudolphi was named in 1810 to the first joint chair of anatomy and physiology at the newly founded University of Berlin.[43] His student, Johannes Müller, was later to praise the professor's keen powers of observation, extensive knowledge and rare gift for teaching.[44] Rudolphi was opposed to vivisection and his main field of endeavor was zoology and comparative physiology. He declared that the essential vital characteristic was excitability, and that a living organism could be induced by stimuli to react, thus expressing its vitality, an idea which Müller developed further. Rudolphi died in 1832 while completing his work entitled *Grundriss der Physiologie.*

The German physiologist *Gottfried R. Treviranus*, on the other hand, was able to conclude his six volume treatise entitled *Biologie oder die*

Philosophie der lebenden Natur für Naturforscher und Ärzte (1802-22). The book will remain a fitting memorial for a scientist who favorably combined proper observations with speculative interpretations. Moreover, Treviranus became one of the first authors to systematically apply the microscope to the study of bodily tissues.

Fig. 54. Karl Friedrich Burdach (1776-1847).

While Treviranus still manifested some strong philosophical beliefs, his contemporary, the anatomist *Karl F. Burdach* (see Fig. No. 54) was even more deeply steeped in "Naturphilosophie."[45] An early adherent of the Brownian medical system and its German adaptation, Röschlaub's "Erregungstheorie," Burdach later shifted his allegiance to Schelling. He first taught at Dorpat, in Livonia, and from 1814 onwards resided at Königsberg. Burdach was extremely productive in many medical fields, including anatomy, embryology, physiology and psychology. Among his published books — he wrote seventeen himself — was the unfinished work entitled *Die Physiologie als Erfahrungswissenschaft* (1826-40), which he published together with other authors such as the embryologist *Karl E. von Baer* (1792-1876), *Martin H. Rathke* (1793-1860), *Adolf Moser* (+ 1810), *Gabriel G. Valentin* (1810-1883), *Rudolph Wagner* (1805-1864), and Johannes Müller.[46]

Burdach carried out superb anatomical investigations concerning the central nervous system and discovered the fasciculus cuneatus or ascending Burdach tract in the spinal cord. His most important publication on this subject was the book entitled *Vom Baue und Leben des Gehirns* (Leipzig, 1819-1826). The experimental approach to physiology propounded by Magendie and Johannes Müller did not impress Burdach greatly, perhaps because of its inherent lack of imagination and prosaic character.[47] In 1827, he delegated his anatomical lectures to a student, von Baer, and continued only his physiological classes. Burdach was the first author to use the word BIOLOGY in his work *Propedeutik zum Studium der gesammten Heilkunst* (Leipzig, 1800), followed by Treviranus' already mentioned book of 1802 and somewhat later the publication *Recherches sur l'organisation des corps vivants* (Paris, 1812) written by the French naturalist *Jean Baptiste Lamarck*

(1744-1829).

In 1796, *Ignaz Döllinger*[48] became professor of medicine at Bamberg after studying in Vienna under Georg Prochaska and in Italy with the public health pioneer *Johann P. Frank* (1745-1821) and the surgeon *Antonio Scarpa* (ca. 1747-1832). Döllinger moved in 1803 to the University of Würzburg, where he taught for the next twenty years. His numerous studies were concerned with a broad range of subjects. He investigated the circulation of the blood, certain secretory functions, blood formation, pulse wave and above all, embryology.[49] He was a master of the vascular injection techniques, and his great knowledge as well as a more Kantian philosophical interpretation of the phenomena attracted many disciples.[50]

In the first decades of the nineteenth century, another group of scholars was active in anatomy and physiology in Germany. On the basis of their objectivity and their methodology, these authors should actually be viewed as the founders of the exact and scientific current of physiology which became dominant later in the century. Among the most prominent researchers was the already mentioned Czech physician *Johannes E. Purkinje* or Purkyně[51] (see Fig. No. 55). Born December 17, 1787 in Bohemia, Purkinje first studied philosophy at Prague, where he fell under the influence of Schelling's "Naturphilosophie."[52] Thanks to a wealthy sponsor, he was later able to study medicine, graduating in 1819 with a remarkable doctoral dissertation entitled "Contributions to the subjective knowledge of vision." The work revealed Purkinje's unusual ability to understand the subjective visual phenomena and even attracted the attention of the German poet Goethe as well as earned him a recommendation from the naturalist Alexander von Humboldt.[53]

Fig. 55. Johannes Evangelista Purkinje (1787-1869).

Thus, Purkinje, who had remained an assistant and prosector in anatomy at Prague after his graduation, was called in 1823 to the University of Breslau. Against the wishes of the Breslauer faculty, he was named to the chair of physiology and pathology, succeeding the anatomist *Ernst D. Bartels* (1778-1838). Although understandably somewhat at odds with his new colleagues, Purkinje was instrumental in adding an hour of experimental demonstrations to his lectures on physiology a year later. We know from one of his memorandums that Purkinje considered an unusually large number of microscopical and physiological exercises necessary for the proper instruction of his students. He illustrated the essential aspects of the blood circulation with the help of a hydraulic apparatus, and showed on living animals the blood pressure elevation with the assistance of the tube designed by Stephan Hales in the eighteenth century. Moreover, Purkinje encouraged his students to use the stethoscope in order to hear the various cardiac sounds, and he demonstrated the mechanics of respiration on certain models while also conducting stimulations on both muscles and nerves.[54]

Purkinje's successor, the German physiologist *Rudolph P.H. Heidenhain* (1834-1897) declared that unfortunately many of Purkinje's original ideas never reached a wider audience. Heidenhain himself found in the Breslau institute a simple spirometer constructed by Purkinje long before the models of the English physiologist *John Hutchinson* (1811-1861) became available. After long years of work, Purkinje was able to construct his own physiology laboratory at Breslau. Although the building was rather small and modest — during Heidenhain's tenure it was only used as a lock-up for unruly students — Purkinje's laboratory was nevertheless an independent facility, a rarity among German universities in 1839.

As mentioned before, Purkinje's building was not the first independent German laboratory of physiology. The first of these experimental institutions had already been erected at the University of Freiburg in 1821 under the direction of the anatomist *Carl A.S. Schultze* (1795-1877), who had been named to the chair of physiology, comparative and pathological anatomy.[55] Schultze's experimental physiology evoked great interest among the attending students, and the course was taught regularly from 1822. In Breslau, however, Purkinje's similar approach was less fortunate. He was not able to stimulate or captivate his audiences with the new demonstrations, which were only sparsely attended in the first years.

Purkinje's most productive years fell between 1820 and 1840, *before* the establishment of the new physiological institute. In those years, he first used a special room in the anatomy building and later furbished his own laboratory at home — he was married to one of the daughters of the anatomist Karl A. Rudolphi. From this private laboratory year after year emerged a series of excellent works, many of them unfortunately printed in publications which were not easily accessible and thus escaping wider attention and recognition.

Purkinje was continuously at work, thinking, experimenting and pondering results. Among his physiological investigations of the period before 1832 — the year he received a good microscope — were important experiences regarding the problem of vision. He examined the objective conditions under which a series of subjective visual impressions occurred[56] and was the first to describe the so-called "Purkinje phenomenon": the color change of light from red to blue at dusk. He also analyzed the ocular spectrum.

After 1832, Purkinje's microscopical research was at the forefront of his activities. He was the first to describe the blastula of a fertilized animal egg, the "Purkinje fibres" in the heart, and able to differentiate within the fibrilar structure of the nerves, as well as the cerebral ganglion cells and cerebellar cells. Purkinje should also be considered a precursor of the cell theory, and he enriched the knowledge regarding the fine structure of bones and teeth. Moreover, he was the first to use the term "protoplasma."[57] The methods of microscopical observation were gradually perfected after Purkinje, and Breslau's institute of physiology became, together with Johannes Müller's laboratory at Berlin, a veritable "cradle of histology."[58] (See Fig. No. 56).

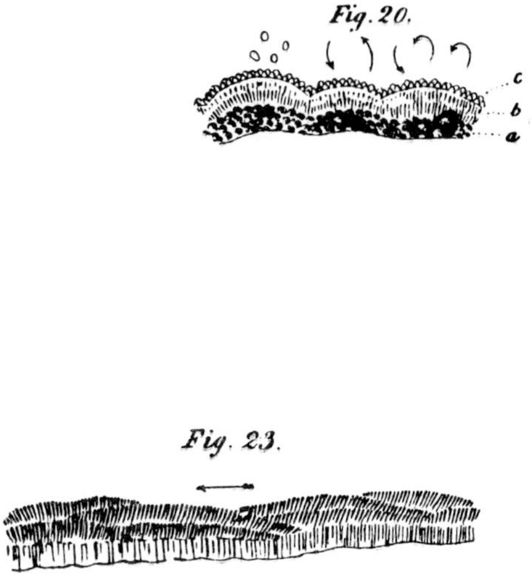

Fig. 56. Microscopical observations of ciliary motion according to Purkinje and Valentin.

After the establishment of the institute at Breslau, Purkinje's steady stream of discoveries began to languish. In 1850, he was called back by the Austrian cultural official Count Thun-Hohenstein in order to teach at the University of Prague. After his return, he transferred his activities to the political arena, becoming interested in the movements connected with the growing Czech nationalism and writing his name ostentatiously as Purkyně. Although several Czech authors have written about him, the sources in other languages are rather scanty.[59,60] More recently, a fair evaluation of his life and work has become available as well as an edition of Purkinje's *Opera Omnia.* Purkinje retired in 1867 and died in Prague, July 28, 1869.

Purkinje did not establish his own school of physiology, although his best student at Breslau became a well-known anatomist: *Gabriel G. Valentin.* At Prague, another physiologist, *Johannes N. Czermak* (1828-1873), studied under Purkinje. Other collaborators were the German zoologist *Carl T.E. von Siebold* (1804-1885), who became the first physiologist in Munich, and *Hermann F. Stannius* (1808-1883), who went to Rostock in 1837 and there founded the first physiological laboratory.

Gabriel G. Valentin (see Fig. No. 57) shared with Purkinje the discovery of the ciliary movements in 1834.[61] The son of a Jewish goldsmith in Breslau,

Fig. 57. Gabriel G. Valentin (1810-1883).

Valentin became an important zoologist, anatomist and physiologist. Valentin's embryological investigations were especially important and in 1835, he won the French "Grand Prix des Sciences Physiques" because of his comprehensive examination of comparative histiogenesis. From 1836 on, he continuously reported his discoveries in the quarterly journal entitled *Repertorium für Anatomie und Physiologie,* which he edited.

Although the University of Dorpat in Livonia offered him an academic position in 1836 if he consented to be baptized, Valentin refused the proposal and went instead to Bern, where he became the first Jewish professor at any German-speaking university. However, the working conditions at Bern were very modest: one room in the anatomical institute and a budget of merely one hundred Swiss francs. In 1839 appeared Valentin's studies of the sympathetic nervous system. In addition, he wrote a well-received textbook and compendium of physiology, and found that the pancreatic juice exhibited an enzymatic activity, a discovery which kindled a series of disputes with French authors about priority.

Fig. 58. The use of Czermak's laryngoscope (1860).

Johannes N. Czermak (1828-1873)[62] was another of Purkinje's assistants, both at Breslau and later at Prague. After a short period in Krakow, he went to Vienna where he studied under the two great physiologists *Ernst W. von Brücke* and *Carl F.W. Ludwig.* Czermak received a two year appointment as professor of physiology at Budapest in 1858-1860, and then went to Jena, Prague and Liepzig in that order. In the latter city he established his own research institute with private funds. Already, in 1858, Czermak described the use of a small mirror in order to visualize directly the larynx. (See Fig. No. 58). This invention of the laryngoscope brought him into a violent dispute with another German physiologist, *Ludwig Türck* (1810-1868) in the so-called "Türkenkrieg." Türck, who resided in Vienna, had been interested in the problem of laryngeal visualization for some time, following in part the suggestions of Carl Ludwig and a Spanish singing teacher named Manuel Garcia (1805-1906). Thus, he claimed to have invented the instrument. In truth, Czermak who had borrowed the first instrument from Türck himself, immediately realized the potential advantages of such an extraordinary method of examination. While Türck was merely concerned with the clinical application of the laryngoscope, Czermak perceived its usefulness for the study of the physiology of voice and speech as well as phonetics.[63] Czermak also devised the well-known vagal effect which can be obtained through pressure applied at the level of the neck, and he also studied some aspects of physiological optics and animal hypnosis.

His adversary, the physiologist Ludwig Türck, had studied the experimental physiological approach of Magendie, Flourens and *François A. Longet* (1811-1817) at Paris, and he subsequently became a well-respected neurologist. Simultaneously with another distinguished French physiologist, *Charles E. Brown-Séquard* (1817-1894), Türck studied the effect of a partial section through the spinal column in 1857. He also discovered the laws concerning the secondary degeneration in the nervous system in 1849, before the publication of the work by the English physiologist *Augustus V. Waller* (1816-1870). Türck had the misfortune of being involved in a series of discoveries which were made almost simultaneously by several authors. Together with Czermak, he received the Monthyon Prize of the Paris Academy of Medicine in 1861.

Another distinguished author was the German physiologist *Julius L. Budge* (1811-1888). Originally from Wetzlar, Budge became director of the anatomical and physiological institute at the University of Greifswald in 1856. He contributed some valuable studies concerning the anatomy and physiology of the vegetative nervous system, discovering the centrum ciliospinale, genitospinale and anospinale. We are also indebted to Budge for the realization that a sympathetic stimulus produces pupillary dilatation while the opposite effect is brought about by the stimulation of the oculomotor nerve. Budge also claimed to have observed a cardiac standstill as a result of vagal stimulation.

Another contemporary of Valentin was the German physician *Theodor L. W. Bischoff* (1807-1882).[64] Bischoff was one of those students who attended the first lectures of a young professor of physiology at Bonn: Johannes Müller. He was primarily interested in embryology, and in 1833 wrote a work concerned with the membranes of the human egg. During his teaching career in anatomy and physiology at the University of Giessen, Bischoff became acquainted with the famous chemist *Justus Liebig* (1803-1873), and his interest shifted thereafter to problems concerning the chemistry of metabolism. He followed Liebig to Munich in 1855 and occupied there the chair of anatomy and physiology at the university.

Bischoff trained two assistants who became distinguished physiologists: *Karl von Voit*[65] and *Conrad Eckhard* (1822-1905). Voit assumed Bischoff's physiological lectures at Munich, and Eckhard, who had also been an assistant to Ludwig in Marburg, became the professor of physiology at Giessen after Bischoff's departure. Eckhard was primarily interested in the physiology of the vegetative nervous system, but also carried out studies on the salivary and gastric secretion.

Other distinguished anatomists and physiologists of the period who did not belong to Müller's circle were *Rudolph Wagner, Hermann Nasse* (1807-1892) and *Jacob Moleschott* (1822-1893). Wagner (see Fig. No. 59), a graduate of Würzburg, studied for some time in Paris with the notable French naturalist *Georges L.C.F. Cuvier* (1769-1832) in Paris. In 1833, he became professor of zoology at the University of Göttingen, and seven years later replaced the famous physician Johann Blumenbach, teaching physiology, comparative anatomy and zoology. In 1847, Wagner demonstrated the genetic link between the ganglion cells in the brain and the peripheral nerve fibres, but his major field of study remained embryology — nucleolus of the ovum — and histology — blood cells, choroid and retina.

Fig. 59 Rudolph Wagner (1805-1864).

Wagner became embroiled in a nasty controversy with his student, the physiologist *Georg Meissner* (1829-1905) regarding the priority in the discovery of the encapsulated receptors for touch located in the connective tissue papillae of the skin. One of Wagner's other great contributions was his editorship of a dictionary of physiology which attempted to depict the state of the science in the years 1842-1853. Among the contributors to the ambitious, six volume publication were Purkinje, Ludwig, Valentin as well as *Alfred W. Volkmann* (1800-1877), *Friedrich H. Bidder* (1810-1894), *Jöns J. Berzelius* (1779-1848), *Ernst H.*

Weber and *Rudolf H. Lotze* (1817-1881).

Finally, Wagner also became well-known for his conservative views on soul and life.[66] This fact provoked the so-called "Materialismusstreit" in the 1850's, a literary controversy concerning the philosophical issues of materialism. One of those who critized Wagner was the famous student of Liebig, the zoologist *Karl Vogt* (1817-1895).[67] In 1855, Vogt published a rather humorous critique of Wagner's anti-materialism entitled "Köhlerglaube und Wissenschaft." Vogt had been a political activist, both as a student during the Restoration period and again during the revolution of 1848. His studies on the nervous system in pythons and other research in zootomy produced valuable contributions to the field. Vogt's subsequent acceptance of Darwinism prompted opponents of the theory to call him "Affenvogt" (Ape-Vogt).

Another prominent author, the Dutch physiologist *Jacob Moleschott,* also participated in the controversies between religion and scientific research. His strong materialistic and anti-Christian bias created difficulties for him at Heidelberg, where he advocated cremation over burial in cemeteries on the grounds that the ashes could be used to fertilize the soil. Moleschott's scientific creed was already expressed in his dissertation of 1845: "Nullam extra physicae studia medicum invenire salutem, immo artem non esse medicinam, nisi a physiologia proficisceretur." A one time student of Jacob Henle, Moleschott went to Turin in 1861, and later transferred to Rome in 1879. He was interested in questions of normal metabolism and nutrition, publishing his results in a work entitled *Untersuchungen* (1857). His well-written autobiography was simply called *Für meine Freunde.*[68]

Moleschott's predecessor at Heidelberg, and a student of Henle was the Swiss physiologist *Rudolf A. von Kölliker* (1817-1-05).[69] After completing his studies at Bonn and Berlin, he became in 1842-1844 prosector in anatomy at the University of Zürich, his hometown. In 1847 Kölliker was appointed professor of physiology and anatomy at the University of Würzburg, a chair which he held until 1865 when he was succeeded by the young physiologist *Albert von Bezold.*[70] Kölliker's primary interests lay in the study of normal histology, but he also carried out several valuable physiological studies.[71] Together with *Heinrich Müller* (1820-1864) he discovered the monophasic current of the heart. Kölliker was instrumental in establishing the prestige of the University of Würzburg during the second half of the nineteenth century, together with the physiologist *Adolf Fick* (1829-1901) and the discoverer of the x-rays, *Wilhelm C. Röntgen* (1845-1923).

All the above named authors were successful both in anatomy and physiology. To some extent, they already belonged to the generation of Johannes Müller, who represents the culmination of this phyiological development. However, in order to complete the contemporary panorama of German physiology, four other distinguished authors should be mentioned who concentrated more on the physical aspects of the fledgling discipline.

Originally from Wittenberg, and sons of a theology professor at the local university, the three brothers Weber represented the new direction in physiology. The oldest of them, *Ernst H. Weber* already became professor of anatomy and physiology at the University of Leipzig at the age of twenty-six (see Fig. No. 60). Completely free of philosophical arguments, Weber spent a long and fruitful career at Leipzig, carrying out a series of physiological studies by employing the physico-mathematical approach, an innovation for the period under discussion. Weber's book, entitled *Wellenlehre, auf Experimente begründet* (Leipzig, 1825), was written in collaboration with his younger brother *Wilhelm E. Weber,* then still a student at Wittenberg. The work laid the groundwork for a more exact analysis of the undulatory movements occurring in fluids, especially when circulating through cylinders, the true physical basis for the motions of the blood. These studies were continued and presented in more detail a decade later, in Ernst H. Weber's publication *De pulsu, tactu et auditu* (Leipzig, 1834).[72]

Fig. 60. Ernst Heinrich Weber (1795-1878).

Ernst H. Weber's investigations concerning the sense of touch — Weber's law — provided the first measurable and quantitative approach to the sensations of locality, pressure and temperature.[73] His collected works were published in 1851 under the title *Annotationes anatomicae et physiologicae programmata collecta.* In 1854, together with his youngest brother, *Eduard F.W. Weber* (1806-1871), Ernst H. Weber demonstrated in Naples his famous experiment about the inhibitory effect of the vagus nerve on the heart.[74]

The third brother, Eduard, who had studied at Halle, became an anatomical prosector to his older brother in Leipzig. Thus Eduard F.W. Weber began an investigative career in which he devoted much time to the mechanics and physiology of the muscles. He also carried out a series of quantitative experiments and was the first to stress the great significance of muscular elasticity. In his article on muscular motion, written in 1846 for Wagner's physiological dictionary, Eduard F.W. Weber unveiled a series of discoveries pertaining to the subject. His physical approach was reflected in the work *Mechanik der menschlichen Gehwerkzeuge* (Göttingen, 1836), which he wrote in collaboration with his brother Wilhelm.

Wilhelm E. Weber, a physicist at Göttingen, contributed substantially to the application of physical principles in physiology. All three brothers should be considered the forerunners of a new exacting, physical current in physiology which only emerged fully in Germany during the second half of the nineteenth century, especially with prominent figures such as *Emil H. Du Bois-Reymond, Hermann von Helmholtz* and *Carl Ludwig* (1816-1895).[75]

Fig. 61. Alfred W. Volkmann (1800-1877).

This historical judgment applies likewise to the German physiologist *Alfred W. Volkmann* (1800-1877) who studied at Leipzig under Ernst H. Weber (see Fig. No. 61). After occupying the chair of physiology at the University of Dorpat in 1837, Volkmann became professor of anatomy and physiology at Halle. His approach to physiology was similar to that of his teacher. He was interested in physiological optics, as revealed in his work *Neue Beiträge zur Physiologie des Gesichtssinnes* (1836) and *Physiologische Untersuchungen im Gebiet der Optik* (1863-64). Volkmann was also active in the field of hemodynamics, in which he exchanged ideas with his teacher Ernst H. Weber and Carl Ludwig. The fruit of his investigations in this area of physiology was a book: *Die Haemodynamik nach Versuchen* (Leipzig, 1850).

Fig. 62. Arnold Adolph Berthold (1803-1861).

Finally, the anatomist *Arnold A. Berthold* (1803-1861) (See Fig. No. 62) should also be included in this account, since he was the founder of the theory concerning internal secretions. Appointed as director of the zoological and anatomical collections of Göttingen in 1840,[76] Berthold also practiced medicine and published many articles. In 1849, he reported certain observations which confirmed the hitherto highly hypothetical notion of an internal secretion by certain bodily organs. Among his skillful experiments was one in which he first eliminated one testicle of a young rooster without any apparent consequences. However, if both testes were surgically removed, the animal lost its crest and tail feathers. Subsequently, after the capon received a testicular implant in the abdominal cavity, the original characteristics of a rooster promptly returned. Since the original nervous connections had not been restored, Berthold concluded that "the interaction in question is mediated through a productive condition existing in the testes, that is by way of their action on the blood, and thereafter through the corresponding influence of the blood on the organism as a whole." Berthold's results failed at first to impress his contemporaries. His description was clumsy and the time not ripe for such ideas of an internal secretory function. In any event, he was the first author to provide experimental proof for a secretion from the testicles into the blood, and thus should be viewed as the founder of the concept of endocrine function.

4.
THE DEVELOPMENT OF BRITISH PHYSIOLOGY 1800-1848

Like France and Germany, Britain also experienced vitalistic and empirical currents in physiology towards the end of the eighteenth century. The vitalists, occasionally strongly motivated by religious considerations, coexisted with the empiricists who proclaimed the supremacy of observation.

Among those who professed a renewed interest in *vitalism* was the anatomist *John Hunter,* who believed in the existence of a "principle of life," an agent such as gravity, inherent in every living particle.[77] Therefore, the composition of all vital organisms could not be conceived in purely chemical nor mechanical terms. Even the body's temperature — Hunter carried out a series of experiments on the subject in 1775-78 — was explained as an effect of the vital principle. In his famous work entitled *A Treatise on the Blood, Inflammation and Gun-shot Wounds,* published in 1794, he considered the blood to be equally alive, and coagulation as well as clot retraction were viewed as terminal actions of a dying, vital fluid.

Hunter's interpretations gave rise to a well-publicized controversy between the surgeon *John Abernethy* (1764-1831), a disciple of Hunter, and another physician, *William Lawrence* (1783-1867) who followed the teachings of Bichat. In 1814, Abernethy gave a series of lectures based on Hunter's vitalistic views in which he declared that the living actions were not a result of the underlying anatomical structures but instead shaped these structures. In this sense, Abernethy accepted the existence of an imponderable matter — the vital principle — which could not possibly be related to electricity, and he strongly defended the vivifying nature of the blood.[78]

Lawrence opposed Abernethy's views in a series of lectures, published in 1819 under the title *Lectures on Comparative Anatomy, Physiology, Zoology and the Natural History of Man.* Like Bichat, Lawrence stressed that the vital activity of the organism was closely related to its structure. "We may smile at these fancies," he wrote, referring to Hunter's vital principle. Lawrence concluded: "If the father of poetry sometimes falls asleep, a physiologist may be allowed to dream a little; but those who are awake need not exhibit another instance of the perverted taste, which led the disciples of an ancient philosopher to drink spinach-juice, that they might look pale like their master."

According to Lawrence, the difference between electrical appearances and vital processes was actually not so great that any analogy between them would be meaningless. All living actions including those of the soul, depended directly upon the structure and organization of the organs. Thus, declared Lawrence, we have to assume that the brain is capable of feeling, remembering, judging and thinking. How these actions took place was difficult to imagine, but it had to be similar to "the secretion of bile by the liver, the contraction of the muscles and any other vital process." Wrote Lawrence: "The same kind of facts, the same reasoning, the same sort of evidence altogether, which show digestion to be the function of the alimentary canal, motion of the muscles, the various secretions of their respective glands, prove that sensation, perception, memory, judgment, reasoning, thought, in a word, all the manifestations called mental or intellectual, are the animal functions of their appropriate organic apparatus, the central organ of the nervous system."

Even if Lawrence himself scoffed at the mounting opposition to his ideas as merely representing "ecclesiastical artillery," one can assume that he had indeed given rise to a veritable storm of adverse comments with his published statements. He was accused of materialism, unlawfulness and friendship towards the French[79] and his book was banned.

Somewhat later, the anatomist *Charles Bell*[80] returned once more to the same subject (see Fig. No. 63). Bell, in a work called *Animal Mechanics* (1828) and other writings, viewed life as an expression of a continuous and Divine act of creation. Life was dominated by a vital principle which expressed itself in the form of sensibility, motion, blood circulation and the turnover of bodily building elements. Thus, Bell emphatically declared that life was not the result, but rather the origin of the bodily organization. Similar thoughts were expressed in 1834 by the physician *William Prout*

Fig. 63. Charles Bell (1774-1842).

(1785-1850), the most prominent and physiologically oriented chemist living in Britain at the time.[81]

Hence, vitalism was the dominant current of physiological thought in Britain until the late 1830's. The phenomenon was probably related to the widely held theories regarding *natural theology*. This position propounded the investigation of nature as a means of proving the marvelous works of God and exposing the inherent perfection of His creation. The attitude established a close connection between religious beliefs and scientific thought, man being viewed as the goal and culmination of Divine creation. Moreover, all creatures were considered to be witnesses of God's infinite wisdom. These concepts were not altogether new; they had already been expressed by Robert Boyle in 1664 and repeated thereafter in the eighteenth century, even on the Continent, by authors such as the German physician *Johann A.H. Reimarus* (1729-1814) in 1760 and the German philosopher and mathematician *Christian von Wolff* (1679-1754) in 1753.

Therefore, it is not surprising to find in nineteenth-century Britain another author echoing these early ideas. In a book published in 1834, the Fullerian professor of physiology at the Royal Institution, *Peter M. Roget* (1770-1869) expressed similar concepts.[82] Entitled *Animal and Vegetable Physiology considered with Reference to Natural Theology*, the publication began with the following introduction: "To investigate the relations which connect Man with his Creator is the noblest exercise of human reason. The being who bestowed on him this faculty cannot but have intended that he should so exercise it, and that he should acquire, through its means, some insight, however limited, into the order and arrangements of creation; some knowledge, however imperfect, of the divine attributes; and a distant, though faint, perception of the transcendent glory with which those attributes are encompassed."

The theological point of view did not hinder scientists in their work, as Boyle's successful career demonstrates. The above mentioned examples were cited in order to prove the existence of a tendency towards generalization and establishment of vital theories in Britain. The movement, which was widely diffused there between 1750 and 1850, was often associated with fundamental religious beliefs.

Next to vitalism, there was a strong *empirical* current of physiological thought in Britain during the early nineteenth century. Since Elizabethan times, and thanks to the philosophers *Francis Bacon* and *Thomas Hobbes,* sensualism and empiricism had become well-accepted traditions in Britain. Even the English philosopher and economist *John Stuart Mill* (1806- 1873), a representative of the empirical viewpoint, had found acceptance both in his own country and on the Continent — he influenced, among others, the German chemist Justus Liebig and the pharmacologist *Friedrich Oesterlen* (1812-1877).[83]

Other examples of the empirical movement were *Alexander Stuart* with his observations regarding the nervous system, and the Scottish physician *Robert Whytt.* Whytt described in 1751 his experiments concerning the central nervous system and pointed out that the destruction of the corpora quadrigemina abolished the pupillary reflex to light. Likewise, he was also interested in the involuntary responses occurring in the body as a result of stimuli — what we today call reflexes. However, he did not ascribe them to the actions of a rational soul, but considered them effects of a "sentient" principle of the mind. Finally, another physician, *Alexander Walker* (1779-1852) of Edinburgh, made valuable contributions to neurophysiology. He claimed to have been the first to demonstrate the different functions of the anterior and posterior roots of the spinal nerves.

Before proceeding with the description of the British developments in neurophysiology, we should mention a highly talented and versatile author, *Thomas Young* (1773-1829).[84] A physician, physicist and Egyptologist, Young's greatest achievement lay in the field of physical optics — the discovery of the phenomenon of light interference due to the undulatory nature of light. Young was also active in physiological optics, having studied Newton's *Principia* and *Optics* at an early age. At the age of twenty, while still a student at London's St. Bartholomew Hospital, Young made a major discovery. During a dissection of a bull's eye there, he correctly interpreted the process of occular accommodation through the bulging and greater convexity of the crystalline lens. Young, however, believed that this phenomenon was actually produced by the muscles in the lens' capsule. Moreover, he also carried out various optical computations concerning the eye and developed an optometer. He was the first to discover the problem of astigmatism, and in 1801 attempted to explain colored vision on the basis of three fundamental color components.[85]

Returning to the British studies of the nervous system, the Scottish physician *Alexander P. W. Philip* (1770-1847) should be listed.[86] Philip was a prominent author who carried out numerous observations and engaged in several controversies. He rejected the dictum of the famous chemist Prout that chemistry was for the physiologist what anatomy meant for surgeons, declaring that chemistry was incapable of explaining a single law concerning the vital functions. In 1815, he expressed his belief in the identity of the living principle and electricity, demonstrating the fact that the gastric secretion could be maintained through electric stimulation of a severed vagus nerve. Following in the steps of the French author Legallois, in 1817-18, Philip examined the dependence of the cardiac function upon the brain. He observed the circulation in the web of a frog's foot after the administration of substances toxic to the cerebrum and concluded that the nervous system was not primarily responsible for the action of the heart and vessels but could influence these functions. Other observations of Philip were directed at the nervous influence on intestinal peristalsis. His most important work was entitled *An Experimental Inquiry into the Laws of the Vital Functions.*

Another important investigator in the field of neurophysiology was the anatomist and surgeon Sir *Charles Bell,* also of Edinburgh. Bell had already acquired some interest in the anatomy of the nerves during his student years, providing excellent drawings of his dissections.[87] In 1811, he printed privately a monograph called *An Idea of a New Anatomy of the Brain, submitted for the Observation of his Friends,* in which the motor functions belonging to the anterior roots of the spinal nerves were described. Later, Bell also claimed, in opposition to Magendie, to have been the first to recognize the sensible functions of the posterior roots. The controversy about the priority of the discovery has not been solved to this day, since Magendie was not always clear about the interpretation of experiments which he conducted.[88,89]

In a work published in 1823, Bell already referred to the doctrine of "specific energies" of the nerves, a phenomenon which was more clearly formulated by Johannes Müller three years later. Although he was very close to the actual discovery of the reflex function inherent in the spinal cord, Bell failed to see the decisive aspects of this phenomenon. His studies about the innervations necessary for respiration produced valuable new insights into the entire process. Bell also postulated that the various areas of the brain exhibited different functions. In 1826, he developed the idea of an actual muscular sense, and proposed the existence of a "circle of nerves" which connected the brain with the muscle. One nerve transmitted the cerebral influence to the muscle while another conducted the muscular sensation back to the brain.

Like Bell, the notable surgeon and anatomist Sir *Benjamin C. Brodie*[90] was also a student of both, Abernethy and Philip. Brodie was similarly interested in the influence of the nervous system on the body's inner organs. He observed that after the destruction of the cervical spinal cord the respiration, cardiac and circulatory action soon came to a halt. However, and in spite of the loss of cerebral influences, both heart and circulation could be maintained through continuous artificial respiration (1811). Brodie, following in the steps of Philip, also studied the action of poisonous substances on the brain and circulatory organs. He was interested in the vagal control over the gastric secretion (1814) and questioned the role played by the brain in thermoregulation. Another interesting observation was made by the Scottish biologist *James Black* (1788-1867) who in 1825 determined the capillary contractibility, which persisted even after the destruction of the spinal cord.

Fig. 64. Marshall Hall (1790-1857).

The most significant neurophysiological contributions of the period were made by the physician *Marshall Hall* (see Fig. No. 64) who almost simultaneously with Johannes Müller dealt with the phenomena of spinal reflexes and claimed priority for his discoveries.[91] Hall, from Basford at Nottingham, had studied at Edinburgh but lived in London since 1826. His enormous number of publications — over 150 — often expressed similar conclusions with some minor variations, making it difficult to assess his proper role in the development of neurophysiology. In the years 1832-33, Hall published for the first time a paper entitled "On the Reflex Function of the Medulla Oblongata and Medulla Spinalis," which first appeared in the *Philosophical Transactions* of the Royal Society of London.

While most authors until then had considered all motions of the skeletal muscles as voluntary, Hall described a series of muscular responses which only occurred involuntarily as a result of the stimulation applied to the sensible spinal nerves. For example, the isolated tail of a salamander responded to a pinprick as long as the spinal cord remained intact. Conversely, the stimulated limbs of a decapitated frog also contracted under those circumstances, and even the severed head of a turtle still showed a reflex closing of the eye lids. All these reactions could be increased by the use of strychnine or abolished with cyanide.

Hall called these nervous phenomena "reflex" functions. In 1827, he designated the so-called reflective system an "excito-motor system," a spinal cord organization distinct from the "cerebral system." Hall divided the spinal cord into a series of segments and ascribed to each of them certain "reflex arcs," as for example, the one concerned with the involuntary closure of sphincters, but he carried his generalizations beyond the available evidence.[92]

In another area of physiological research, Hall clearly differentiated for the first time between arterioles, capillaries and venules. He observed the pulse in the arteries but could not find it in the veins, discovering also that the arterioles actually branched out into capillaries of similar size. In his studies concerned with capillary circulation, published in 1831, Hall stressed his approach: "The sources of our knowledge in physiology as well as in natural science are *observation* and *experiment.*" Unfortunately, he restricted the usefulness of his method by declaring that experimentation was not necessary if the plain observations were satisfactory, since one had to avoid unnecessary cruelty during those procedures involving experimental animals.

Hall's neurophysiological observations were supplemented by the work of another anatomist and physiologist, *Richard D. Grainger* (1801-1865) who confirmed in 1837 the role played by the spinal cord in the reflex arc. In retrospect, it is clear that most of these British studies exhibited a great similarity, both in method and topic of research.

Finally, the first half of the nineteenth century also witnessed the important works of the physician *William Prout* (1785-1850) and the physiologist Sir *William Bowman* (1816-1892). Prout, who had taken his medical degree at Edinburgh, became the first British physiological chemist of any stature. In 1815, he established the so-called "Prout hypothesis" which postulated that the atoms of all elements were composed of hydrogen in various and changing combinations. In 1824, Prout discovered the existence of free hydrochloric acid in the stomach of a rabbit. Examining in the same year the foodstuffs in order to discover the nutritive elements in them, he divided these substances into three basic groups: sacharina, oleosa and albuminosa which correspond roughly to our modern carbohydrates, fats and proteins.

Prout also declared that our body was formed by the same chemical elements present in the food. Therefore, he analyzed the urine and urinary calculi in order to determine their chemical composition. As a physician, Prout was concerned with the specific gravity and daily volume of urinary excretion, as well as with the minerals present in the sediment.[94]

More will be said of William Bowman in one of the subsequent sections. However, reference should be made here to the five volume textbook of anatomy and physiology which he composed with the assistance of the physician *Robert B. Todd* (1809-1860), professor of anatomy at King's College Hospital, London. There were several of these textbooks in England during this period and most of them also included pathology, such as the works of the Scottish physician *William Pulteney Alison* (1790-1859) and the naturalist *William B. Carpenter* (1813-1885). Also to be mentioned are the physician *John Davy* (1791-1860), author of the work entitled *Physiological and Anatomical Researches* (London, 1839), Sir *Astley Cooper* (1768-1841) a well-known surgeon and investigator of the thymus gland, *John Hutchinson* of Newcastle-on Tyne, who constructed a spirometer in 1844, and finally, the Scottish physician *Thomas Wharton Jones* (1808-1891) who differentiated granular from non-granular leucocytes in 1846.

5.
AMERICAN PHYSIOLOGY FROM ITS BEGINNINGS UNTIL 1860

The development of physiology in the United States followed essentially the European antecedents and especially the British model during the eighteenth and early nineteenth centuries. Both currents, vitalism and empiricism, dominated the various physiological works carried out during this period. In general, American physiology remained in its *first, pre-experimental phase* until 1860, as an object of study outside of the laboratory. These investigations were carried out sporadically by physicians who were primarily clinicians, among them *William Beaumont* (1785-1853), *John Richardson Young* (1782-1804), *Robley Dunglison* (1798-1869) and others.[95]

A *second, transitional period* in American physiology covered the decade 1860-1870, in which new interest for physiological problems emerged. In addition to some new teaching positions in physiology, some interested physicians of this time also began to carry out some experimental work in a few privately endowed laboratories. Gradually, it was realized that the United States had still much to learn from European science. Among the most prominent authors of the period were *John W. Draper* (1811-1882), *John C. Dalton, Jr.* (1825-1889), *Austin Flint* (1836-1915) and *Silas Weir Mitchell* (1829-1914).

The *third phase* began after 1870, when it was realized that without laboratories and experimental research, American physiology could not readily parallel or duplicate the European developments. Thus, in 1871, *Henry P. Bowditch* (1840-1911), a student of Carl Ludwig, opened the first physiological laboratory at the Harvard Medical School in Boston.[96]

With this broad sketch in mind, let us consider here merely the first phase. The physician John Richardson Young can be truly considered one of the pioneers of American physiology. In 1803, he carried out a series of experiments concerned with the action of gastric juice on meat in Philadelphia. Two other authors, *D. H. Spencer* in 1803, and *Thomas Ewell* (1785-1826) were concerned with similar questions. Ewell published the work *Notes on the stomach and secretion* (Philadelphia, 1803) and performed the first American vivisection experiment on an animal

In 1833, physiologists throughout the world focused their attention on America for the first time. In that year, *William Beaumont*[97] (see Fig. No.

Fig. 65. William Beaumont (1785-1853).

65) published his remarkable results in a book entitled *Experiments and Observations on the Gastric Juice, and the Physiology of Digestion.* Beaumont had the fortune of treating and investigating a Canadian voyager, Alexis St. Martin, who had sustained a gunshot wound resulting in a gastric fistula. Under those circumstances, he carried out over seventy experiments, partly in vivo, in the interior of St. Martin's stomach, and the rest with the patient's isolated gastric juice. In this fashion, he came to a series of important conclusions about gastric motility and digestion, especially the effect on meat.

Beaumont received some advice from *Robley Dunglison,* who had obtained his diploma from the German university of Erlangen, although he was originally from England. Dunglison, who taught first at the University of Virginia and then at Jefferson Medical College of Philadelphia, wrote a two volume textbook entitled *Practice of Medicine* (1832-42). Dunglison's work was strongly oriented towards anatomy and histology, since he was a professor of the Institutes of Medicine and anatomy. For him, digestion was not the effect of a vital power but rather a process of chemical dissolution. Apparently, Beaumont did not acknowledge Dunglison's contributions to the subject.[98] In any event, Beaumont's observations gave rise to further experiments carried out by Johannes Müller and his student Theodor Schwann in Berlin.[99]

Dunglison was also the teacher of *Silas Weir Mitchell* at Philadelphia. Mitchell also spent a year with the famous French physiologist Claude Bernard and the microscopist *Charles Robin* (1821-1885) at Paris, where he became interested in experimental work and studied the formation of uric acid, hemoglobin and cerebellar functions. Concerning the latter, Mitchell emphasized the existence of cerebellar coordinating functions which had been first proposed by Flourens. Mitchell was also one of the founders of the American Physiological Society, and as a clinician published a series of excellent works in neurology — one was entitled *On Rest in the Treatment of Nervous Disease* (1875) — in addition to poems and novels.[100]

John W. Draper came from England and studied physics, chemistry and medicine at the University of Pennsylvania. He was especially interested in the exchange which occurred at the level of the capillaries and the cellular tissue, such as osmosis, endosmosis and capillary attraction. In 1838, Draper became professor of chemistry. He was a pioneer in both photography and microphotography. In 1856, he published his work *Human Physiology, Statical and Dynamical,* in which he denied the existence and action of special vital forces in addition to the purely physicochemical phenomena.[101]

Both *John C. Dalton* and *Austin Flint* already belonged to a generation of American physicians who went abroad in order to complete their education. Both men studied under Claude Bernard at Paris, Dalton in the 1850's and Flint in 1860-61. Dalton, originally from Massachusetts, graduated in 1847 from Harvard, and has been considered the first true physiologist of the United States, who devoted his time exclusively to that discipline.[102] He was active in experimental physiology and enriched his lectures with a series of animal experiments. Initially Professor of Physiology in Buffalo (1851-54) and Vermont (1854-56), Dalton was appointed to the Chair of Physiology and Microscopical Anatomy at the College of Physicians and Surgeons of New York. Not only did he write the celebrated work *Treatise on Human Physiology,* but he also attacked convincingly the anti-vivisectionists in the publication *The Experimental Method in Medical Science* (1882). Dalton was interested in the formation of sugar in the liver, and in charge of the first private American laboratory of physiology. Throughout his work, the influence of Claude Bernard was clearly visible.

During the 1870's, the French influence gradually receded in America. In 1865, the first American to study in Berlin under the direction of Du Bois-Reymond was the physician *Charles E. Morgan* (1832-1867). Not only did Morgan acquire a more systematic education in Germany, but he also was instrumental in translating essential portions of Du Bois-Reymond's work on animal electricity into English: *Electro-physiology and Therapeutics* (New York, 1868).[103]

Next to Dalton, *Austin Flint* (see Fig. No. 66) has been considered the inaugurator of laboratory research in America. After graduating from Jefferson Medical College in 1857, Flint studied at the University of Vermont under Dalton. After a year in Paris under the direction of

Fig. 66. Austin Flint (1853-1915).

Bernard and Robin, his interests shifted towards histology. Flint became for many years professor of physiology at Buffalo, and then moved to Bellevue Hospital Medical College in New York City.[104] His five volume textbook, entitled *The Physiology of Man; Designed to Represent the Existing State of Physiological Science* (New York, 1866-1874) was histologically oriented and described relatively few physiological instruments. It was dedicated to his French teacher, Charles Robin.[105]

REFERENCES

[1]J. Schiller, "Physiology's struggle for independence in the first half of the 19th century", *Hist. Sci.* 7 (1968), pp. 64-89.

[2]G.J. Goodfield, *The Growth of Scientific Physiology,* London, 1960.

[3]K.E. Rothschuh, "J. Müller und C. Ludwig". *Deutsch. med. Wschr.,* (1953), H. 2.

[4]H.H. Eulner, *Die Entwicklung der medizinischen Spezialfächer an den Universitäten des deutschen Sprachgebietes.* Stuttgart, 1970.

[5]E.T. Nauck, "Bem. zur Geschichte des physiol. Instituts Freiburg i.Br.". *Ber. naturforsch. Ges. Freiburg* 40 (1950), p. 147.

[6]K.E. Rothschuh, "Carl August Sigmund Schultze (1795-1877) und seine Vorlesungen über Experimentalphysiologie in Freiburg (1830)." *Sudhoffs Archiv* 47 (1963) pp. 347-359.

[7]H.F. Kilian, *Die Universitäten Deutschlands in med.-naturwiss. Hinsicht betrachtet,* Heidelberg Leipzig, 1828.

[8]R. Herrlinger and J. Krupp, *Albert von Bezold. 1836-1868. Ein Pionier der Kardiologie,* Stuttgart, 1964.

[9]W. Trendelenburg, "60 Jahre Berliner Physiologische Gesellschaft". *Klin. Wschr.* (1936), p. 311.

[10]K. Sudhoff, "Das medizinische Zeitschriftenwesen in Deutschland bis zur Mitte des 19. Jahrhunderts", *Münch. med. Wschr.* 50 (1903), pp. 455-463.

[11]G. Rosen, "Romantic medicine: a problem in historical periodization", *Bull. Hist. Med.* 25 (1951), pp. 149-158.

[12]See P. Diepgen, "Alte und neue Romantik in der Medizin", *Klin. Wschr.* (1932), p. 28. and I. Galdston, "The romantic period in medicine", *Bull. N. Y. Acad. Med.* 32 (1956), pp. 346-362.

[13]H.A.M. Sneless, "The influence of the dualistic system of Jacob Joseph Winterl (1732-1809) on the German romantic era", *Isis* 61 (1970), pp. 231-240.

[14]E. Hirschfeld, "Romantische Medizin" *Kyklos* 3 (1930), pp. 1-89.

[15]J. Neubauer, "Novalis und die Ursprünge der romantischen Bewegung in der Medizin", *Sudhoffs Arch.* 53 (1969), pp. 160-170.

[16]John Neubauer, *Bifocal Vision, Novalis' Philosophy of Nature and Disease.* Chapel Hill, 1971.

[17]K. Bürker, "Über den Werdegang der Physiologie u. das neue physiologische Institut an der Landesuniversität Giessen", *Nachr. der Giessener Hochschulgesellsch.* 11 (1937), p. 19.

[18]K.E. Rothschuh, "Ansteckende Ideen in der Wissenschaftsgeschichte, gezeigt an der Entstehung und Ausbreitung der romantischen Physiologie", *Deutsch. med. Wschr.* 86 (1961), pp. 396-402.

[19]G. Risse, "Kant, Schelling, and the early search for a philosophical 'science' of medicine in Germany, *J. Hist. Med.* 27 (1972), pp. 145-158.

[20]Richard H. Shryock, "Science in a romantic age 1800-1850", in *The Development of Modern Medicine,* New York, 1947, ch. VI, pp. 109-119.

[21]Friedrich W.J. von Schelling, *On University Studies,* transl. by E.S. Morgan, Athens, O., 1966.

[22]K.E. Rothschuh, "J. Görres und die romantische Physiologie". *Med. Msch.* 5 (1951), p. 128.

[23]S.F. Mason, "German Nature-philosophy", in *Main Currents of Scientific Thought,* New York, 1953, ch. 29, pp. 280-290.

[24]O. Temkin, "Basic science, medicine and the romantic era", *Bull. Hist. Med.* 37 (1963), pp. 97-129.

[25]Werner Leibbrand, *Romantische Medizin,* Hamburg & Leipzig, 1937.

[26]J.H. Oesterreicher, *Versuch einer Darstellung der Lehre vom Kreislauf des Blutes,* Nürnberg, 1826.

[27]Paul Diepgen, *Deutsche Medizin vor 100 Jahren,* Freiburg, 1923.

[28]J.J. Wagner, *Von der Natur der Dinge,* 3 vols., Leipzig, 1803.

[29]A. Winkelmann, *Einführung in die dynamische Physiologie,* Göttingen, 1802.

[30]Johann B. Wilbrand, *Lehre vom Kreislauf in den mit Blut versehenen Tieren nebst weiterer Nachweisung, dass eine Blutzirkulation weder in d. Beob. noch wissenschaftlich begründet ist u. sich mit d. sonstigen Verhalten d. Natur nicht vereinbaren lässt,* Frankfurt, 1826.

[31]Johann B. Wilbrand, *Was ist Physiologie und wie ist diese Wissenschaft zu behandeln?* Frankfurt, 1827.

[32]Johann B. Wilbrand, *Allg. Physiologie, insb. vergleichende Physiologie der Pflanzen und Tiere,* Heidelberg, 1833.

[33]K.E. Rothschuh, "J. Görres und die romantische Physiologie", *Med. Mschr.* 5 (1951), p. 128.

[34]For a biography see G. Canguilhem, in *Dict. Sci. Biogr.* 2 (1970), pp. 122-123.

[35]Xavier Bichat, *General Anatomy,* transl. from French by G. Hayward, Boston, 1823.

[36]P. Lain-Entralgo, "Sensualism and vitalism in Bichat's Anatomie générale", *J. Hist. Med.,* 3 (1948), pp. 47-64.

[37]K.E. Rothschuh, "Deutsche Biedermeiermedizin. Epoche zwischen Romantik und Naturalismus", *Gesnerus* 25 (1968), pp. 167-187.

[38]Marie J.P. Flourens, *Expériences sur le Système Nerveux,* Paris, 1825.

[39]Claude Bernard, *François Magendie,* Paris, 1856.

[40]O. Temkin, "The philosophical background of Magendie's physiology", *Bull. Hist. Med.* 20 (1946), pp. 10-35.

[41]François Magendie, *An Elementary Compendium of Physiology; for the Use of Students.* Transl. from the French, Philadelphia, 1824.

[42]James Olmstead, *François Magendie, Pioneer in Experimental Physiology and Scientific Medicine in XIX Century France.* With a preface by John F. Fulton, New York, Schuman, 1944. and
R. Langmann, *Das Werk des französischen Physiologen François Magendie,* Düsseldorf, 1937.

[43]M. Dittrich, "Die Bedeutung von Karl Asmund Rudolphi (1771-1832) für die Entwicklung der Medizin und der Naturwiss. im 19. Jhdt." *Wiss. Z. d. Ernst Moritz Arndt-Univ. Greifswald,* 16 (1967), pp. 249-277.

[44]J. Müller, "Gedächtnisrede auf Carl Asmund Rudolphi 1835". *Abh. Akad. Wiss.* Berlin, (1837), p. 23.

[45]See biography by A.S. Kay in *Dict. Sci. Biogr.* 2 (1970), pp. 594-597.

[46]Arthur W. Meyer, *Human Generation: Conclusions of Burdach, Döllinger and von Baer,* Stanford, 1956.

[47]H.B. Picard, "Philosophie und Forschung bei K.F. Burdach", *Med. Mschr.* 5 (1951), p. 125.

[48]See biography by G. Risse in *Dict. Sci. Biogr.* 4 (1971), pp. 146-147.

[49]B.H. Willier — J.M. Oppenheimer, eds., *Foundations of Experimental Embryology,* Englewood Cliffs, N.J., 1964.

[50]See his following works: Ignaz Döllinger, *Grundriss der Naturlehre des menschlichen Organismus,* Bamberg, 1805.
Ibid., Von den Fortschritten, welche die Physiologie seit Haller gemacht hat. Eine Rede. München, 1824.
Ibid., Grundzüge der Physiologie. 2 vols. Regensburg, 1835.

[51]Henry J. John, *Jan Evangelista Purkyně: Czech Scientist and Patroit* (1787-1869), Philadelphia, 1959.

[52]R. Toellner, "Naturphilosophische Elemente im Denken Purkyněs", in *Centenary Symposium,* ed. by Vl. Kruta, Brno, 1971, pp. 35-41.

[53]Vl. Kruta, "J.W. Goethe's and J.E. Purkyně's studies on vision", *Actes XIe Cong. Int. Hist. Sci.* 5 (1965), pp. 58-63.

[54]Vladislav Kruta, *J.E. Purkyně (1787-1869) Physiologist,* Prague, 1969.

[55]H. Schreiber, *Chronik der Albert-Ludwig-Universität zu Freiburg im Breisgau vom Sommerhalbjahr 1824 bis dahin 1829,* Freiburg i. Br., 1829.

[56]E. Thomsen, "Über Johannes Evangelista Purkinje und seine Werke", *Skand. Arch. Physiol.,* 37 (1919), p. 1.

[57]P.K. Studnička, "Einiges über das Wort Protoplasma", *Protoplasma* 27 (1937), p. 619.

[58]M. Teich, "Purkyně and Valentin on ciliary motion: an early investigation in morphological physiology", *Brit. J. Hist. Sci.* 5 (1970), pp. 168-177.

[59]Armin von Tschermak-Seysenegg, *J.E. Purkyně,* Prague, 1937, and
Armin von Tschermak-Seysenegg, "Erprobung der Methode des Augenspiegelns nach Joh. Ev. Purkinje (1823)", *Klin. Mbl. Augenheilk.,* 107 (1941), p. 85.

[60]A. Schröter V. Kristelli, "J.E. Purkyně", *Almanach d. kais. Akad. d. Wiss.* Wien, 20 (1870), p. 182, and
E. Ebstein, "Purkinje, der Begründer der physiologischen Institute in Breslau und Prag", *Hippokrates* 3 (1931), p. 508.

[61]B. Kisch, "G.G. Valentin, the life of a Jewish pioneer of modern medicine", in *Victor Robinson Memorial,* New York, 1948.

[62]See biography by Vl. Kruta in *Dict. Sci. Biogr.* 3 (1971), pp. 530-531.

[63]E. Lesky, "Ludwig Türck (1810-1868). Neuroanatom und Neurophysiologe", in *Von Boerhaave bis Berger,* ed. K. Rothschuh, Stuttgart, 1964, pp. 129-133.

[64]See biography by K.E. Rothschuh in *Dict. Sci. Biogr.* 2 (1970), pp. 160-162.

[65]G. Lusk, "Carl von Voit, master and friend", *Ann. Med. Hist.* 3 (1931), pp. 583-594.

[66]Rudolph Wagner, *Über Wissen und Glauben,* Göttingen, 1854.

[67]W. May, "Karl Vogt zu seinem 100. Geburtstag", *Naturwiss.* 21 (1917), p. 449.

[68]J. Moleschott, *Für meine Freunde,* Giessen, 1894.

[69]G.R. Cameron, "Rudolf A. von Koelliker (1817-1905)", *Ann. Sci.* 2 (1955), pp. 166-172.

[70]See biography by J. Steudel in *Dict. Sci. Biogr.* 2 (1970), pp. 110-111.

[71]Rudolf A. Kölliker, *Erinnerungen aus meinem Leben,* Leipzig, 1899.

[72]P Hoffmann, "Ernst Heinrich Webers 'Annotationes anatomicae et physiologicae'", *Med. Klin.* (1934), p. 1250.

[73]Carl Ludwig, *Rede zum Gedächtnis an Ernst Heinrich Weber,* Leipzig, 1878.

[74]H.E. Hoff, "Vagal stimulation before the Webers", *Ann. Med. Hist.* 8 (1936), pp. 138-144.

[75]E. Mendelsohn, "Physical models and physiological concepts: explanation in 19th century biology", *Brit. J. Hist. Sci.* 2 (1965), pp. 201-219.

[76]G.G. Gruber, "Arnold Adolph Berthold (1803-1861)" (=Niedersächsische Lebensibilder, Bd. 4), Hildesheim, 1960, pp. 23-30.

[77]Jessie Dobson, *John Hunter,* Edinburgh, 1969.

[78]J. Goodfield-Toulmin, "Some aspects of English physiology: 1780-1840", *J. Hist. Biol.* 2 (1969), pp. 283-320.

[79]P.G. Mudford, "William Lawrence and 'The Natural History of Man'", *J. Hist. Ideas* 29 (1968), pp. 430-436.

[80]See biography by P. Amacher in *Dict. Sci. Biogr.* 1 (1970), pp. 583-584.

[81]W.H. Brock, "The life and work of William Prout", *Med. Hist.* 9 (1965), pp. 101-126.

[82]Donald L. Emblen, *Peter Mark Roget, the World and the Man,* New York, 1970.

[83]K.E. Rothschuh, "Friedrich Oesterlen (1812-1877) und die Methodologie der Medizin", *Sudhoffs Arch.* 52 (1968), pp. 97-129.

[84]Alexander Wood, *Thomas Young, Natural Philosopher, 1773-1829,* New York, 1954.

[85]E.C. Millington, "History of the Young-Helmholz theory of colour vision", *Ann. Sci.* 5 (1942), pp. 167-176.

[86]W.H. McMenemey, "Alexander Philips Wilson Philip (1770-1847), physiologist and physician", *J. Hist. Med.* 13 (1958), pp. 289-328.

[87]G. Gordon Taylor — E. Walls, *Sir Charles Bell, his Life and Times,* Baltimore, 1958.

[88]Max Neuburger, *Die hist. Entwicklung der experimentellen Gehirn- und Rückenmarksphysiologie vor Flourens,* Stuttgart, 1897.

[89]M. Brazier, "Rise of neurophysiology in the 19th century", *J. Neurophysiol.* 20 (1957), pp. 212-226, and
M. Brazier, "The historical development of neurophysiology" in *Handbook of Physiology,* Washington, D.C., 1959, I, pp. 1-58.

[90]See biography by G.J. Goodfield in *Dict. Sci. Biogr.* 2 (1970), pp. 482-484.

[91]G. Canguilhem, "Le Concept de réflexe au XIXe Siècle", in *Von Boerhaave bis Berger.* ed. K.E. Rothschuh, Stuttgart, 1964, pp. 157-167.

[92]G. Jefferson, "Marshall Hall, the grasp reflex and the diastaltic spinal cord", in *Science, Medicine and History,* London, 1953, II, pp. 303-320.

[93]A.M. Kasich, "William Prout and the discovery of hydrochloric acid in the gastric juice", *Bull. Hist. Med.* 20 (1946), pp. 340-358.

[94]H. Kausch, "William Prout — der Übergang von der Uroskopie zur Urochemie", (Med. diss., Münster, 1968).

[95]W.J. Meek, "The beginnings of American physiology", *Ann. Med. Hist.* 10 (1928), pp. 111-125.

[96]W.B. Cannon, "The first American laboratory of physiology", *Science* 78 (1933), p. 365.

[97]See biography by G. Rosen in *Dict. Sci. Biogr.* 1 (1970), pp. 542-545.

[98]J.J. Bylebyl, "William Beaumont, Robley Dunglison and the 'Philadelphia Physiologists'", *J. Hist. Med.* 25 (1970), pp. 3-21.

[99]George Rosen, *The Reception of William Beaumont's Discovery in Europe,* New York, 1942.

[100]Richard D. Walter, *S. Weir Mitchell, M.D., Neurologist. A Medical Biography,* Springfield, Ill., 1970.

[101]Donald Fleming, *John William Draper and the Religion of Science,* Philadelphia, 1950.

[102]H. Sewall, "The beginnings of physiological research in America," *Science* 58 (1923), p. 187, and
John C. Dalton, *A Treatise on Human Physiology,* Philadelphia, 1859.

[103]P.F. Cranefield, "Charles E. Morgan's 'Electrophysiology and Therapeutics': an unknown English version of Du Bois-Reymond's 'Thierische Elektricität'", *Bull. Hist. Med.* 31 (1957), pp. 172-181.

[104]A.S. Evans, "Austin Flint and his contributions to medicine", *Bull. Hist. Med.* 32 (1958), pp. 224-241.

[105]Austin Flint, *Collected Essays and Articles on Physiology and Medicine 1855-1902,* New York, 1903.

CHAPTER VI

Johannes Müller, Carl Ludwig and their Circle of Students

1. JOHANNES MÜLLER

The birth of *Johannes Petrus Müller* (see Fig. No. 67) was registered in the city of Koblenz "on the 25th of Messidor, 9th year of the Franconian Republic" (July 14, 1801). Koblenz, then boasting a population of about eight thousand people, was the capital of the Rhine and Moselle Departments, both under French administration since the peace of Luneville.[1]

Fig. 67. Johannes Müller (1801-1858).

Thus, the young Müller grew up in politically unstable times, and was also exposed to the contemporary flowering of romanticism and Schelling's "Naturphilosophie." His father sent the talented boy to the local "École secondaire" at the age of ten, and later he went to the "Gymnasium" where the already mentioned poet, Joseph Görres, taught between 1808 and 1811. Strangely enough, Müller never made any reference to this famous romantic author in his writings.

Although the classes were often interrupted because of the prevailing political unrest, Müller already exhibited a great desire for knowledge. His mathematical as well as linguistic abilities clearly distinguished him from his other classmates. Like many other early students of nature, he collected animals and plants during his childhood, dissecting insects and acquiring knowledge quite independently from his deficient instruction at school.

After completing his military obligations, Müller initiated his medical studies at the newly created university in Bonn. Among his professors, he was especially impressed by the physiologist and surgeon *Philipp F. von*

Walther, whose philosophy of nature fascinated the young medical student. In addition, the faculty at the University of Bonn included the chemist *Karl W.G. Kastner* (1783-1857), teacher and sponsor of Justus Liebig, a man also deeply influenced by "Naturphilosophie," and the physician *August Brandis*. Brandis was also a philosopher of nature and an earnest follower of Aristotle, a teacher who infused Müller with a great admiration for the famous Greek philosopher and biologist.

In spite of his activities concerning student affairs, Müller plunged into his scientific studies with great enthusiasm. Barely a sophomore, he competed in 1820 for a prize-essay instituted by the university concerned with fetal respiration. After carrying out fifty-seven experiments on living animals of all types, such as stray cats which he personally caught in the surrounding villages, Müller attempted to demonstrate the phenomenon of fetal breathing. His work, entitled *De respiratione foetus* (Leipzig, 1823), was successful in the contest and won him the prize.

Müller's doctoral thesis was completed at age twenty one. It was concerned with the laws and quantitative relationships governing the motions of various animals, including the milleped: "Diss. inaug. physiol. sistens commentarios de phoronomia animalium, Bonnae, 1822." The influence of the prevailing philosophical spirit was clearly visible in the preface to Müller's thesis. Muscular flexion was interpreted as a motion with circular tendencies, while extension simulated a straight line, all in accordance with the poles of an electrical pile. Among several other concepts, Müller defended the dictum "Psychologus nemo, nisi physiologus," a subject which he dealt with repeatedly.

Thanks to the favorable impression which he caused at the University of Bonn, and the support of the local curator, P.J. von Rohfüs, Müller received a stipend to study in Berlin and receive further training for a career in academic medicine. Between 1823-1824, he spent one and a half years at Berlin, especially in the anatomical institute there under the direction of Carl A. Rudolphi. This short sojourn actually clarified his approach towards the study of nature, also contributing to the solidification of his love and devotion for anatomy. Rudolphi was an excellent teacher and investigator who greatly stimulated the students because of his zeal for the truth and his opposition to the ideas of "Naturphilosophie" in both anatomical and physiological research.

In an eulogy of Rudolphi, given in 1837, Müller described his former teacher in the following way: "Every time he had the opportunity, Rudolphi would express himself in rather strong terms against a study of nature based on a series of misinterpreted philosophical principles. To be sure, Rudolphi's criticism was somewhat pretentious, since it addressed itself to vague generalities because of a lack of precise methods... But the presumptuous and irresponsible way to philosophize about natural subjects then in vogue could not comfort the sober-minded observer, particularly in the face of the frauds which were being perpetuated in the name of animal magnetism. Even in Berlin, usually a focus of respectable scientific efforts, there was no want of credulous authors. In those circumstances, Rudolphi continued to hamper the philosophical current with his forceful opposition, and we owe much to his voice for having regained physicians from the field of medical miracles."[2]

During his stay at Berlin, Müller was forever healed from his previous tendencies towards a romantic interpretation of nature. However, he never denied the value of a philosophical contemplation of nature nor did he consider such an effort superfluous. To the contrary, Müller always continued to express the idea that no scientific investigation was meaningful unless the experiences could be eventually included in an ideally conceived picture of nature. The sojourn at Berlin also provided him with numerous suggestions for the field of comparative anatomy.

Shortly after his return to Bonn, Müller received permission to teach, and he held his inaugural lecture entitled "Von dem Bedürfnis der Physiologie nach einer philosophischen Naturbetrachtung" (On the need for a philosophical contemplation of nature in physiology) on October 19, 1824. The carefully prepared speech contained a complete program of action, and characterized a line of approach which Müller maintained throughout his career. After depicting the erroneously advancing "Naturphilosophie" as leading everybody astray, he referred to those ideas and methods truly necessary in the natural sciences and especially in physiology.[3]

One can clearly sense in Müller's words the influence of Goethe towards observations and experiments. In fact, Müller almost used the same words which Goethe had expressed in 1792, when he declared: "An observation should be carried out in a plain, candid and industrious manner without preconceived opinions. In turn, the experiment is artificial, fidgety, expeditious, digressive, enthusiastic and unreliable. One sees daily the accumulation of one experiment upon another, one fiction supporting the next one, all by men who cannot bring intellectual excellence nor enough self-denial and honesty to these pursuits. Nothing is easier than to carry out a series of so-called experiments. Assuredly, one can always forcibly interrogate nature and she will give us an answer in her distress. But nothing is more difficult than the correct interpretation of nature, nothing tougher than to plan a truly valid experiment. We consider it the foremost objective of contemporary physiology to show and clearly grasp these obstacles. I have already attempted to show that research into nature has something religious about it; in other words, it is really a cult. One can add to this statement, I believe, that the investigation of nature also has a permanent priesthood."

Müller then continued: "We have experiences which can only be shaped from ideas. In turn, new ideas are derived in many imaginable ways from these experiences. The unassuming, simple contemplation of nature, is being able to recognize the right thing, the truth of the appearances should be the relevant goal for both the naturalist and the physiologist. Under those circumstances, let your mind experience whatever it wants, since it will feel more than what is apparent to the senses in the objects under scrutiny. Since experiences and observations are derived from preconceived ideas, so these sensible phenomena will form other ideas. In truth, experience becomes the generating ferment of the mind. Abstract thought about nature is not the proper area for physiological endeavor, since the physiologist must experience nature in order to develop thoughts about her."

Seldom in the history of physiology has the relationship between experience, observation or experimentation and thought been as clearly expressed as in Müller's above statement. The idea which prompts a series of observations or motivates the performance of an experiment connects with those conclusions derived from the new data. Thus, thought necessarily envelops the scientist on both the beginning and the end of any given experience, and is incorporated into his conceptual framework.

"Physiology is not a science unless it combines intimately with philosophy." This, a somewhat strange sounding characterization of experimentation, has its roots in Müller's *vitalistic outlook*. He was deeply convinced of the absolute distinction between the inorganic and organic realms. Only the action of a "Lebenskraft" made it possible for the living being to survive and adapt to the various influences it was subjected to and effect the necessary restorations when structure and function were both disturbed.

Two reasons made Müller's acceptance of a vital force unavoidable. In the first place, the French chemist *Antoine F. de Fourcroy* (1755-1809) had already stressed the binary structure of inorganic elements in contrast with the tertiary and quaternary one displayed by living substances. Therefore, it was assumed that a special force was needed to achieve such an organic chemical structure. Secondly, the organism existed "according to the laws of a rational, goal oriented plan," and therefore the "Lebenskraft" had to preside over the formation of tissues and organs. In truth, a rational and creative force was already present in the embryo. Moreover, life was characterized as a reaction to stimuli, a response triggered by a stimulus and dependent on the nature of the living part which had been excited, its so-called "specific energy." Clearly, Haller's concept of irritability and Brown's so-called "excitability" were still influential. According to Müller, the oxygen which passed into the blood acted as a stimulus to facilitate decomposition at the tissue level.

Müller's objections to certain modalities of animal experimentation were based on the above mentioned ideas. How, in Müller's view, could an experiment provide meaningful answers when only partial aspects of an organismic totality were actually subjected to artificial experimental manipulations? Above all, Müller could not sympathize with a purely physical or chemical interpretation of the vital processes. This attitude prevailed even throughout the period in which Carl Ludwig achieved new and important insights into the lawfulness of the living appearances by adopting a physico-chemical approach.[4]

Another of Müller's students, Du Bois-Reymond, recalled later that he had been unable to convince his teacher of the strict validity of certain physical laws in living organisms. Du Bois-Reymond characterized Müller's position as follows: "The organisms are susceptible to physical and chemical actions, but the way these living entities react to the inorganic influences is markedly different. Physical agents transmit to one another a certain motion, and chemical elements interact with one another to form a third compound. By contrast, when these physico-chemical stimuli excite the organism, they can only provoke the appearance of an organic property: the living energy."[5] Müller expressed his ideas about the methods and goals of physiology and the essence of vitality in several other writings. Especially important was the prologue to his work *Bildungsgeschichte der Genitalien* (1830) and the famous *Handbuch der Physiologie* (1834), but the above remarks must suffice.[6]

Müller's teaching years at Bonn between 1824 and 1833 were extremely busy. In addition to his many lectures on comparative anatomy, human and animal physiology, medical methodology, general pathology as well as ocular and cerebral diseases, Müller worked endlessly on major and lesser anatomical and physiological problems. Stimulated by Goethe, he conducted *subjective physiological examinations* in the years 1824-1827, publishing two works in 1826: *Zur vergleichenden Physiologie des Gesichtssinnes der Menschen und Tiere* (Comparative physiology of vision in man and animals) and *Über die phantastischen Gesichtserscheinungen.*

In the first of these publications, Müller expressed his now famous law of the "specific energies" of the nerves,[7] while in the second he described a series of subjective visual effects which occurred in a dark room. From the physiology of the senses, Müller attempted to gather some insights into the nature of the soul. He declared that "the soul is only a special form of life among many other vital manifestations which are subjects for physiological investigation. Thus, I am also convinced that research in physiology will deal with psychological matters in the last analysis." Later, in his *Handbuch,* Müller expressed similar views about the relationship between life and soul.

In 1827 Müller's small text *Grundriss der Vorlesungen über die Physiologie* appeared, based on his physiological lectures. Interestingly enough, the work had been written in accordance with the ideas of Philipp von Walther and other romantic authors. It divided the organic functions into sensibility, irritability and a reproductive force. A year before, Müller had been appointed assistant professor, but in the spring of 1827 he suffered a nervous breakdown from which he recovered following extensive travel. A combination of factors seemed to have been responsible for the episode. In the first place, Müller used to consume excessive amounts of coffee and ate irregularly in order to provoke his subjective optical experiences. Secondly, his teaching load was proving to be excessive. Perhaps, the mental collapse was just the first of a series of depressive episodes which Müller was to suffer in the course of his life.[8] In any event, the temporary mental disability terminated the physiologist's subjective and philosophical life period.

Thereafter, Müller devoted himself more to *observations, anatomical dissections and experimentation.* Above all, and still at Bonn, he studied microscopical structures intensively. Müller was convinced that knowledge about the finer organic composition would contribute substantially to the solution of physiological problems. The first of these endeavors were two pioneering works, published in 1830: *De glandularum secernentium structura. .* and *Bildungsgeschichte der Genitalien.* The first book was concerned with glandular function, and eradicated some old preconceptions. The result of an incredible amount of observations on representatives of virtually all classes of vertebrates, Müller's publication prompted the following remark from the German physiologist *Rudolf P.H. Heidenhain* in Hermann's *Handbuch der Physiologie,* published in 1883. Wrote Heidenhain: "In contrast to the concepts available from the preceding century, which postulated that blood vessels were secretory structures, a new explanation is now at hand. Secretion actually occurs from the walls of closed glandular spaces, which are surrounded by capillary networks. Thus, the glands represent a vast surface, a living lining which initiates the secretory function."

During the years 1827-1833, Müller also carried out numerous single studies, such as those which confirmed Bell's laws on the frog, the discovery of the frog's lymphatic heart, as well as a great number of observations concerned with blood and lymph. In the process of these investigations, Müller duplicated and confirmed, independently from Hewson, the work concerning blood coagulation while adding more knowledge to this rather important subject. The results of such a research were published in the fifth volume of Burdach's *Physiologie als Erfahrungswissenschaft* (1832). Müller's versatility was reflected in another publication entitled *Über die Diffusion durch dünne, thierische Häute.* Between 1833 and 1840 a series of superb microscopical investigations were carried out in Müller's institute, and the microscope became the favorite research tool for physiologists.

A trip to Paris in September, 1831, brought Müller in contact with Cuvier, the French physiologist *René J.H. Dutrochet* (1776-1847), the Belgian zoologist *Henri Milne-Edwards* (1800-1885) and Alexander von Humboldt, who were deeply impressed by the presentations of the physiologist. In a series of animal experiments, Müller severed the spinal cord in cats and rabbits, stimulated and achieved the contraction of the ductus thoracicus in a goat, and produced a vagal irritation in dogs. Moreover, he achieved a stimulation of the splanchnic nerve in dogs and rabbits, and proved the influence of the renal nerves on urinary secretion.

In 1833, Müller made the most important decision of his life: he accepted the call to succeed Rudolphi in the chair of anatomy and physiology at Berlin. A reflection of Müller's growing self-confidence was the fact that he actually proposed his own candidacy to the vacant position in a letter to the Prussian minister Stein von Altenstein. After negotiations with the anatomist and physiologist *Friedrich Tiedemann* of Heidelberg, and Carl G. Carus of Dresden broke down, Müller was called to Berlin in early 1833. In the summer semester of that year, he took over the institute with its excellent collections, housed however in rather deficient physical facilities. The old building was located behind the "Garnisonkirche" and had insufficient room, light and ventilation. According to Du Bois-Reymond, the dissecting rooms resembled foul-smelling caves in which about two hundred students carried out their work on the cadavers. Some of Müller's collaborators, like the pathologist *Jakob Henle* (1809-1855) and the anatomist *Theodor Schwann* (1810-1882), from 1835 onwards had to work in their private living quarters, since there was no additional space in those narrow rooms, cluttered with boxes, preparations, living and dead frogs, books and even eatables.

The ancillary resources necessary for research and physiological work were sparse at best. The physician *Heinrich F. Bidder* (1810-1894), later a physiologist at Dorpat, spent a few months at the Berlin Institute in 1834, and described the existing conditions in his unpublished reminiscenses.[9] Müller's so-called "Kabinett" consisted of two small and simply furnished rooms. In the first one worked his prosector, Henle, while Müller occupied the second one with Schwann. He also shared the only existing and borrowed microscope of the entire institute with Schwann.

Müller's lectures were given almost without any experiments, although he brought ample material to his classes on comparative anatomy. His delivery was enthusiastic, the content of his discourses clear, relevant and scholarly, reflecting his complete control over the subject matter. Virchow declared that Müller often acted more like a preacher[10] and was very sensitive about interferences which occurred while he was lecturing. On occasion, he would gaze intently at the undesirable listener until the culprit finally left the classroom. Wrote Virchow: "What a contrast, when the usually cold and stern face suddenly changed into an expression of kind benevolence, and the eyes laughed more than his face, like warm sunrays piercing through the clouds. In those moments, Müller was a captivating individual and one became conscious of the spiritual greatness of the man. Physically, one was also aware of the contrast between the great, wonderful head and a body whose only prominence was the wide shoulders."

Müller's transfer to Berlin coincided with the publication of the first section belonging to the *Handbuch der Physiologie.* The second section of this first volume appeared in 1834 and the second volume later, in 1837. With this work, Müller set a lasting monument for himself. According to Du Bois-Reymond, the book had a significance for the nineteenth century akin to that of Haller's physiological work a century earlier. Everywhere in the text, Müller's prodigious knowledge and own experiences came to the fore. His rigorous methodology, balanced judgment and complete knowledge characterized the entire book. Included in Müller's work was an enormous amount of material, critically sifted and presented according to its importance. When dealing with observations and experiments, his narrative was purely inductive. Only where general views about vitality, soul and nature were offered to the reader, Müller ventured to present carefully contrived philosophical considerations. He exhibited a masterful, theoretical incisiveness matched only by the sharp powers of observation used for the empirical subjects.

Many still extant questions were alluded to in Müller's *Handbuch,* and possible answers or avenues of progress suggested. Almost all areas of physiology were enriched by new observations. In spite of its unpretentious exterior and the total lack of illustrations, Müller's book was extraordinarily successful. It expounded a brand new physiology, compared with similar previously published texts.[11] Zoologists, anatomists, physiologists and physicians became unmistakably aware of the great success which a careful observation of nature was able to provide. This achievement was in direct contrast to the deceptive romantic form of interpretation which had dominated in Germany during the preceding decades.

To be sure, Müller's approach was still largely qualitative in nature, more a higher kind of "animated anatomy" rather than a measuring enterprise searching for causes and effective forces. The physico-chemical considerations still remained in the background: their time had yet to come. Müller was more of a morphologist than an experimentalist. He viewed more with distaste than disgust the direction taken by the French physiology under Magendie with its vivisection experiments. Müller's preference for morphology, comparative anatomy and animal systematization soon re-emerged. The anatomical museum in Berlin offered great possibilities for further study, and Müller himself brought home a wealth of material from his various travels. After studying the Myxinoides in 1834, he worked on cartilaginous fishes and a complete system of fishes (1844). Thus, sea-urchins, starfishes, infusoria and primitive animal forms became subjects of Müller's untiring research, an activity which was far removed from physiology and should only be mentioned here peripherally.

In 1839, Müller published his last strictly physiological work: *Über die Kompensation der Kräfte am menschlichen Stimmorgan,* which furnished important contributions to the nature of the human voice. Thereafter, the great physiologist moved away from the rapidly expanding experimental physiology to the extent that he was unwilling to rewrite his *Handbuch* for a planned revised edition in 1855. The final edition of Müller's original text was printed in 1841-1844.

Müller's entire scientific production encompassed twenty independent writings and over two hundred smaller and diverse publications, a truly immense accomplishment and proof of genius and industry. A great number of honors were bestowed on Müller. He became a member of numerous scientific organizations in his own country and abroad. Twice he was elevated to the deanship, the last time during the unfortunate year 1848, when political turmoil engulfed the lecturing halls. Virchow, who was deeply involved in those episodes, relates the events and great inconveniences suffered by Müller as the presiding dean.

During the years in Berlin, many men who later became famous worked with Müller at the Institute. Henle and Schwann have already been mentioned, but there was also *Karl B. Reichert* who in 1840 assumed Henle's position as prosector, Ernst Brücke and Hermann von Helmholtz. During his days as a student, Du Bois-Reymond was in contact with Müller's institute through his friendship with *Eduard Hallmann,* one of Müller's assistants. Then, after his graduation as a physician under Müller's deanship, Du Bois-Reymond became Hallmann's successor and a teaching assistant at the Institute in 1846.[12]

As mentioned, Rudolf Virchow was another student of Müller who received from his teacher valuable suggestions regarding the use of the cell theory in pathological anatomy. Müller had already been interested in questions of pathological histology in the years 1837-1838, and some of his work was published under the title *Über den feineren Bau der krankhaften Geschwülste* (1838), although the entire work was never completed by Müller himself. Another author, the physiologist *Friedrich Miescher* (1844-1895) was urged by Müller to work on the microscopical changes observable in inflammations of the bones. Thus, the great physiologist seems to have greatly influenced his vast circle of assistants and students to carry out research in a variety of fields. He truly became the point of departure for a great school of famous anatomists and physiologists.

Starting in 1854, Müller suffered frequent episodes of mental exhaustion and insomnia. In addition, he began to experience dizzy spells, especially when carrying out microscopical observations, a fact which prompted him to travel in search of relief. In September, 1855, he barely but luckily survived a maritime disaster, and was fished from the sea clinging to a piece of wood. Suffering from no obvious acute disease, Müller died suddenly on April 28, 1858, and perhaps he committed suicide. Among the students who carried his coffin was the biologist *Ernst H. Haeckel* (1834-1919). An enormous crowd bid him farewell and Alexander von Humboldt, then eighty-nine years old, delivered the eulogy at the graveside.[13] The citizens of Koblenz have erected a monument in his honour as a tribute to a truly great son.

2.
CARL LUDWIG (1816-1895)

Under the impact of the successes achieved by prominent authors such as Magendie and Müller, physiology gradually abandoned the strongly speculative view of nature which had prevailed earlier in the nineteenth century. Thus, physiologists such as Purkinje, Valentin, the brothers Weber, Volkmann, Wagner and others in Germany as well as Claude Bernard in France, began to carry out empirical research of isolated phenomena, an approach which greatly enriched physiology. The majority of these investigators used a method which may be called *experimental, biological* and *vivisectional.* Their goal was to clarify the importance and usefulness of the various parts of the organism.

For this purpose, morphological research went hand in hand with the physiological one. Year after year, numerous new findings were made in the areas of gross and microscopic anatomy, forcing time and again corrections and expansions of the prevalent physiological notions. The kind of questioning inherent in this approach, however, was more a "what is it?" rather than "how does it actually happen?". In other words, the performance of individual organs and tissues was of more interest than the determination of the physico-chemical forces responsible for these functions. Therefore, a whole generation of authors and a particular direction of physiological research ensued which has been designated "biological".

Many of these so-called "biological" physiologists were still adherents of a moderate vitalism, in the sense that they postulated the need for a "Lebenskraft" or vital force in order to explain the organic purposefulness and harmony of the bodily actions.

In the meantime, physics and chemistry had made great strides. Among the most notable authors were the physicist *Ernst F.F. Chladni* (1756-1827), the chemist *Hans C. Oersted* (1777-1851), the physicist *Joseph von Fraunhofer* (1787-1826), the mathematician *Carl F. Gauss* (1777-1855), the chemist *Robert W. Bunsen* (1811-1899), the physicist *Wilhelm E. Weber*, and the chemists *Friedrich Wöhler* (1800-1882) and *Justus von Liebig* (1803-1873), all in Germany. We should mention the French chemists *Joseph L. Lagrange, Pierre S. de Laplace* (1749-1827), the physicists *Jean B. Biot* (1774-1862) and *Nicolas L.S. Carnot* (1796-1832). In England, also, there were several prominent authors such as *Thomas Young, Michael Faraday* (1791-1867), *James C. Maxwell* (1831-1879) and *Thomas Thomson* (1773-1852).

To be sure, Weber, Purkinje and Volkmann, among others, had already used physical principles in order to interpret vital processes. However, this approach only became a generalized program of action around the middle of the nineteenth century, especially in the hands of Du Bois-Reymond, Helmholtz and above all Carl Ludwig. The two volume *Lehrbuch der Physiologie,* published by Carl Ludwig between 1852 and 1856, constituted a veritable turning point in the history of physiology, giving rise to the so-called *physico-chemical* approach, which went on to dominate in the second half of the century.[14]

The first introductory sentence in Ludwig's *Lehrbuch* reflects the crucial change in direction: "The task of scientific physiology is to determine the performances of the animal body and *derive these functions necessarily from the elementary conditions inherent in the organism itself*". Therefore, physiology should be a physics and chemistry applied to the living organism. Declared Ludwig: "We are obviously still far removed from a complete application of such an approach in science, but we should, nevertheless, continue to advance towards such a goal. The above interpretation is, obviously, not the one which had hitherto prevailed. Rather, it is a new direction which may be called *physical* in order to contrast it with the *vital* approach. Such a notion demands that any object contains within itself the causes for its actions and effects, in accordance with the law of causality which must be upheld in order to be able to think properly. Moreover, this idea is also in harmony with the often expressed axioms for empirical knowledge which include only the proven mediate or immediate facts into the basis of the conclusions. Hence, such an approach justifiably rejects the acceptance of hypothetical entities such as the nerve-ether or others, and will not hesitate to accept any new fundamental condition for consideration, provided that such an entity has a real existence".

The program enunciated by Ludwig drew sharp opposition and criticisms. In 1854, Wagner wrote in the *Zeitschrift für rationelle Medizin,* that such an approach, although praiseworthy, was still premature, and should not be used in textbooks of physiology. He felt that Ludwig was decades or perhaps centuries ahead of his time in proposing such a direction. Wagner was probably not entirely incorrect, since Ludwig's proposed goal at the outset was merely a program of action. Therefore, Ludwig himself became the "Fahnenträger der Schule" — the standard-bearer of a new school of physiology, as Du Bois-Reymond called him in their correspondence. In truth, Ludwig was the unquestionable founder of the precise, quantitative, physico-chemical direction in physiology, whose results were destined to place medicine on an entirely new basis.[15]

Carl Friedrich Wilhelm Ludwig (see Fig. No. 68) was born on December 29, 1816, in Witzenhausen.[16] According to his own account, Ludwig received a very unexciting education in Witzenhausen and later Hanau, perhaps because of its emphasis on classical-humanistic instruction, and the total absence of topics related to the natural sciences. However, this situation was experienced by almost all of the famous scientists of the nineteenth century.

Fig. 68. Carl Ludwig (1816-1895).

Ludwig studied medicine at the University of Marburg, graduating there in 1840 with an obscure and unimportant dissertation. A year later, he became the anatomical prosector to *Franz Ludwig Fick* (1813-1858), and was allowed to teach after writing the work *De viribus physicis secretionem urinae adjuvantibus* in 1842. This essay, dealing with urine formation, contained the program for his later physicochemical approach. Because of his unpleasant relationship with the physiologist Hermann Nasse and some financial difficulties, Ludwig remained at Marburg only until 1849, when he assumed the chair of anatomy and physiology at the University of Zűrich.

Two years prior to this date, Ludwig had met in Berlin his later friends Du Bois-Reymond, Helmholtz and Brűcke. All four were united in their enthusiasm for a truly physical direction in physiology, whose foundations Ludwig had already laid at Marburg with his studies about hydromechanics and blood pressure. Among other innovations, Ludwig constructed the first *kymograph* in 1846, an instrument for measuring and recording variations in fluid pressure (see Fig. No. 69), and the first graphic method in experimental physiology.[17]

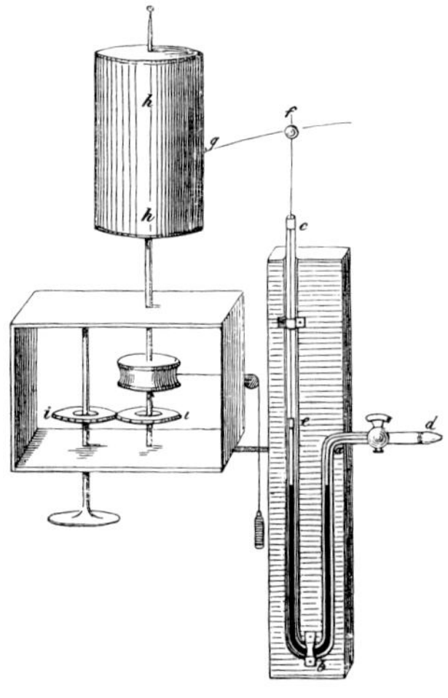
Fig. 69 Ludwig's Kymograph of 1846.

In Zűrich, Ludwig continued his studies on the phenomena of diffusion, endosmosis and secretion. In 1850, he discovered the existence of a genuine secretory nerve in the salivary gland. Since the reactionary government of Prussia hesitated to offer him a teaching position because of his liberal and antireligious views, Ludwig accepted a call to become the physiologist at the medico-surgical section of the Viennese Military Academy in 1855. His friend, Ernst Brücke, was active in the same city as physiologist at the university's Institute, while Ludwig had only scanty quarters available to him at the Josephinum.[18] Nevertheless, Ludwig's decade in Vienna yielded a series of valuable contributions to the physiology of respiration, blood and the innervation of blood vessels.

Ludwig fully developed his great investigative qualities, organizational skills and teaching talents at the University of Leipzig, where he went in 1865 to replace Ernst H. Weber and occupy the chair of physiology. In 1869, Ludwig was able to move into the *Neue Physiologische Anstalt,* the institute which became the model for many similar institutions erected by some of his students throughout the world in the second half of the nineteenth century. Ludwig directed the Institute for almost thirty years until his death, in 1895. It became the workshop for an extraordinary amount of experimental work and a theoretical as well as practical training ground for an enormous number of students from around the globe. Thus, in the span of only a few years, Ludwig's institute achieved world-wide renown, and outstanding discoveries in all areas of physiology were made there. Some of this work was concerned with the vegetative functions such as vasomotor centers, the various gases in the blood, the metabolism of isolated organs and the action of cardiac nerves. Among Ludwig's most popular instruments was the so-called "Stromuhr" which was capable of measuring the blood flow rate.[19] His discoveries have been listed elsewhere.

In these decades it was almost an obligation for every young and ambitious physiologist to spend at least a year studying with Ludwig at the Institute in Leipzig. Among the great number of students, easily exceeding two hundred, one could find next to many Germans a large number of English, American, Scandinavian and Russian physiologists who later became famous in their own lands. The family tree of Ludwig's large school of physiology provides an impressive and graphic picture of the entire group. Moreover, Ludwig's collaborator of many years, *Hugo Kronecker* (1839-1914), was in charge of maintaining personal contact with this international association of scholars. Ludwig looked after each of his students, advised them personally, directed their planning and execution of experiments, and often helped to write their results in a publishable form. Yet, almost all of these works executed under his immediate supervision carried only the names of his disciples, although his own contribution usually was substantial and decisive. This attitude cast an especially favorable light on the extraordinary German physiologist and his proverbial modesty. All his students lauded Ludwig for both his great kindness and helpfulness. Nevertheless, a certain dignity, "something majestic issued from the middle-sized, slender man with the sharp facial features wearing glasses".[20]

The physician *Warren P. Lombard* (1855-1939) nicely described Ludwig's relationship with his students: "Every morning Ludwig would come to the tables where the various students were doing their work, and discuss with them the next steps needed to continue their investigations. Often he would set up a timetable with the students in order to be able to participate with them in certain crucial phases of their research, and he was always punctual. At times, Ludwig would take some of his disciples to his own workroom and discuss critically with them the methodology employed as well as check the notes made and the graphs obtained during the previous experiments. Ludwig would then ponder about the alternatives and directions for the subsequent investigations, and arrive at appropriate conclusions. However, not all these activities were arbitrarily performed on the spur of the moment. Each evening, when he left the laboratory, Ludwig would take home with him a series of notes and protocols concerning some of the experiments in progress, in order to study and review them carefully".[21] The great physiologist died on April 23, 1895, the victim of a bronchitis complicated by heart failure.

Ludwig's accomplishments in physiology were numerous and varied.[22] His work of 1842 on renal function already reflected the direction of his subsequent thought, since he considered the acceptance of vital, secretory forces as superfluous. Rather, Ludwig believed that the process of renal secretion could be envisioned as a phenomenon in which certain substances existing in the blood were pressed out of the vessels at the level of the glomeruli, thanks to the existing blood pressure. This activity was followed by a resorption of water in the adjacent renal tubuli, an explanation which Ludwig owed in part to his careful experimental representation of the renal vessels.

During his years in Zurich, Ludwig wrote most of his two volume *Lehrbuch,* characterized by his physico-chemical approach. The first section of the initial volume was entitled "physiology of the atoms", reflecting Ludwig's concern: to depict and explain the living substance from the elementary particles of matter. Subjects such as diffusion and osmosis, animal electricity, vision, the mechanics of joint movement and hemodynamics were all topics of medical physics, which came to the fore through Ludwig's research at the expense of morphological studies. Whenever the 'fundamental actions' could not be apprehended, the subject was soon dropped by Ludwig. He was the first to carry out in hemodynamics the differentiation between measurements of lateral versus retrograde pressure.

Ludwig was also the first to measure the arterial pressure in the pulmonary artery. In 1867, he introduced the already mentioned "Stromuhr", an instrument which could measure the blood flow of an entire organ (see Fig. No. 70). Ludwig observed the rise of vascular tone with the increase of blood pressure when the spinal cord was stimulated, and in 1864, he discovered the sympathetic innervation of the vascular muscle layers. Finally, in 1871, he determined the approximate position of the vasomotor center in the spinal cord.

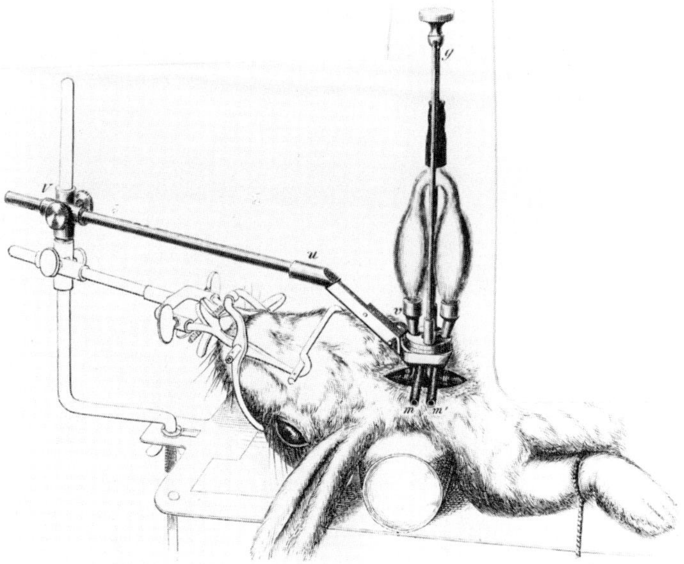

Fig. 70. Ludwig's "Stromuhr" for the measurement of blood flow.

Thanks to a rather ingenious arrangement, Ludwig was able to determine in 1867 with the Lithuanian physiologist Johannes Dogiel that the first cardiac sound was muscular rather than valvular in origin. With another collaborator, the Russian physiologist *Elie de Cyon* (1843-1912), Ludwig examined in 1866 the effect of temperature on cardiac action in an isolated heart. Moreover, in 1871, he studied the all-or-nothing law of the heart with his American disciple *Henry Pickering Bowditch.* Finally, the discovery of the Nervus depressor with Cyon, in 1866, clarified the role of the afferent nerves in the visceral nervous system.

Ludwig also spent much time analyzing the lymphatic vascular system. In 1858, before Pflüger's work, he constructed a mercurial pump for the gases of the blood, and in 1870, he formulated a theory regarding the forces operative in gas exchange. Ludwig should also be remembered for a method used to preserve the vitality of organs such as the heart, kidneys and muscles of warm-blooded animals.[23]

As already mentioned, over two hundred collaborators worked actively with Ludwig at Leipzig. No other physiological laboratory in the world made as many fundamental and lasting discoveries as Ludwig's institute. Among those who worked actively with Ludwig were the British physiologists *Walter H. Gaskell* (1847-1914), *John S. Burdon-Sanderson* (1828-1905), *Thomas Lauder-Brunton* (1844-1916), *William Stirling* (1851-1932), and *Edward Sharpey-Schafer* (1850-1935). From America came *Henry P. Bowditch, Warren P. Lombard, William H. Welch* (1850-1934) and *George R. Minot* (1885-1950), while from Scandinavia came *Alarik F. Holmgren* (1831-1897), *Olof Hammarsten* (1841-1932), *Christian Bohr* (1855-1911) and *Robert A.A. Tigerstedt* (1853-1923). Two prominent figures came from Italy: *Luigi Luciani* (1840-1919) and *Angelo Mosso* (1846-1910). (See Fig. No. 71.)

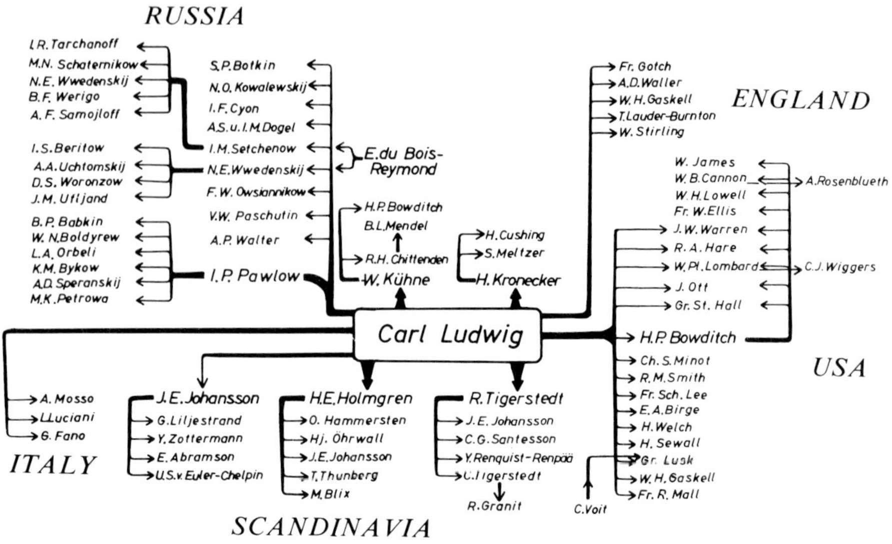

Fig. 71. Ludwig's most prominent foreign collaborators.

Ludwig was a convinced antivitalist like his friends Du Bois-Reymond, Helmholtz and Brücke. This position truly encouraged a whole generation to analyse the living processes for their elementary physico-chemical components, unencumbered by "Naturphilosophie" and vitalism.[24] There is no question that these authors were only partially successful in that task, and that they concentrated primarily on the area of organ physiology. Studying middle-sized organs, Ludwig's fundamentally mechanical orientation was supported by the help of ingenious devices. Hence, measurement and a quantitative, numerical tendency emerged in physiology, an approach which was gradually and successfully adopted throughout the world. Whether one cares to call it "sterile scientism"[25] or "an upside down philosophy of nature" does not matter. Without the reduction of the complex organs into their active components, it was impossible to understand and influence medically the entire organism.[26]

After the narrative of the life, work and influence of Germany's two greatest physiologists of the nineteenth century, Müller and Ludwig, one feels the need to compare them and attempt a historical evaluation of their relative significance for physiology in general.[27] According to a widely accepted interpretation, Müller was indeed the founder of the classical period of German physiology in the last century. His name needs no introduction, his bust can be found in laboratories, and there are several good biographies of his life. By contrast, who knows anything about Ludwig's life and work outside of the older generation of scientists? A German biography of Ludwig has only recently appeared together with a presentation of his work.[28]

Thus it appears that the correction of a widespread historical error is in order concerning the influence of both, Johannes Müller and Carl Ludwig, on the development of physiology. There is no question that Müller made essential contributions towards a veritable rebirth of empirical physiological studies in Germany, although he was not the originator of such an approach. However, only with Ludwig did the experimental, causal and analytical method clearly come to the fore, and his quantitative, physico-chemical approach to physiology spread throughout the world. By contrast, Müller showed relatively little interest in such a pursuit, and he often even failed to recognize the great significance of the results being achieved with the new direction, such as the discovery of the ophthalmoscope. In general, Müller remained rather skeptical and aloof about the newer analytical and numerical viewpoint, as Du Bois-Reymond recalled in his letters and speeches.

Concerning the discovery of new physiological facts, Müller undoubtedly made many remarkable observations, but did he achieve any really new and fundamental insights in experimental physiology except for the law of the specific nerve energies? Du Bois-Reymond formulated this question, and not without reason provided a negative answer for it.

Therefore, it is fair to say that Müller was an excellent *precursor* and a veritable trail-blazer, but not the great inaugurator and *discoverer* like Ludwig to whom we owe a great number of radically new physiological facts. Müller may have been superior to Ludwig in intellectual capacity, education, versatility and the universality of his efforts, but this is not relevant to the questions and issues raised here. When one is concerned with the scientist who was of greater *significance,* who had greater influence on the development and the direction which physiology took after him, Ludwig is really that figure. Müller's attitude towards the organism was one of reverence, and his research geared more to observe passively rather than to intervene aggressively and experimentally. Thus, his thought can be more adequately described as biological. Müller was more concerned with the total relationship of the appearances rather than the mechanism of a partial function. He was truly the supreme representative of a more biologically oriented physiology which, perhaps unjustly, has faded into the background. Against this viewpoint, Carl Ludwig emphasized with rigid partiality his pure physico-chemical direction in physiology, an approach which shaped the style of the discipline for the last hundred years.

3.
HERMANN VON HELMHOLTZ AND HIS DISCIPLES

Between 1830 and 1850, the reputation and significance of physiology as one of the basic medical sciences greatly increased. New physiological laboratories were founded in all German universities, and young physiologists appointed to teach the subject. The most famous institution in Germany capable of training these promising new scientists was Müller's Institute in Berlin. To be sure, it was not the only one, since Purkinje's institution at Breslau served the same purpose. However, the various students of Müller at Berlin not only received instruction in a new discipline - physiology - but were also well schooled in anatomy. This was particularly true of Du Bois-Reymond, Helmholtz, Brücke and later Henle, Schwann and even Virchow. In addition, there was a numerous group of younger people who had worked as students under the direction of Müller, and thus proudly considered themselves as belonging to the so-called *Müller school of physiology* (see table). A common bond uniting them was the strong feeling of affection which they exhibited for their beloved teacher as well as for each other.

Fig. 72. Hermann von Helmholtz (1821-1894).

Among Müller's students, Hermann von Helmholtz (see Fig. No. 72) clearly was the most outstanding figure, not only in talent and in his capacity for work, but because of his logical and incisive mind. Born August 31, 1821, at Potsdam, the son of a college teacher, Helmholtz was in his early years a sickly child. Although he received a classical education in his home-town, the young man early developed well-defined interests in scientific subjects: mathematics and physics. In 1838, Helmholtz decided on a career in medicine, enrolling in the *Königlich-Medizinisch-Chirurgisches Friedrich-Wilhelms-Institut* — informally called the "Pépinière." Education in the government-supported institution was free, and as for Rudolf Virchow, it was his only means to afford such an instruction.

At about this time, Helmholtz was strongly attracted by Müller's lectures, and he graduated in 1842 with a dissertation inspired and dedicated to Müller, entitled "De fabrica systematis nervosi Evertebratorum". In this thesis, which dealt with leeches, Helmholtz was able to demonstrate that the nervous fibres were connected to the ganglion cells discovered by the German physiologist *Christian G. Ehrenberg* (1795-1876) in 1833.

Helmholtz's life and achievements have been adequately depicted in a series of publications.[29,30] At the beginning of his medical career he was a squadron surgeon with the Prussian Hussar Guards at Potsdam. However, Helmholtz was able to install in the barracks a small physicomedical laboratory, where he and his friends Du Bois-Reymond and Brücke often met and worked. At the time, his main interest resided in the relationship between muscular work and heat, a problem of muscular energy. For this purpose, Helmholtz was able to build a highly sensitive *thermomultiplier.* Eventually, this line of research led him to the more general and fundamental physical problems connected with energetics.

Moreover, Helmholtz was prompted to apply some mathematical and physical thinking about the quantitative relationships existing between forces during the conversion of one type of energy into another. Barely twenty-six years old, the fledgling military surgeon read in 1847 a paper at the Berlin Physical Society entitled "Über die Erhaltung der Kraft". The success of this epochmaking but purely theoretical communication was initially rather meager. Poggendorf, the editor of the famous *Annalen,* actually declined to publish it. However, it received warm praises from the higher military authorities, which lauded the important and practical direction of Helmholtz's studies. In doing so, they were obviously confusing the kind of force Helmholtz was talking about with another of more prosaic character which also interested the layman.[31]

The ensuing controversy over the priority of this law of thermodynamics, which was also enunciated by the Heilbronn physician and physicist *Julius Robert Mayer* (1814-1878), and the deep impact which the principle of the conservation of energy had on physics and physiology, cannot be recounted here in detail.

After a short period as an assistant to Müller, Helmholtz applied and was appointed in 1848 a teacher of anatomy at the Berlin Academy of Arts, thanks to the personal recommendation of Müller who viewed his former student as "one of those rare great talents". Moreover, the great naturalist Alexander von Humboldt was instrumental in having Helmholtz released from his military obligations. Helmholtz only occupied his new civil post for a year. When his friend, Ernst Brücke, left Königsberg for Vienna in 1849, he was appointed Brücke's successor there, both in physiology and general pathology.

After his arrival in Königsberg, many of Helmholtz's significant scientific works came gradually to fruition. Among them was the determination of the velocity of nervous transmission, the form and duration of muscular contraction, the existence of a latent period and other topics. With the simplest of auxiliary devices, Helmholtz was able to create the proper experimental conditions for solving his physiological problems. His measurement of the velocity of the nerve-impulse destroyed the belief that such phenomena occurred at immeasurably great speeds, or were, in fact, produced by a nervous spirit.[32] Instead, Helmholtz defined nervous transmission as a material process capable of measurement.[33]

On December 6, 1850, Helmholtz communicated to the Berlin Physical Society one of his inventions: *the ophthalmoscope*. The significance of this discovery does not have to be emphasized. It truly placed ophthalmology on a completely new basis. Nevertheless, the value of the instrument was not immediately recognized by everyone. In his famous speech entitled "Das Denken in der Medizin", which Helmholtz delivered in 1877, he recalled: "In connection with the ophthalmoscope, one of my very famous surgical colleagues told me that he would never use such an instrument, since in his opinion it was far too dangerous to allow the entrance of such bright light into the eye. Another declared that the ophthalmoscope could perhaps be useful to those physicians with poor eyesight, but since his own vision was normal, he had absolutely no use for such a gadget".[34]

During his years at Königsberg, and especially after 1853, Helmholtz also became interested in questions related to sensory physiology. Topics such as accommodation, ophthalmometry and the colors of the spectrum attracted his penetrating and philosophical mind, always probing the most essential problems of any given subject. Likewise, Helmholtz was interested in the formation of electricity in various animal organs, a subject which Du Bois-Reymond had considered in a two volume work published in 1848-49. In dealing with such a topic, Helmholtz departed, as always, from the pure physiological facts in his search for the general physical laws which would explain the various phenomena.

In 1855, Helmholtz left Königsberg to assume the chair of anatomy and physiology at the University of Bonn. He remained there only for three years, moving on to Heidelberg, where he became professor of physiology. In those years he worked simultaneously on problems related to *physiological optics* and *physiological acoustics*. He examined both fields broadly and was able to integrate and harmonize the various problems and results. The combination of physical, mathematical, anatomical and physiological data was a reflection of Helmholtz's versatility, and he pursued the topics further through new lines of questioning. These approaches varied somewhat in character, and some were purely epistemological, philosophical, psychological or even aesthetical.

Helmholtz concerned himself experimentally with the physical analysis of sound and the application of acoustics to the physiology of the ear and the organs of speech. This led him to become preoccupied with the physiological causes of musical harmony in 1857. The first of his two great works dealt with the physiology of sense perception: *Die Lehre von den Tonempfindungen als Grundlage für die Theorie der Musik,* published in 1863. The fruit of eight long years of work, Helmholtz's book contained not only acoustical information, but a physiology and psychology of hearing, and even a theory of musical harmony. Many new findings were presented there for the first time.[35]

Somewhat later, Helmholtz completed his more extensive work, the *Handbuch der physiologischen Optik* (1856-66). The publication was divided into three parts, dealing with dioptrics, visual sensations and perceptions on the basis of an "empirical" theory. As with his other works, Helmholtz's book remains today an unequaled standard work on sense physiology, thanks to the scope, clarity and accuracy of the physico-mathematical analysis and its richness in new discoveries.[36] Due to the author's tendency toward general issues, the epistemological problems as well as the relationships between sensory stimulations and perceptions were thoroughly dealt with. He clearly evaluated such sensory experiences both in thought and judgement. What really dominates all these investigations, however, are ideas belonging to an empirical world view, which Helmholtz accepted in lieu of the already discarded philosophical approach.[37]

Beginning with his work on the conservation of energy, Helmholtz everywhere presupposed the concept of nature as something that could be understood. He wrote: "The final goal of the theoretical natural sciences is, therefore, to discover the ultimate and unchangeable causes of natural phenomena. Whether all the processes of nature be actually referable to such, or whether nature is capable of being completely comprehended, or whether changes occur which are not subject to the laws of necessary causation, but spring from spontaneity or freedom, this is not the place to decide; it is at all events clear that the science whose object it is to comprehend nature must proceed from the assumption that it is comprehensible, and, in accordance with this assumption investigate and conclude until, perhaps, she is at length admonished by irrefragable facts that there are limits beyond which she cannot proceed."[38]

During the last years at Heidelberg, Helmholtz was gradually drawn more into the field of pure physics, especially electrodynamics, induction, polarization and others. Thus, in 1871, he was appointed to the chair of physics at the University of Berlin. From that time on, Helmholtz's works in physics predominated, and he studied problems of aero-, electro- and thermodynamics, electromagnetism and several other related fields, publishing important works concerned with these subjects.[39] However, his interest in biological and medical questions remained, evidenced by his occasional speeches and lectures, which are still delightful to read. Helmholtz received numerous honors, becoming not only a member of many important scientific academies, but also elevated to the nobility by the German emperor in 1882. For his contributions to ophthamology, he was awarded the Graefe-Medal in 1855. Finally, on the occasion of his seventieth birthday, he received numerous testimonials acknowledging his numerous and distinguished contributions. He died three years later, in 1894, the victim of a stroke, and with his passing an unusually bright star faded from the sky of the natural sciences.

Characteristic for Helmholtz's method of work was the fact that he always used the investigation of relevant details in order to solve more fundamental questions underlying them. At all times his goal was to discover general laws of nature, and for this purpose he used his tremendous physico-mathematical knowledge. Helmholtz was indeed a tireless worker. "Those who have had the good fortune to see Helmholtz carry out his experiments will never forget the way in which this superior intellect was capable of overcoming innumerable difficulties with his methodical actions. With the simplest accessories, built of cork, glass sticks, wood and carton boxes, Helmholtz would construct models of rather suggestive contrivances, long before they were given to the mechanic for final elaboration. No misfortune was able to shake his temperament, characterized by an admirable composure and poise."

Helmholtz remained modest in spite of his superiority, and always felt the great responsibility which actually befalls those who have the fortune of being talented. These special obligations and inexorable demands were imposed on the physician at the bedside as well as on the researcher in the laboratory. In the already mentioned speech, "Das Denken in der Medizin", Helmholtz said the following: "Perhaps one must have witnessed the eyes of a dying patient as well as the grief and desperation which befalls the affected families in order to ask oneself the crucial question: was everything possible done to avoid the tragedy? Did, indeed, science furnish all the possible knowledge and aid in such a situation? Therefore, the theoretical questions can acquire dimensions and practical implications which go far beyond the merely methodological issues. The theoretical investigator may smile glibly and remain detached, his imagination and pride may even flourish for a time in the isolation of his untroubled laboratory. He may even find the older prejudices interesting and excusable, labeling them as mere reflections of poetical romanticism and youthful ecstasy. Those, however, who have to struggle with the real adverse powers, cannot be indifferent nor romantic. They always have to test critically what they know and are able to do, and use only the harsh light of facts, rather than entertaining pleasant illusions".

In spite of his long career as a physiologist which extended for over twenty years from 1849 to 1871, Helmholtz had only a handful of disciples, much less than one would have expected from such an outstanding figure. Perhaps only a few students were confident enough to follow in the footsteps of such a towering spirit. Among those who worked with Helmholtz at Heidelberg for some time was the physiologist *Julius Bernstein* (1839-1917), while another student, the Austrian physiologist *Sigmund Exner* (1846-1926) remained there only for a brief period. Two other prominent followers who studied under Helmholtz in Berlin were *Arthur König* (1856-1901) and *Johannes von Kries* (1853-1928).

Born in Berlin, *Julius Bernstein* was attracted to physiology during his studies at Breslau under the direction of Rudolph Heidenhain.[40] When Bernstein completed his education at Berlin, one of his friends, Ludimar Hermann, introduced him to Du Bois-Reymond, in whose laboratory he worked from 1860 to 1864. In the latter year, Bernstein came to Heidelberg, highly recommended by his former teacher, and he remained there until 1871. After that date he was named professor of physiology at the University of Halle, remaining active there for forty-six years until his death. Bernstein, in turn, had also a series of distinguished students and collaborators, among them Julius Steiner, Paul Jensen, Armin von Tschermak-Seysenegg, Ernst Laqueur and Fritz Verzar.

In his laboratory, Bernstein was an example of carefulness and exactitude when carrying out his investigations. He always criticized in kind terms without deterring his students from further work, and was a master at improvising from his rather modest resources, displaying a great deal of methodological originality. He allowed his more advanced students a great deal of freedom and was not bent on forming a large group of immediate followers around him.

Bernstein was a master of *biophysics,* demonstrated by his works concerned with muscle and neurophysiology, especially bioelectricity. These works actually began during his first years in Du Bois-Reymond's laboratory, where he perfected an instrument to interrupt stimuli. Bernstein observed, among other phenomena, the increase and diminution of a negative oscillation in the catalectrotonic or anelectrotonic fields. In 1867, he demonstrated that the negative depolarization wave in the nerve had the same velocity as Helmholtz's nerve conduction, thus relating one phenomenon to the other. A similar relationship was discovered between the wave of contraction in the skeletal muscle and its electrical potential. Bernstein's greater work, entitled *Untersuchungen über den Erregungsvorgang im Nerven- und Muskelsystem* (Heidelberg, 1871) also contained valuable observations regarding the subject.

Finally, Bernstein was also the founder of the so-called "membrane theory", a new doctrine which explained the bioelectrical potentials through differences in ionic concentrations across a hypothetical membrane, a working theory which proved to be extraordinarily useful. (See Fig. No. 73) Many other discoveries, especially in the fields of cardiac and sensory physiology, form part of the 135 publications which make up Bernstein's life work. This original, highly skilled and successful physiologist died on February 6, 1917.

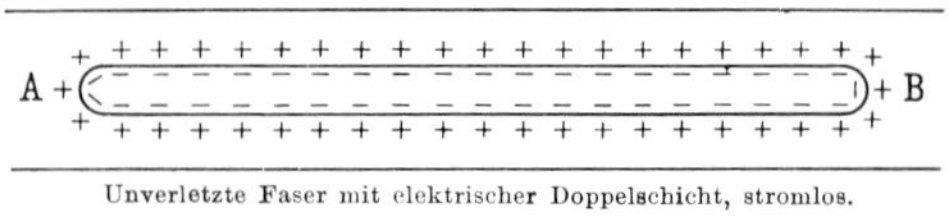

Fig. 73. Bernstein's "membrane theory" (1912). Uninjured fibre with double electrical layer.

Johannes von Kries (see Fig. No. 74), another of Helmholtz's disciples, has left us an interesting autobiography[41] in which he vividly depicts his career with its triumphs and disappointments. Born October 6, 1853, near Graudenz, von Kries studied at the University of Halle, where he was deeply influenced by the "grand old man", the physiologist Alfred W. Volkmann. After his medical graduation at the University of Leipzig, von Kries went to work with Helmholtz at Berlin for one year in 1876, and received from his teacher lasting suggestions and ideas.[42]

Fig. 74. Johannes von Kries (1853-1928).

In 1877, von Kries started to work at Carl Ludwig's Institute in Leipzig, where he was deeply impressed by the systematic work being conducted there at the time, and the discussions and consultations he was able to have with Ludwig himself. Later, in 1880, he was appointed to the chair in physiology at the University of Freiburg, a position which he kept until his death in 1928 in spite of several tempting offers from other universities.

Von Kries concentrated primarily on sense physiology after some preliminary work on the mechanics of muscular contraction, the theory of stimulation and the cardiac conduction system with its anomalies. Together with *Willibald A. Nagel* (1870-1911) he distinguished between red and green color blindness, introducing also the concept of daylight and twilight vision (1896-97). In 1894, he was an important co-founder of the so-called "duplicity theory", although the idea had already been expressed before. Especially during the last decades of his life, von Kries turned to more general and fundamental philosophical questions, such as epistemology, logic and the psychology of judgement. He studied Kantian philosophy in great detail, and thus changed from an experimental physiologist into a successful philosopher.

Almost all of von Kries' students also exhibited great interest for the physiology of the senses and the functions of the nervous system. Among them were *Rudolf Metzner* (1858-1935), *Wilhelm Trendelenburg* (1877-1946), *Emil von Skramlik* (*1886), *Hans Piper* (1877-1915) and the already mentioned Willibald A. Nagel. With these men did Helmholtz's original suggestions and ideas come into our own times.

Originally from Leipzig, *Rudolf Metzner* had previously been an assistant of Carl Ludwig's in 1889-90. He worked with von Kries for four years during 1890-94 and then became professor of physiology at the University of Basel. Metzner was especially interested in the histological changes which occurred in a secreting gland, as well as in problems of urine formation and the hearing damages caused by excessive noise.

Another disciple of von Kries, *Willibald A. Nagel,* of Tübingen, studied medicine and natural sciences at Freiburg, and in 1920 assumed the direction of the section devoted to sensory perception at the Berlin Institute. Nagel was primarily known for his work in physiological optics, a field which he enriched with some new instruments such as the "Anomaloskop", an apparatus for the recognition of red and green color blindness. He also published a five volume textbook of physiology between 1904-10.

Hans Piper, of Altona, was another of von Kries' best students, an investigator who worked for some time with the notable English physiologist *Charles S. Sherrington.* In 1908, Piper was appointed as section chief of the Physiological Institute in Berlin. He was concerned with muscular electrophysiology and the production of stimuli in the nervous system — the so-called "Piper rhythm". In addition, Piper was involved in studies regarding the electrophysiology of the hearing organ and the retina, an area of research which was subsequently developed by one of his students, the physiologist *Arndt Kohlrausch.* Another disciple, *Paul Hoffman* (1884-1962), also began his work on reflexes under the direction of Piper.

Finally, the already mentioned *Emil von Skramlik* (Berlin, Humboldt University), *Ernst Mangold* (Berlin, Landwirtschaftliche Hochschule) and *Wilhelm Trendelenburg* also belonged to the circle of von Kries' students. Trendelenburg,who started his medical studies at Freiburg in 1895, was attracted to physiology by the lectures of von Kries.[43] After a short sojourn at the Hering Institute in Leipzig, he became an assistant of von Kries in 1902, later teaching also at Innsbruck, Giessen, Tübingen and finally Berlin.

About Trendelenburg's extensive scientific career only a few highlights can be mentioned here. He worked on the inhibitory character of certain areas within the nervous system, and carried out some experimental research concerning epilepsy. Moreover, Trendelenburg devised a method to freeze nerves, and made an analysis of the visual color disturbances. He created some glasses designed for adaptation to the Röntgen rays, studied X-ray stereoscopy and was interested in voice physiology using modern acoustical methods. Finally, Trendelenburg employed bipolar thoracic electrodes in electrocardiography and examined the dependence of the bleaching velocity of visual purple in relation to the light's wave length. Many monographs and review articles are testimonies of this physiologist's extraordinary industry. According to his collaborators, Trendelenburg not only had a stimulating personality, but was also an excellent teacher and investigator.

To conclude the short list of Helmholtz's disciples, we have to mention briefly *Arthur König*. He worked with Helmholtz in Berlin in the 1880's, where he also edited his teacher's second edition of the textbook concerned with physiological optics. König was later appointed director of the physical section of the Physiological Institute in Berlin, then under the direction of Du Bois-Reymond. He greatly advanced our knowledge concerning Weber's law, the mixture of colors, the distribution of brightness along the color spectrum, visual purpura and acuity. He was the co-founder of the journal *Zeitschrift für Psychologie und Physiologie der Sinnesorgane.*[44]

4.
EMIL DU BOIS-REYMOND AND HIS DISCIPLES

Emil Du Bois-Reymond (see Fig. No. 75), a somewhat younger and life-long friend of Helmholtz, was born November 7, 1818, in Berlin.[45] His father was originally from Neuchâtel, Switzerland and his mother the daughter of a preacher in the French community living in Berlin. Thus, French was the language primarily spoken at the future physiologist's home. He also went to a French school where he received some mathematical and scientific instruction in addition to the common humanistic education of the times.[46]

After graduation from college, and without any definite professional goal in mind, Du Bois-Reymond took courses in theology, literature, natural sciences and mathematics until his older friend, Eduard Hallmann, directed him towards the study of medicine. Hallmann, who had worked with Johannes Müller since 1834, introduced his young friend to both Müller and Henle, and thus, Du Bois-Reymond was able to learn microscopy and morphology.[47]

After the Italian physiologist *Carlo Matteucci* (1811-1868) published his work entitled *Essai sur les phénomènes électriques des animaux* in

1840, Müller encouraged Du Bois-Reymond to verify Matteucci's experiments, aware of the budding physiologist's interest in physics. Müller's suggestion truly became Du Bois-Reymond's lifelong professional cue. With great expectations he plunged into his experimental work, and in spite of licensing examinations, military service and graduation, he published his first article on the subject in 1843.

Fig. 75. Emil Du Bois-Reymond (1818-1896).

The publication, entitled "Vorläufiger Abriss einer Untersuchung über den sogenannten Froschstrom und über die elektromotorischen Fische", appeared in Poggendorf's *Annalen,* and firmly established his reputation as a researcher. Du Bois-Reymond arrived at the somewhat one-sided conclusion that many of Matteucci's statements were "wrong, his methods sloppy and his writings fraudulent." However, the Italian author had already demonstrated the electrical negativity of a cross-section of muscle and, in 1840, had described the electric oscillations in a tetanized muscle.

Du Bois-Reymond published the first volume of his most important work, the *Untersuchungen über tierische Elektrizität,* in 1848, while parts of the second volume appeared in 1849 and 1860 and the supplements as late as 1884. With this book the young investigator took his rightful place among the most prominent physiologists of his time, and it led to his nomination to a membership in the Prussian Academy of Sciences at the age of thirty-three. Later another of Du Bois-Reymond's important works appeared: *Gesammelte Abhandlungen zur allgemeinen Muskel-und Nervenphysik* (1875-77) which helped greatly in making his research readily available to others.[48]

As expected, Du Bois-Reymond's harsh criticism of Matteucci was greatly exaggerated, a fact which has come especially to light in the course of some recent studies concerned with the Italian author.[49] Nevertheless, Du Bois-Reymond should be credited with the application of physical methods for the study of animal electricity, such as the use of certain sensitive measuring devices, multiplicators, non-polarizable electrodes, and an improved method of compensating the native animal electricity of the tissues with the experimentally applied current, and many others (see Fig. No. 76)

The tendency and ability to apply these physical procedures when deal-

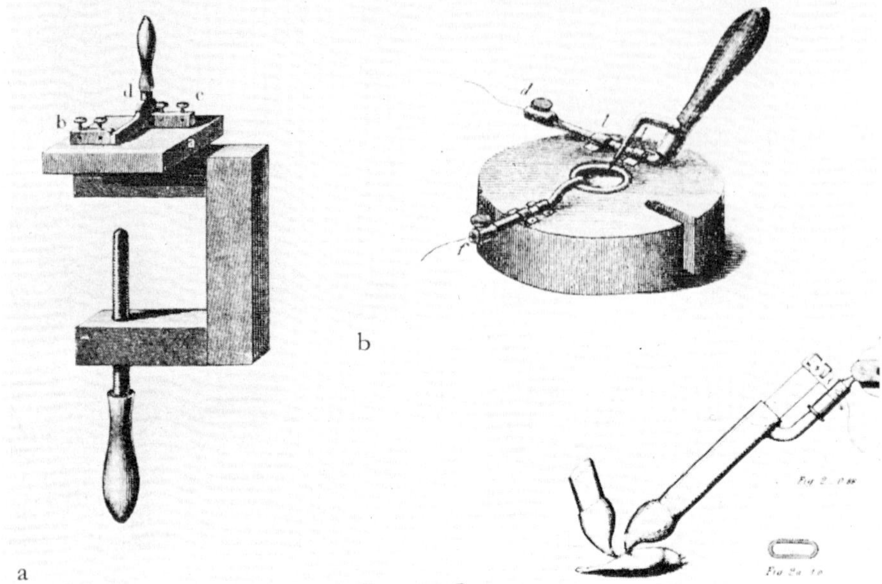

Fig. 76. Some of Du Bois-Reymond's measuring devices:
a) simple switch
b) mercury switch
c) non-polarizable electrodes

ing with specifically physiological problems, made Du Bois-Reymond one of the leaders of the so-called physical approach, together with Ludwig, Helmholtz and Brücke. With some other fellow physicians, including Helmholtz, he was the founder of the Berlin Physical Society. Since the concern with animal electricity always pointed back to general issues in physics, it was natural that Du Bois-Reymond attempted to solve some of these underlying questions. For example, he studied polarization, such as the one generated at the boundaries between different electrolytes, electro-endosmosis, cataphoresis, diffusion and thermal currents. Moreover, he introduced for the first time an induction device which allowed a precise gradation of current, useful both in physics and physiology.

Among the various electrophysiological manifestations, Du Bois-Reymond discovered — simultaneously with Matteucci — the so-called "injury current" and the "negative oscillations" in the muscle. The similarity in electrical reactivity between the muscles and a magnet led him to establish a theory, the famous "Molekulartheorie", which was fashioned according to Ampere's conception of the magnetic constitution.[50] Du Bois-Reymond never got over the refutation of this hypothesis by his own student, *Ludimar Hermann* (1838-1914). Unfortunately, he was always highly critical of others, sometimes extremely prejudiced in his judgment, but little amenable himself to criticisms of his own work. Therefore, he was very conservative in maintaining the content and form of his lectures and demonstrations, especially after Müller's chair was divided into a separate anatomical and physiological position after 1858. The unchanging charts, the verbatim repetition of certain expressions, anecdotes and even jokes brought Du Bois-Reymond the reproach that he was still in 1890 teaching the physiology of 1868.

Testimony of Du Bois-Reymond's extraordinary oratorical talents are the two published volumes of his speeches: *Reden von Emil Du Bois-Reymond*[51] with an introduction written by one of his students, Julius I. Rosenthal (1912). The work also provides an outline of Du Bois-Reymond's broader interests, and his capacity to probe deeply into the fundamental scientific questions of his day. Among the topics discussed were his orations "Über die Lebenskraft" (1848), "Über die Grenzen des Naturerkennens" (1872), "Ignoramus-ignorabimus", and the commemoration speech for Johannes Müller (1858), and Hermann von Helmholtz (1895), both the best biographical sources for these distinguished physiologists.

Historically valuable also were Du Bois-Reymond's addresses delivered at the Berlin Academy of Sciences, an institution which had appointed him permanent secretary since 1868. The speech which Reymond delivered in 1877 at the formal opening of the newly erected Physiological Institute in Berlin is of special value, inasmuch as it reflects the historical development of physiology. He vividly described the external obstacles which stood in the path of experimental research as it was carried out in Du Bois-Reymond's early years as a physiological investigator. "If, at that time, a young man wanted to carry out physiological experiments," he declared, "he had to do them in his own living quarters. Thus, using frogs and rabbits as the subjects for these trials - we did not dare to employ dogs under those circumstances - the budding scientist often got into trouble with his landlord. Hence, many investigations were impossible to perform in those conditions, while others faced almost insurmountable difficulties. There were no eager assistants whose criticism could be valuable, nor accessible specialized libraries and collections of necessary instruments."

Du Bois-Reymond continued his reminiscences: "The young physiologist of the day had to purchase books, chemicals, tools and all the other research material out of his own meager funds, often designing and constructing his own devices. Moreover, we all had to solder our own elements, and glue our rubber tubes together which were not as yet commercially available. We sawed, planed and drilled, filed, shaped and polished. The need for advice and help regarding mechanical matters drove us into the workshops themselves, where we obtained many useful tips from talented craftsmen. Hence, we learned to understand the construction of an instrument to the last nut and bolt like the anatomy of an animal. If a valuable apparatus was entrusted to us through the kindness of a teacher, we took advantage of the situation, studying its mannerisms and above all, keeping it clean!"

Du Bois-Reymond settled his dispute with the adherents of the vital force or "Lebenskraft" in a brilliant essay, published in 1848 as a prologue to his first volume on animal electricity. Adequately entitled "Über die Lebenskraft", Du Bois-Reymond's work should be considered, together with Lotze's article "Leben, Lebenskraft" which appeared in Wagner's handbook in 1842, as the most significant and effective indictment of vitalism. Following these writings, the assumption that there were certain metaphysical peculiarities characteristic of the living beings became an obsolete concept.

During his lifetime, Du Bois-Reymond received numerous honors, and at his death many tributes were paid to him by prominent scientists. Many interesting details regarding the lives and personal relationships of the *four friends,* Du Bois-Reymond, Ludwig, Helmholtz and Brücke have been uncovered in the correspondence between Ludwig and Du Bois-Reymond, edited in 1927.[52]

One of the physiologist's sons, René Du Bois-Reymond, was for many years in charge of the physiological section of the Berlin Institute. In fact, the Physiological Institute was actually divided into four relatively independent physiological sections: chemical, physical, histological and experimental. The building was always filled with a profusion of instruments and assistants. Among the latter were *Albert von Bezold* (1836-1868), *Eduard F.W. Pflüger* (1829-1910), *Ludimar Hermann* (1838-1914), *Isidor Rosenthal* (1836-1915), *Hermann Munk* (1839-1912), *Franz C. Boll* (1849-1879), *Karl S. Sachs* (1853-1878) and *Johannes Gad* (1842-1926). A fair number of Russian students could always be found with Du Bois-Reymond.[53,54]

One of the most prominent disciples of Du Bois-Reymond's rather numerous students was *Albert von Bezold,* who had a brilliant but rather short career. The son of an Ansbach physician, Bezold was a highly gifted youth who already had gone to study in Munich at the age of seventeen. Unfortunately, he was stricken with rheumatic fever there, which left him with the sequelae of a cardiac valvular lesion. After further studies in Würzburg, Bezold published the results of some early research carried out at the local Chemical Institute which was headed by Scherer, as well as the Anatomical Institute under the direction of Kölliker. Then, in 1857, Bezold went to work with Du Bois-Reymond at Berlin, where he studied the action of curare and the phenomenon of the electrotonus.

In 1859, barely twenty-three years old, Bezold was appointed professor of physiology at the University of Jena, where he published the results of several important investigations related to the innervation of the heart

Fig. 77. Eduard F.W. Pflüger (1829-1910).

(1863). Six years later, he was called to fill the newly created chair in physiology at the University of Würzburg, where he published several good articles on topics of cardiac and circulatory physiology. Bezold's early death occurred in 1868, the result of his cardiac ailment.[55]

From the questions which occupied Bezold during his stay in Berlin with Du Bois-Reymond, one can also see the influence exerted by the master on another student: *Eduard F.W. Pflüger* (see Fig. No. 77). Pflüger was likewise concerned with the subject of electrotonus as well as the problem of intestinal innervation. Born in Hanau, he started studying at Marburg in 1850 and later studied at Berlin where he graduated and became a professor in 1858. Pflüger's first lecture, delivered in 1855, dealt with the inhibitory effects of the N. splanchnici on the gut, his first important discovery. Subsequently, there were violent discussions about his studies concerning the reflex actions of the spinal cord (1853).[56]

During the next few years, Pflüger was constantly experimenting in his rented apartment, dealing mainly with the effects of electricity on irritability and the frog nerves' response to stimuli. The results of these investigations were published in 1859 under the title *Untersuchungen über die Physiologie des Elektrotonus.* They reflected Pflüger's exceptional experimental skills, his patience and precise analytical procedures.

A year after the publication of this monograph on electrotonus, Pflüger, then barely twenty-nine years old, was appointed to the chair of physiology at the University of Bonn. The physical facilities at his disposal were no longer the old anatomical building in Bonn's so-called "Hofgarten", but a temporary pavilion at the University. However, only in 1878 was Pflüger able to occupy a well equipped and roomy laboratory in Bonn-Poppelsdorf.

Between his first important work on electrotonus and the last famous publication concerned with the glycogen in 1903, Pflüger's working methods remained essentially unchanged. Noteworthy were both his capacity to define physiological questions as well as the clarity with which he formulated the various experimental conditions. His choice and creation of new methods for the execution of numerous experiments, as well as his critical review of personal and contemporary scientific results must be stressed. Pflüger's cautious use of conclusions for the establishment of new hypotheses and theories completes the picture of this remarkable scientist.

Another fundamental trait of Pflüger's thought was his teleological formulation, which was designed to correct and supplement the causal and analytical treatment of the vital phenomena. Pflüger had already expressed in his first youthful work that "the organic form is not answering the purpose because it exists; rather, the structures can only be viable because they serve a purpose". Twenty-five years later, he reached back to stress this viewpoint again. Recognizing the fruitfulness of teleology as a leading thought in physiological investigations, Pflüger characterized organic phenomena as "teleological mechanics". He also declared that "the cause of every need in a living creature is at the same time the inducement to satisfy this need". He used numerous examples to illustrate this dictum, such as the relationships between light intensity and pupillary diameter, lack of nourishment and hunger-food ingestion; intestinal distention and emptying; lack of oxygen and cellular absorption of oxygen; and muscular exertion hypertrophy; lack of water and thirst.

One of Pflüger's important studies, carried out over a number of years, concerned itself with problems of respiration and the gases in the blood. About this subject he finally published a work in 1875, entitled *Über die physiologische Verbrennung in den lebendigen Organismen.* Pflüger postulated that the living cells determined and regulated their own oxygen consumption, but had no influence over the oxygen content in the blood and other factors.[57]

In more than two hundred articles, Pflüger dealt with virtually all areas of physiology and physiological chemistry, and was greatly opposed to any separation of these two disciplines. His criticism was often unsparing, unusually harsh, and many polemics in his journal were the results of such an attitude. Pflüger justified his conduct by declaring that "criticism is the most important impulse for progress, and therefore I practice it". His life mirrored his rather restless and creative activities, both in research and teaching. He kept the chair of physiology at the University of Bonn for fifty-one years, until he died on March 16, 1910, at the age of eighty-one, the victim of an hepatic cancer.[58]

Beginning in 1868, Pflüger edited the journal *Archiv für die gesammte Physiologie des Menschen und der Tiere.* Its pages reflect the tremendous progress which German physiology made in those years. In the eighteenth volume, published in 1878, the following statement was printed: "A physician without physiological knowledge is like a watchmaker who is supposed to correct the abnormal action of a clockwork, but is ignorant of the normal operating conditions of the machine he is called to repair".

Pflüger's scientific accomplishments have been extensively described by one of his own students.[59]

Another disciple of Pflüger, *Nathan Zuntz* (1847-1920), already worked in the laboratory before his graduation. After a short period of rural medical practice, Zuntz returned to Bonn and became Pflüger's assistant in 1870, receiving a professorial appointment four years later at the Agricultural Academy of Bonn. In 1880, Zuntz transferred to the newly established Agricultural College in Berlin, where he died in 1920.

Zuntz had already concluded in his dissertation that the combination between carbon dioxide and hemoglobin was dissociable instead of fixed. Together with Joseph von Mering he examined the effects of food absorption on the metabolism, coining the word "Verdauungsarbeit" (digestive work) for the increased metabolic rate. Zuntz also demonstrated that intravenously administered foodstuffs could produce a similar metabolic increase.

Fig. 78. Nathan Zuntz (1847-1920) and his portable breathing apparatus.

A tremendous methodological advance was the introduction of the so-called Zuntz-Geppert breathing machine. This portable apparatus facilitated short metabolic investigations in the laboratory as well as studies during the actual performance of certain physical activities (see Fig. No. 78). Thus, Zuntz became involved in those physiological events which were associated with sports, hiking and high altitude, subjects on which he published a series of monographs with his collaborators. He was a popular lecturer who could present in simple terms a series of rather complex biological phenomena, a gift which made him contribute to the dissemination of physiological knowledge to the persons interested in sports activities.

Zuntz studied extensively the various gases in the blood, and established the carbonic acid method for the measurement of blood volume as well as the formula of oxygen and caloric consumption. Many foreign students were attracted to Zuntz's laboratory, and among his better known collaborators were Adolf Loewy, Adalbert Magnus-Levy, A.J. Geppert, W. Caspari and Arnold Durig.

Another student of Pflüger was *Max Bleibtreu* (1861-1939), who came to Pflüger's laboratory in Bonn after studying science and medicine. His investigations were primarily directed towards chemico-physiological questions, and Bleibtreu observed the increase of the respiratory quotient during food intake. In 1903, he succeeded another well known physiologist, *Leonard Landois* (1837-1902), at Greifswald.

Rudolf Rosemann (1870-1943) was a disciple of both Landois and Pflüger, who re-edited Landois' successful textbook, a publication which went through thirteen editions in twenty-eight years. Rosemann's special topics of interest concerned the physiology of chloride metabolism and the gastric secretions. His short sojourn at Pflüger's laboratory occurred in 1903, and he was appointed, thereafter, professor at the University of Münster. Rosemann held this position between 1904 and 1937, a kind person of broad interests and great erudition as well as a popular speaker.[60]

Fig. 79. Ludimar Hermann (1838-1914) (right) together with Martin Gildemeister (middle) and Otto Weiss (left) in the Physiological Institute.

Although the German physiologist *Ludimar Hermann* (see Fig. No. 79) called himself an autodidact in his posthumously published memoirs or *Erinnerungen,*[61] he had been greatly inspired in his work by Du Bois-Reymond's investigations of animal electricity. Of Jewish ancestry, Hermann was born on October 21, 1838 in Berlin. He studied in his hometown and came early to Du Bois-Reymond's laboratory as an amanuensis, precisely at a time when the Berlin Institute was the center of physiological research in Germany. Hermann was especially fascinated by chemistry, and he concerned himself with questions of muscular physiology and chemistry between 1859 and 1861. Characteristically, Hermann tended to deal with his physiological subjects in a *mathematical* fashion.

After his graduation in 1859 and promotion to teach six years later, Hermann became the successor of Adolf Fick, accepting the chair of physiology at Zürich in 1868. During the ensuing sixteen years, he investigated predominantly those questions concerned with animal electricity. In a publication which appeared in Berlin during 1867, Hermann demonstrated that the muscle was capable of releasing sufficient energy for the contraction even in the absence of oxygen through anaerobic processes.

In the following year, Hermann published another work which sought to demonstrate the lack of electrical potentials in an intact and resting muscle.

This assertion, which was in direct opposition to Du Bois-Reymond's theories, caused Hermann to become embroiled in a long personal and professional conflict with his former teacher. According to his views, the electrical potentials observed after the muscle sustained an injury were expressions of an altered living substance — the so-called "Alterationstheorie." Hermann believed that these currents reflected a gradual dying of the living muscle instead of being expressions of preexisting electrical tensions.

Hermann had presented his novel views first to Du Bois-Reymond himself in a private conversation, and was allowed to carry out the needed experiments to verify such an hypothesis at the Berlin Institute. However, since the early results of such a research were not in accordance with Du Bois-Reymond's own conceptions, Hermann was told that he was no longer welcome to experiment there. An attempt to resume his investigations under the direction of Willy Kühne at Virchow's institute was thwarted by Virchow himself. Thus, Hermann installed a small laboratory in his parent's printing press, where he strenuously worked with borrowed instruments and materials. He was finally able to prove the lack of electrical potentials in intact muscles and established the laws of extrapolar nervous currents.

Hermann subsequently published numerous and excellent papers on action potentials, polarization, mechanisms concerning the conduction of the stimulation — the so-called "Strömchentheorie" — and electrotonus. These investigations, carried out during his tenure at the University of Zürich, significantly increased the available knowledge of the various manifestations related to animal electricity. Other work was devoted to questions of respiration, metabolism and energy conversion as well as the physiology of the blood. According to Hermann, digestion consisted of a series of hydrolytic breakdowns which facilitated resorption. In turn, the products of these chemical separations enabled the body to make new and multiple syntheses, in a fashion "like letters of dissolving sentences in a book could be reassembled in order to make new ones". (1868)

In 1884, Hermann went to teach at the University of Königsberg, where he concentrated on physiological problems related to the organs of sense perception, primarily vision and hearing. Moreover, he worked especially on the physiology and physics of the sounds connected with speech. In those years, Hermann also edited the six volumes of a textbook, the *Handbuch der Physiologie* (Leipzig, 1879-83) in which he himself wrote the sections pertaining to muscular and neurophysiology after Du Bois-Reymond's refusal to cooperate. Among the excellent co-authors of this ambitious project were a galaxy of German physiologists, such as Fick, Heidenhain, Zuntz, Engelmann, Grützner, Rollet, Rosenthal, Voit, Drechsel, Maly, Wittich, Mayer and Hensen. Hermann died on June 5, 1914 at the age of seventy-five.

Among Hermann's students in Zürich was *Johann B. Luchsinger* (1849-1886). Beginning in 1878, Luchsinger taught physiology at the veterinary school in Bern, transferring to a chair at the University of Zürich in 1884. Three disciples of Hermann at the physiological institute of the University of Königsberg became well known: *Oscar Langendorff* (1853-1909), *Otto Weiss* (1871-1943) and *Martin Gildemeister* (1876-1943).

Originally from Breslau, Langendorff[62] was an assistant of Wittich and Hermann between 1875 and 1880. In 1892, he accepted an offer to teach physiology at the University of Rostock, succeeding *Hermann Aubert* (1826-1892), who had made some important advances in optical physiology. Langendorff remained in Rostock until his death in 1909, and among his studies were some investigations concerning the location and function of the respiratory center. Moreover, he improved considerably the method to isolate the mammalian heart in 1895, a technique pioneered by the British physiologist *Henry Newell-Martin* (1848-1896).

Langendorff also searched for the nature of the cardiac automatism, preferring the neuroganglionar explanation over the myogenic hypothesis. Other experiments were related to the role and importance of digestive ferments, both in the stomach and the pancreas, as well as the function of ganglions in the pathways of nerves. During the latter studies, Langendorff demonstrated in 1898 that there was no need for an excitation of the cells in the spinal ganglions for the transmission of impulses.

Otto Weiss, of Vilsen near Hannover, spent almost his entire life in Königsberg. Versatile in his research, technically skilled, Weiss concentrated on the development of newer methods for the study of physiology. Among his instruments was a phonoscope to register weak sound waves, and he also promoted the application of cinematography. Regarding his numerous areas of research, Weiss dealt with cardiac sounds, blood circulation, neuromuscular phenomena, vision, hearing, voice and language. Weiss was an excellent teacher and he carried out rather difficult physiological experiments during his lectures. His unusually broad education and a great love and understanding for nature contributed towards the great esteem in which he was held.

The third prominent disciple of Hermann was *Martin Gildemeister,* born in Thorn and an assistant at Königsberg from 1899 until 1907. Gildemeister then went to work with Richard Ewald at Strassburg and became section chief at the Physiological Institute of Berlin. In 1924 he was named professor of physiology at the University of Leipzig, and died, like his friend and colleague, Otto Weiss, in 1943. Gildemeister concerned himself with physiological topics which could be measured and hopefully expressed in mathematical terms, rather than dealing with qualitative observations and differences. He was interested in the relationship between stimulus and excitement, hoping to establish a general theory of excitability. Among

Gildemeister's important investigations was the study of the so-called psychogalvanic reflex of the skin. He was endowed with a rather pleasant personality and possessed great technical proficiency, furnishing new approaches to the burgeoning physiological methodology of his time.

During those years in which both Pflüger and Hermann were active at Du Bois-Reymond's laboratory, another physiologist was working there: *Rudolf Heidenhain* (see Fig. No. 80). The son of a physician, Heidenhain was born on January 29, 1834, in Marienwerder. His early interests in collecting plants and animals already pointed to a career in science, and Heidenhain began his studies of the natural sciences at the University of Königsberg. Thereafter, he received his medical education in Halle, where he became interested in physiology thanks to the influence of Alfred W. Volkmann. Heidenhain completed his medical career at Berlin, where he graduated in 1854. His dissertation was based on some research suggested to him by Du Bois-Reymond and purported to refute the ideas of Moritz Schiff about the vagus nerves as causes of cardiac activity. Instead, Heidenhain proposed that the heart's automatism was actually maintained by the cardiac ganglia, and the vagus simply regulated the process.

Fig. 80. Rudolf Heidenhain (1834-1897).

For several semesters, Heidenhain worked as an assistant of Du Bois-Reymond, returning finally to Halle in 1856. For awhile he cooperated with Volkmann, whose daughter he married. The thesis which qualified him to teach in 1857 dealt with blood volume. Two years later, at the early age of twenty-five, he was appointed to the chair of physiology at the University of Breslau, succeeding Karl B. Reichert. In spite of initial student strikes there against the alien "greenhorn" professor, Heidenhain was capable of restoring order and gaining the respect of those he taught. He remained at Breslau until his death in 1897, developing an unusually versatile investigative career. Short of stature, vivacious and active, Heidenhain exhibited a tremendous endurance and unflinching capacity for work. Instead of being a fighter, he was a more critical and conservative personality.[63]

Heidenhain's own capacities and ideas directed him more towards a biological standpoint rather than an approach to the living phenomena from a physico-mathematical viewpoint. He especially rejected the hasty and purely physical interpretations of glandular activity, stressing the peculiarities of the living actions without paying lip service to vitalistic theories. Heidenhain carried out a series of investigations in the 1870's, searching for the nerves responsible for the processes of vasoconstriction and dilatation. During this research, he confirmed some of Schiff's earlier experiments with the vasodilating neural pathways of the sciatic nerve in the dog. Moreover, he justified Schiff's conclusions which in his view had been derived from work "in the mainstream of contemporary research".

In 1864, Heidenhain published an important book on muscular physiology entitled *Mechanische Leistung, Wärmeentwicklung und Stoffumsatz bei der Muskeltätigkeit,* which was dedicated to Du Bois-Reymond. In this work, Heidenhain stressed the rather economical operation of the muscles and their own control of the type of energy consumption needed in relation to the amount of muscular activity. He demonstrated that as a muscle gradually became more fatigued, it tended to operate more economically.

Heidenhain's most important investigations were concerned with the physiology of glandular secretions. Beginning in 1867, he carried out some pertinent experiments on this subject for almost thirty years until his death, achieving in the process a great technical mastery. Among his valuable techniques was the use of a small gastric cul-de-sac (1875), and the histological examination of those morphological changes occurring during periods of secretion and rest. Heidenhain also observed in detail the salivary and gastric glands, studying their structure, secretions, response to innervation, etc. Other organs which came under his scrutiny were the pancreas, liver and kidneys.

In 1888, Heidenhain published his well-known inquiries concerning the absorption of hypotonic solutions in the small bowel, initiated a few years earlier. These studies, together with the investigations about the formation of lymphatic fluid, provided Heidenhain with a lasting place among the great physiologists of the period. He died on October 13, 1897, after suffering from a duodenal ulcer.[64]

A series of rather gifted students received valuable suggestions from Heidenhain. They were *Paul Grützner* (1847-1919), *Karl Hürthle* (1860-1945), *Richard Gscheidlen* (1842-1889), *Franz W. Schweiger-Seidel* (1795-1838), *Wilhelm Waldeyer* (1836-1921), Ernest Starling and of course Ivan P. Pavlov.

Paul Grützner, of Festentag in Silesia, studied in Breslau, Berlin and Vienna, graduating at the University of Berlin in 1869. For eleven years he remained an assistant of Heidenhain until his appointment to a professorial chair at the University of Bern and later Tübingen. Grützner was well-known for his work in muscular and neurophysiology, circulation and urinary secretion. He also wrote an excellent monograph on voice and speech which appeared in a contemporary handbook. One of Grützner's disciples was *Karl Bürker* (1872-1957), known for his investigations in myothermia (since 1899) and hematology: hemoglobin, clotting time, counting chamber, etc. From 1917 on, he taught physiology at the University of Giessen.

Another of Heidenhain's students, *Karl Hürthle,* worked at the Breslau Institute between 1898-1927 and became his teacher's successor there. Hürthle made valuable contributions to the histology of muscular fibres and studied circulatory phenomena, especially in the small peripheral vessels to which he ascribed an independent role in the total circulation of the blood.

Finally, three other students of Du Bois-Reymond advanced his fields of electro- and muscular physiology. They were *Isidor Rosenthal, Johannes Gad* and *Jean-François Heymans.* Isidor Rosenthal started to teach physiology at the University of Berlin in 1862 and became full professor at Erlangen ten years later. His most important work was related to respiratory physiology — regulatory mechanisms — which he published in Hermann's textbook. Rosenthal also investigated the general physiology of muscles and nerves.

The second author mentioned was *Johannes Gad,* originally from Posen, who studied at the University of Berlin, where he became assistant to Du Bois-Reymond. After some years with Adolf Fick at Würzburg, Gad returned to Berlin and was appointed director of the experimental section of the Physiological Institute. In this capacity, he influenced, together with his mentor, Du Bois-Reymond, several native and foreign physiologists and clinicians such as *Alfred Goldscheider* (1858-1935), *George Marinesco* (1864-1938), *E. Flateau, William T. Porter* and *Otto Kohnstamm.* In 1887, Gad became one of the founders of the journal *Centralblatt für Physiologie.* He concentrated on questions dealing with respiration, electro-physiology, digestion of lipids and functions of the spinal cord. After 1885, he occupied the chair of physiology at the University of Prague. "Gad's bovine heart" is still a popular object today for physiological demonstrations. One of Gad's students was *Richard H. Kahn* (*1876) of Prague, the other was *Jean-François Heymans* who should be considered the founder of Belgian experimental pharmacology. Heymans worked with Du Bois-Reymond and Gad at Berlin between 1887 and 1890.

Before concluding this long list of disciples who followed Du Bois-Reymond, *Hermann Munk* of Posen, also a student both of Müller and Virchow, should be mentioned.[65] Munk became a physiologist for the Berlin Veterinary Institute in 1876, and studied the physiology of nerves, cerebral cortex, thyroid and other organs. One of Munk's most important discoveries concerned the localization of the center of vision in the occipital lobe of a dog. In general, our knowledge of cerebral locations was greatly enhanced by Munk's research. His brother, *Immanuel Munk* (1852-1903) was later a departmental chief in the Physiological Institute of Berlin, and a successful worker in the fields of metabolism, nutrition, renal function and muscular activity in connection with protein catabolism. He was a long time collaborator of Nathan Zuntz.

5. ERNST BRÜCKE AND HIS DISCIPLES

The third author from the circle of friends and disciples of Johannes Müller was *Ernst Wilhelm Ritter von Brücke* (see Fig. No. 81). Born in Berlin on June 6, 1819, Brücke studied both at Berlin and Heidelberg, and in 1844 was granted academic teaching privileges under Müller. Three years later, following the death of Burdach, he was called to Königsberg in order to fill the chair of physiology and general pathology at the local university. However, Brücke remained there for only two years, moving on to the University of Vienna in 1849. This transfer allowed Helmholtz to fill Brücke's position at Königsberg.[66]

Fig. 81. Ernst Brücke (1819-1892).

Unfortunately, the working conditions which Brücke encountered in Vienna were less than ideal. According to his grandson, who wrote a biographical account of the physiologist,[67] the equipment available at the University Institute merely consisted of one microscope, an incubator and a dynamometer. On the brighter side, Brücke encountered a brilliant group of men at the medical school in Vienna such as the clinician *Josef Skoda* (1805-1881), the pathologist *Carl von Rokitanski* (1804-1878), the dermatologist *Ferdinand von Hebra* (1816-1880) and the anatomist *Joseph Hyrtl* (1810-1894). The latter got into an argument with Brücke in 1853, and the ensuing controversy between the two scholars gave rise to a series of humorous anecdotes.

During his stay at Königsberg, Brücke studied the physiology of sense organs, predominantly the anatomy and physiology of the ocular ciliary muscles. Later, in Vienna, he devoted his interest to problems of digestive physiology, such as pepsin, salivary action, the breakdown of fats, intestinal motility and other subjects.

After the arrival of Carl Ludwig in Vienna during 1855, both Brücke and Ludwig became personal friends, and they shared the same ideals in physiological research. Since his days in Berlin, Brücke had been an exponent of the physico-chemical direction like Ludwig. In this connection, Du Bois-Reymond wrote to Hallmann: "Brücke and I, we have both sworn to expose the truth, namely that there are no other forces operating in the organism except those physico-chemical ones". However, in another paragraph of the same letter, Du Bois-Reymond continued: "Teleology is a kind of lady whom no biologist is able to live without, but he is rather reluctant to be seen with her in public".

During his years at Vienna, Brücke returned to one of his favorite ideas: the so-called "Pasigraphie". He wrote a book in which he attempted to represent with specific signs the sounds used in various languages. In this fashion, Brücke hoped to establish a unified phonetic writing method and be able to read every foreign language without any prior preparation.

Brücke was a versatile individual, not only concerned with the problems of his chosen field and the natural sciences, but also conversant in topics such as painting, sculpture, poetry and artistic handicrafts. These interests were reflected in several of Brücke's publications, such as *Die Physiologie der Farben für die Zwecke der Kunstgewerbe bearbeitet* (Leipzig, 1866) — a physiological explanation of colored vision for arts and crafts — *Über die physiologischen Grundlagen der neuhochdeutschen Verskunst* (Vienna, 1871) — about physiology and poetry. Moreover, Brücke published two works related to problems of anatomy and art, entitled *Die Beckenlinie antiker Statuen* (1888) and *Schönheit und Fehler der menschlichen Gestalt* (1891). Each one of these writings was directly based on his physiological considerations.

For over forty years Brücke remained the most prominent physiologist in Vienna. His work *Vorlesungen über Physiologie* went through several editions. For a long time he was one of the most important representatives of the Viennese scientific community, also enjoying a great reputation in royal circles. He was honored for his work by the Austrian emperor, and elevated into the ranks of the aristocracy. After his death on January 7, 1892, another extensive work was published: *Wie behütet man Leben und Gesundheit seiner Kinder* — how to protect the life and health of children (Vienna and Leipzig, 1892).

For the first time since the activities of Georg Prochaska, Vienna acquired a series of top-notch scholars in physiology aroung the middle of the nineteenth century. Under the direction of Brücke at the University and Ludwig at the Josephinum, the teaching of physiology resumed under the best of auspices. Brücke became the starting point and founder of a new *Austrian school of physiology* which included his students *Alexander Rollet* (1834-1903), *Ernst Fleischl von Marxow* (1846-1891) and *Sigmund Exner*.

A second generation of disciples which trained under the above named physiologists consisted of *Oscar Zoth* (1864-1933) and *Arnold Durig* (1872-1961). Among those who formed the third generation of the Austrian school were *Ludwig Haberlandt* (1885-1932), *Leopold Löhner* (1884-1958) and *Ferdinand Scheminsky* (*1899). Moreover, the founder of psychoanalysis, *Sigmund Freud* (1856-1939), also studied initially under Brücke, a fact which was reflected in his manner of working. Likewise, one of Freud's early collaborators, *Joseph Breuer* (1842-1925), who studied vestibular functions, also received his physiological training under Brücke, and was in turn the teacher of another prominent student: *Alois Kreidl* (1864-1928). Some of these Austrian disciples of Brücke will be presented here in more detail, inasmuch as their work has already become historical.

Alexander Rollet (see Fig. No. 82) was born on July 14, 1834, in Baden, near Vienna. During his Viennese studies with scholars such as Hyrtl, Oppolzer, Rettenbacher, Brücke and Ludwig, Rollet was especially attracted to physiology, and went to work in Brücke's laboratory. In 1857, after the departure of *Maximilian R. von Vintschgau* (1832-1902), another of Brücke's students, Rollet, became one of Brücke's assistants and remained in the University Institute for seven years. His close circle of collaborators and colleagues during that time included other prominent figures such as Elie de Cyon, Ivan M. Sechenov, Willy Kühne, Wilhelm Preyer and Johannes N. Czermak, a tightly knit group surrounding both Brücke and Ludwig. In those years Rollet carried out his studies on muscular physiology.

Recommended by Brücke and barely twenty-nine years old, Rollet was appointed full professor at the University of Graz in 1863. Although he was short of funds, with a newly founded institute still housed in makeshift quarters, Rollet successfully organized both teaching and research. He built many of the necessary instruments for work himself, using cork, strings and sealing wax. Fortunately, in 1872 a new and well-equipped anatomical-physiological institute was inaugurated in Graz, offering Rollet better possibilities to carry out physiological research. Thus, he was able to develop an extensive and fruitful activity until his death in 1903.

Rollet's growing reputation attracted many students to Graz, among them figures such as *Viktor von Ebner* (1842-1924), *Rudolf Klemensiewicz* (1848-1922), *Julius Glax* (1846-1922), *Oscar Zoth* and numerous Russians. Indeed the Grazer Institute virtually became a Russian colony, composed of men such as *Ivan M. Sechenov,* (1829-1905), *W.N. Boldyreff* (1872-1946), *Dobroslavin, Kistiakovsky, Golubev* and *Rynek*. The spirit of Rollet's school was characterized by thoroughness, precision, reliability, modesty and clearness, reflecting the outstanding personality of its founder. Rollet's unusual capacity for work, his sharp thinking and admirable memory were coupled with a systematic approach to his activities and personal modesty. His disciple, Oscar Zoth, has preserved for posterity many of Rollet's outstanding characteristics.[68]

Rollet's primary areas of investigation were the physiology of the blood, muscular histology and physiology, and physiological optics. His inquiries into hemoglobin, hemolysis, blood analysis, specific resistance and other topics coined him the term "Blut-Rollet". From his days as a student, he remained interested in the problems of muscular structure and function, especially between 1874-1898. With unbelievable patience and perseverance, he examined histologically all sorts of muscular structures, among them the muscles of three hundred varieties of beetles. Rollet was thus able to observe under the microscope a wave of contraction passing through a living muscle. He also carried out careful studies on muscular work, fatigue and recovery.

Fig. 82. Alexander Rollet (1834-1903).

Rollet's dislike for premature hypotheses in physiological areas not fully explored was reflected in 1896 towards the end of his work on muscles. Instead of establishing a theory, he enumerated no less than nineteen questions which actually had remained unanswered and demanded further experimental elucidation. Like Brücke before him, Rollet was also very interested in several questions concerning the physiology of the sensory organs. Among other phenomena, he explained spatial vision and the sensation of depth through the convergent effort of the ocular muscles (1861). Other works dealt with the problems of contrasting colors and the fading color perception. In 1899, he described and analyzed his own olfactory sensations after being affected by a severe and artificially induced anosmia.

Finally, Rollet's importance as a researcher was matched by his reputation as a revered academic teacher. As chancellor of the university, and in other numerous capacities, he placed his talents at the service of the public in general, and received many honors and early recognition for his accomplishments. He died of pneumonia on October 1, 1903, shortly before his seventieth birthday.

Among Rollet's Russian disciples were *Ivan M. Sechenov,* later professor of physiology in Odessa, St. Petersburg and Moscow, *W.N. Boldyreff,* professor of special therapeutics at the University of Kasan, and *Karl A. Spiro* (1867-1932), professor of comparative physiology in Odessa. Other students of Rollet included *Viktor von Ebner* (1842-1924) who worked at the Graz Institute from 1868 until 1871 and was later appointed professor of histology and embryology in Graz and Vienna. In addition, there were *Richard Maly* (1839-1894), who first occupied the chair of physiological chemistry at Innsbruck and then received an appointment in chemistry at Graz and Prague, and *Rudolf Klemensiewicz* who became professor of general and special pathology at the University of Graz.

In the 1880's, the physiologist *Karl Laker* worked in Rollet's institute, where, together with *Georges Hayem* (1841-1933) and *Giulio Bizzozero* (1846-1901) he discovered that the thrombocites were independent elements within the blood — the so-called "Blutscheiben." Laker, who was clinically interested in otorhinolaryngology, later spent a great deal of time studying the physiological and acoustic aspects of music, developing a method for the "visualization of music" with the help of some instruments. Laker eventually became professor of physiological acoustics at Graz in 1923, and retired in 1930.

The most prominent disciple of Rollet was *Oscar Zoth.* A native of Austrian controlled Padua, Zoth became an assistant, instructor and collaborator of Rollet during the years 1885-1902. After a short professional appointment at Innsbruck in 1902, Zoth succeeded Rollet at Graz in 1904. He studied blood and circulatory physiology, and was concerned with problems of physiological optics and muscular motion. Zoth not only published a series of original works and monographs, but was also responsible for a series of technical innovations and the construction of certain instruments. A scholar of extraordinary modesty and technical acumen, Zoth retired in 1926 and died in 1933 at the age of seventy.[69]

Zoth was succeeded at the University of Graz by a disciple, *Leopold Löhner* (1884-1958), whose work dealt with almost all areas of physiological research. Another of Zoth's students was *Ludwig Haberlandt,* who started teaching physiology at Innsbruck in 1913. Haberlandt was well-known for his research concerning a cardiac hormone and hormonal sterilization. Finally, another of Zoth's students in Graz, *Robert Stigler* (*1878), should be mentioned. Stigler was active in Vienna at the School of Agriculture.

A native of Tyrol, *Maximilian Ritter von Vintschgau* was another disciple of Ernst Brücke. The predecessor of Rollet at the Viennese University Institute in 1856-57, Vintschgau became professor at the University of Padua under Austrian rule in 1857, remaining there until 1866. After three years in Prague, he was appointed to the chair at Innsbruck, where he died at the age of seventy. Vintschgau was renowned for his investigations of the physiology of taste (1879), sense of color and the reaction time of facial sensations during temperature variations.

Among Vintschgau's students was *Arnold Durig,* who followed Cyon's methodology regarding physiological techniques. Durig, who worked in Vintschgau's laboratory from 1893 to 1900, was allowed to use the available laboratory hardware for teaching purposes only. To compensate, he carried out, in addition, a series of experiments at night and during vacation time, a fact which provoked the ire of his teacher after his surreptitious activites were discovered. In 1900, Durig went to work with Sigmund Exner in Vienna, after a short sojourn in Oxford where he studied under *John Burdon-Sanderson,* and a period in Berlin under Nathan Zuntz. Later, he was called to Vienna, where he received, in 1905, an academic appointment to teach physiology at the Viennese "Hochschule für Bodenkultur," eventually becoming professor at the University in 1918. Durig's work in physiology was of an unusually wide scope. He studied the effects of high altitude climate, nutrition, and the physiology of physical work, as well as some aspects of fatigue and respiration, contributing significant new knowledge in all these fields of endeavor. Among his collaborators were *Richard Wagner* (1893-1970) and *Ferdinand Scheminzky* who later occupied prominent academic positions.

Fig. 83. Willy Kühne (1837-1900).

The physiologist *Willy Kühne* (1837-1900) (see Fig. No. 83) was another disciple of Ernst Brücke, and also trained under the French physician and scientist Claude Bernard. Both Brücke and Bernard exerted great influence over the subsequent career of Kühne.[70] Even as a student, Kühne demonstrated a marked interest in the natural sciences, especially chemistry. Therefore, he went to work in the laboratory of the chemist Friedrich Wöhler at Göttingen at the age of seventeen. After some studies with Rudolf Wagner, Kühne transferred to Berlin where he carried out research in muscular physiology at Du Bois-Reymond's institute. He demonstrated the direct muscular excitability on the M. sartorius, the dual capacity of nervous transmission — the so-called "Zweizipfelversuch" — and discovered the terminal motor plate.

Kühne spent two years with Claude Bernard learning physiological techniques before coming to work with both Brücke and Ludwig at Vienna, where they provided him with valuable suggestions for further research. In 1868, he was appointed professor of physiology in Amsterdam, and three years later he settled in Heidelberg as the successor of Helmholtz, an academic position which he held until his untimely death in 1900. His friends and colleagues depicted him as an unusually spirited and lively personality with a solid preparation in his field, in addition to an excellent general culture. Artists, musicians and other interesting persons formed part of his selected group of friends.

Kühne was a very gifted man, both in attributes and opportunities, and his life work can be divided into three separate fields of investigation: muscles, visual substances and digestion. In all of them, Kühne applied his chemico-physiological method. In the muscle, he discovered and named the liquid and contractile muscular plasma *myosin.* One day, he observed a nematode worm swimming in a living muscular fibre. In its movements, the parasite clearly passed through the striated part of the muscle which closed again behind the nematode's tail.

Finally, Kühne also coined the name *enzyme* to designate the formless precursors of ferments. He named the pancreatic enzyme capable of digesting albumin, *trypsin,* discovering a series of intermediate steps in the protein metabolism. Moreover, he observed living pancreatic cells under the microscope.

The discovery of visual purple by *Franz C. Boll* prompted Kühne to carry out a series of investigations in this field of physiology. His so-called "Optogramm" or retinal image was described as follows: "The retina (with its epithelium) acts not as a photographic plate, but actually more like a complete photographic laboratory, in which workers continuously supply new highly photosensitive material to the plate, erasing the older images" (1876). Kühne was fortunate enough to fix, from the retina of a rabbit, the image or "Optogramm" of a window with the help of an alum solution. He also tried to place Helmholtz's image on a rabbit's retina, but only the white collar of the scientist's shirt became visible. Kühne recognized early that vision could take place without visual purpura. The stimulation of the cones actually mediated the sensation of color while the rods conveyed light or darkness.[71] He died on June 10, 1900 at the age of sixty-three, and should be considered one of the founders of the *German chemical physiology,* a topic on which he wrote a textbook in 1868.

A large number of famous future physiologists worked with Kühne at the University of Heidelberg, such as *Hugo Kronecker, Albrecht Bethe* (1872-1954), *Otto Cohnheim-Kestner* (1873-1953), *Jacob J. von Uexküll* (1864-1944), as well as the Swedish author *Alarik F. Holmgren,* the Dutch physiologist *Hendrik Zwaardemaker* (1857-1930), the British physiologist

Ernest H. Starling and the American *Russell H. Chittenden* (1856-1943). Two of Kühne's students should be mentioned here, while the rest will be considered in subsequent sections.

Born in Estonia, *Jacob J. von Uexküll* (see Fig. No. 84) studied zoology at Dorpat and then worked together with Albrecht Bethe, Otto Cohnheim-Kestner and others at Kühne's laboratory. After 1925, Uexküll became the founder and director of the "Institut für Umweltforschung" — an institute for environmental studies — in Hamburg. He was a highly original scientist whose ideas and intelligent research proved fruitful for physiology in general. Uexküll's studies about the relationships between the environment, bodily structures, anatomy of the sensory organs and instinct in animals, provided essential ingredients for a comprehensive biological "Weltanschauung" or worldview. To that effect, he published a series of impressive works, among them *Theoretische Biologie* (Berlin, 1920). Uexküll died on the island of Capri in 1944.

His colleague, *Otto Cohnheim-Kestner,* was born in Breslau as the son of the well-known pathologist, Julius Cohnheim. From 1897 to 1912 he studied at the physiological institute of the University of Heidelberg, first with Willy Kühne and then Albrecht Kossel. After spending some weeks with Ivan Pavlov in St. Petersburg, he worked several months at the zoological station in Naples. Later, in 1918, Cohnheim-Kestner founded the physiological institute of the hospital in Hamburg-Eppendorf, and became professor of physiology there. He published several important monographs in the fields of digestive and nutritional physiology, the chemical aspects of albumin and metabolism. The author gratefully remembers his excellent and stimulating lectures, which were enlivened by numerous and impressive experiments. He died on February 21, 1953.

Fig. 84. Jacob Johann von Uexküll (1864-1944).

Another prominent disciple of Brücke was *Alois Kreidl,* a native of Gratzen in Bohemia. Kreidl was formally allowed to teach in 1897 at Vienna, after being an assistant to Brücke and Exner at the physiological institute for several years. In 1918, Kreidl became section chief and director at the Viennese Institute for General and Comparative Physiology. He was a master in various operative techniques and carried out excellent investigations of the midbrain, the sympathetic center, the vestibular apparatus and the sense of hearing.

Another physiologist, *Sigmund Mayer* (1842-1910), also trained under Brücke in physiology. Thereafter, he went with Ewald Hering to Prague, and eventually assumed the direction of the Histologic Institute at Leipzig.

Brücke's successor, the physiologist *Sigmund Exner*, was also a product of Brücke's school where he was active as a student. After some training with Helmholtz at Heidelberg in 1868, Exner became Brücke's assistant in 1871. He combined Brücke's interest in sensory physiology with Helmholtz's concerns for physiological optics, and he worked on questions related to the central nervous system and the physiology of the senses.

Exner was one of the founders of the journal entitled *Zeitschrift für Psychologie und Physiologie der Sinnesorgane* (journal of the psychology and physiology of the sensory organs) together with such other prominent authors as Hermann Aubert, Heinrich Hering, Arthur König, Hermann von Helmholtz, Johannes von Kries, Georg E. Müller, Wilhelm Preyer and C. Stumpf. The relationship between psychic manifestations and the activity of the central nervous system occupied him time and again, as evidenced by his work on cerebral localization (1881). Concepts such as reaction time, facilitation and inhibition were coined and then explained by Exner. He was also attracted by the physiological aspects of optics, and studied hallucinations, the regeneration of the retina, color contrasts, facet-eye, and other topics. Moreover, Exner was interested in the questions of sound transmission in the field of physiological acoustics. Both Arnold Durig and Otto von Fürth worked in his institute.

6.
OTHER DISCIPLES OF JOHANNES MÜLLER

With a certain degree of justification the physiologist *Karl (von) Vierordt* (1818-1884) can be considered a student of Johannes Müller.[73] Vierordt, originally from Baden, attended Müller's lectures during the years 1839-40 and was deeply influenced by the great teacher of physiology. He initially went into medical practice after his graduation from medical school, but was able to devote some time to scientific pursuits thanks to his position as a military physician. Vierordt's interests lay in the area of measurable, quantitative physiological phenomena, especially the gas exchange in respiration and specifically the elimination of carbonic acid. His pertinent ideas were published in Wagner's handbook in 1846. Through his valuable investigations, he soon became famous and was appointed to teach theoretical medicine at Tübingen in 1849. Following the departure of Arnold, physiology became an independent discipline there in 1853, quite distinct from anatomy, and thus Vierordt was appointed professor of physiology. Later, in 1868, he had the good furtune to receive a new physiological institute at the University of Tűbingen.

Vierordt carried out a series of excellent and skillfully executed investigations. He enriched the physiology of the blood with his method of counting and measuring the volume of red cells. His sphygmograph, presented in 1853 at the meeting of natural scientists in Tűbingen, soon became widely known as the first pulse-recording device, in spite of its initial imperfections. Vierordt wrote a popular work, entitled *Grundriss der Physiologie,* which was translated into several foreign languages. In his last years he concerned himself with topics such as the blood velocity and the application of spectrophotometry in physiology and chemistry. He died from a cardiac ailment at the age of sixty-six. A posthumously published book also contained his biography and bibliography.[74]

Fig. 85. Georg Meissner (1829-1905).

Vierordt's son, *Hermann von Vierordt* (1853-1944) was also interested in physiological subjects — cardiac sounds, the mechanics of walking — in addition to clinical medicine. Both father and son displayed a great interest in the history of medicine reflected in Hermann von Vierordt's book entitled *Medizingeschichtliches Hilfsbuch mit besonderer Berűcksichtigung der Entdeckungsgeschichte und Biographie* (Tűbingen, 1916)

Another physiologist who was deeply influenced by Johannes Műller was *Georg Meissner.* Meissner, (see Fig. No. 85) a contemporary of Adolf Fick and Friedrich Bidder, was attracted by the physician Rudolf Wagner to study the natural sciences during his student years at the University of Gőttingen. He even accompanied Wagner on his zoological study trip to Trieste.[75] In 1853 Meissner discovered the tactile corpuscles in the skin which still carry his name, a feat which made him instantly famous but also provoked protracted quarrels regarding the priority of the disclosure with Wagner himself.[76] Meissner, however, did not forget to write in a dedicatory note to his previous teacher: "Thanks to you my accidental discovery has received the necessary meaning and significance."[77]

Meissner conceived the stimulation of these tactile corpuscles as pressure changes on the skin, which in turn triggered neural responses. After the retirement of Wagner, Meissner became his successor at Göttingen in 1860. He was a great friend and collaborator of the anatomist *Jakob Henle* (see Fig. No. 86) who was also active at the university. Between the years 1850-1872, Meissner wrote a series of excellent papers. Thereafter, until his death in 1905, he ceased publishing his work as a result of violent attacks by Ernst Brücke, Willy Kühne and especially Du Bois-Reymond.

Fig. 86. Jacob Henle (1809-1885).

Meissner not only discovered the tactile corpuscles but also the internal myenteric plexus. For years he was concerned with the physiology of proteins, and was known as a versatile and prominent biologist. Among his students were *Otto Weiss, Ludwig Thiry* (1817-1897) and *Heinrich Boruttau* (1869-1923). The latter wrote his so-called "Habilitationsschrift" or thesis which allowed him to begin his university teaching career in 1894 at Göttingen under the direction of Meissner. In 1907, Boruttau, who was primarily interested in medical physics — electro-diagnosis and electrotherapy, went to Berlin. He was also concerned with the history of physiology which he enriched with various articles and the first comprehensive narrative in the field: *Geschichte der Physiologie in ihrer Anwendung auf die Medizin bis zum Ende des 19. Jahrhunderts* (Jena, 1903), a work which was published in the historical textbooks of Puschmann, Neuburger and Pagel.

Heinrich Bidder, a native of Livonia, was already mentioned as a disciple of Müller. He was sent from the university of Dorpat to study under Müller in 1834-35. On his return, he was appointed professor of anatomy instead of Jakob Henle, a position which he later held for anatomy and physiology until 1869. Bidder's name is forever linked with that of the physiologist *Carl Schmidt* (1822-1894) through their well-known work about the chemical aspects of digestion and metabolism entitled *Die Verdauungssäfte und der Stoffwechsel, eine physiologisch-chemische Untersuchung* (Leipzig, 1852). Here, Bidder concluded that the excessive albumin was promptly metabolized in the organism, a process which he called "Luxuskonsumption".

In 1842, Bidder carried out some studies with Alfred W. Volkmann regarding the autonomy of the sympathetic nervous system, a work which belongs to the most significant publications in neurophysiology. Besides, he carried out several valuable microscopic observations about the anatomy of the retina, epithelial tissue and the structure of the bones. Finally, Bidder's description of his 1834 study trip to Germany and specifically Berlin is still a rich source of information on Müller's institute and the work being carried out there at the time.[78]

Reflecting his earlier predominantly morphological interests, Johannes Müller had a series of students which became famous for their anatomical work. In this regard, two very prominent figures should be mentioned here rather briefly: *Jakob Henle* and *Theodor Schwann,* both friends and collaborators of Müller for many years. Henle had met Müller in the house of the Zeiller family[79] and soon the talented student became a helper — "Gehülfe" — of the great physiologist. After a brilliant final examination for his medical degree, Henle accompanied Müller on a trip to Paris. In 1834, at Berlin, Henle was officially appointed Müller's prosector, a position he held while being persecuted because of his political beliefs and involvement in the German fraternities of students or "Burschenschaften". These activities also foiled his academic appointment at the University of Dorpat. Beginning in 1837, Henle actively carried out microscopical studies regarding the anatomy of epithelia, discovering, like Remak, the neuraxon.

Henle's friend Theodor Schwann[80] was even more fortunate in those years. Schwann discovered pepsin in 1836, and described a device to determine quantitatively the muscular force. He also considered the problem of spontaneous generation. In 1839, Schwann reached the apex of his anatomical career with the publication of his famous work *Mikroskopische Untersuchungen über die Übereinstimmung in der Struktur und dem Wachstum der Thiere und Pflanzen.* In this book he clearly expressed and demonstrated for the first time the idea that the cell was the fundamental unit of plants and animals, a concept previously alluded to by the French naturalist René J.H. Dutrochet in 1820.[81]

Schwann's great religious piety, reflected in his autobiography,[82] gave rise to a situation unique in the annals of the history of science: Schwann refused to publish his work until he had obtained the official approval or "imprimatur" from the pertinent ecclesiastical authorities. In 1839, Schwann went as an anatomist and physiologist to Löwen, followed by a call to teach physiology at Liège in 1858. In those years, he only published one work about the effects of bile on the digestive process. Many other investigations simply remained unpublished. A series of accidents in the Belgian mines prompted Schwann to design a portable apparatus for artificial respiration which could be used on the unfortunate workers. He spent his last years attempting to improve the device.[83]

In 1840, Henle also left Berlin in order to assume the anatomist *Philipp F. Arnold's* (1803-1890) position in Zürich,[84] followed by R.A. von Kölliker, who became Henle's prosector there. Henle wrote his important work entitled *Allgemeine Anatomie* in Zürich, in which he successfully related histology to the cell-theory. Some years later, in 1844, Henle transferred to Heidelberg where he became a great attraction for many students together with his friend, the physician *Karl von Pfeuffer* (1806-1869), an editor of the journal *Zeitschrift für rationelle Medizin.*

At Heidelberg, Henle began his work on the textbook entitled *Handbuch der rationellen Pathologie,* which he later completed at the University of Göttingen in 1855. The book was actually the first systematic attempt to write a genuine physiopathology, and the publication contributed to better relationships between physiology and clinical medicine, adding greatly to the fame of its author. In 1882, Henle celebrated his fiftieth anniversary as a physician amid great expressions of sympathy and recognition. He died three years later at the age of seventy-six.[85]

From the prominent Schultze family came another of Müller's disciples, the physiologist *Max J.S. Schultze* (1825-1874), a son of the anatomist Carl A.S. Schultze of Freiburg and Greifswald. After studying for some years with his father in Greifswald, Max Schultze went to Halle and Bonn as an anatomist, becoming one of the early researchers in general gross, and microscopical anatomy, and one of the founders of the cell-theory. Schultze's physiological investigations encompassed the first description of the blood platelets, the introduction of the concept of physiological body fluids, as well as the discovery that the retina of diurnal animals had a predominance of cones while the nocturnal ones had more retinal rods.

The anatomist *Robert Remak* (1815-1865) also spent a few years with Müller at the Berlin Institute. The son of a Jewish merchant in Posen[86] and a student of Schwann, Remak received in Berlin a series of valuable technical suggestions and acquired a great love for microscopical anatomy. His observations on the structure of the nerve-fibres, the axis-cylinder or "Primitivband", and the intestinal neural pathways were of great significance. In 1837, Remak demonstrated that the neurofibres originated in the ganglionar cells, proving experimentally that there was no extracellular origin for new cells, an aspect which made him a forerunner of Virchow. Remak also initiated the concept of three germinal layers in embryological development.

Another well-known physician who belonged to this generation of students of Müller was *Rudolf Virchow* (1821-1902) who acquired world fame with his dictum "omnis cellula e cellula" - every cell derives from another cell - and applied the cell-theory to pathology. After studying with Helmholtz at the Pépinière in Berlin, Virchow worked as an assistant to the anatomist *Robert Froriep* (1804-1861) and was later a prosector at the Charité Hospital. He was dismissed from this post in 1849 because of his participation in the revolutionary events of the preceding year, receiving an appointment as professor of pathological anatomy at the University of Würzburg.[87]

The following years, until 1856, were for Virchow the most fruitful years of his scientific career. The result of his efforts was the publication of a book concerned with cellular pathology, in 1858, which truly began a new chapter in the history of pathology. Virchow was also one of the founders of the famous Physico-Medical Society of Würzburg, which celebrated its 100th anniversary in 1950. Virchow also edited a journal, the *Archiv für patholoische Anatomie, Physiologie und für klinische Medizin,* which became the setting for the new developments in cellular pathology. Many of his articles were concerned with physio-pathological questions, reflecting his growing recognition that physiology was the basis for pathology as well as clinical medicine.[88]

The reason for the largely morphological character of Virchow's cellular pathology should be attributed to the lack of a proper contemporary cellular physiology.[89] Although the decade between 1830 and 1840 had seen notable improvements both in microscopic methodology and morphology, this new knowledge only gradually acquired more significance after 1840, when it started to be used for the elucidation of functional questions. Thus, the English physiologist *William Bowman* described for the first time in 1840-41 the minute structure and functions of the striated, voluntary muscle, discovering in 1842 the relationships between the glomeruli and the urinary tubules. Conversely, Ernst Brücke observed the actual contraction of intestinal villi only in 1853, and the English author Thomas Wharton Jones the ameboid movement of the leucocytes in 1853.

In addition, the Italian physiologist *Alfonso Corti* (1822-1888) discovered the cells in the inner ear which carry his name, while *Paul Langerhans* (1847-1888) disclosed the existence of the pancreatic islet-cells in 1869, and Heidenhain the gastric cells. However, no investigator actually dealt with living cells, a necessary requirement for a genuine study of cellular physiology. Only with the advent of the physiologist *Max Verworn* (1863 1923) did such a direction in physiology come into being.

Virchow also expressed his opinion about the crucial contemporary problems of mechanism and materialism in a series of speeches and articles dealing with general topics. This was especially true after he became the target of many attacks such as the article by the German botanist *Matthias Schleiden* (1804-1881) about the materialism prevalent in German science: *Über den Materialismus der neuen deutschen Naturwissenschaft, sein Wesen und seine Geschichte* (Leipzig, 1863). Virchow's response to these accusations has been the subject of a careful analysis.[90]

Another author who should be included in this review is *Ernst Haeckel* who deeply influenced scientists and laymen alike in the second half of the last century with his materialistic monism and the work entitled *Welträtsel* (1899).[91] Haeckel was one of Müller's students who greatly venerated his teacher but departed greatly from the philosophical position espoused by Müller. His excellent zoological work had already given him a good name in scientific circles at an early age, and he was in some instances directly following Müller's valuable suggestions. Haeckel was also one of the first renowned authors on the continent to publicize Darwin's ideas on the origin of the species and apply these concepts "biogenetically" in the works *Generelle Morphologie der Organismen* (Berlin, 1866) and *Natürliche Schöpfungsgeschichte* (Berlin, 1868). The establishment of ancestral lines in the gradual evolution of life, including that of man, evoked both great enthusiasm as well as violent opposition. Pertinent and irrelevant criticisms were especially evident after Haeckel's general work *Welträtsel,* in which he unveiled rather strong anti-Christian tendencies in his system of monistic philosophy. However, Haeckel himself realized the limitations and weaknesses of his philosophical arguments, and thus one cannot deny his basic honesty and respect his efforts to achieve a higher conceptual synthesis. Max Verworn, who called himself a student of Haeckel,[92] followed many of Haeckel's suggestions and expressed clearly his teacher's irreproachable opinions.

7.
THE SCHOOL OF CARL LUDWIG

During the years 1865-1895, Carl Ludwig's Institute in Leipzig became the meeting place and training center of a young generation of physiologists from all parts of the globe. Such an international group of distinguished students was only duplicated to some extent by Claude Bernard and his disciples in Paris.

Fig. 87. Adolf Fick (1829-1901).

One of Ludwig's most prominent disciples was *Adolf Fick* (See Fig. No. 87). Born on September 3, 1829 in Kassel, Fick was the offspring of a highly talented family.[93] After completing his basic schooling, he went to study mathematics at the University of Marburg, where he soon changed to medicine. Thanks to his mathematical abilities, he was ideally suited for solving certain physiological questions, such as muscular mechanics and physiological optics. At Marburg, Adolf Fick soon became a friend of the young professor Carl Ludwig, then a prosector to Fick's older brother. Ludwig recognized the potential capacities of the younger Fick who, initially at least, seemed more predisposed to loafing than working. During his stay in Berlin, Fick came into contact with both Müller and Du Bois-Reymond. Although Müller impressed him greatly, the dean of German physiology was unable to stimulate significantly Fick's physico-mathematical approach.

After returning to Marburg (Ludwig had already left for Zürich), Fick graduated with a physiological thesis entitled *Tractatus de errore optico,* and became a prosector for his brother Ludwig Fick. In 1852, he followed Ludwig to Zürich where he began his teaching career. After Ludwig's departure from that city, he worked under Moleschott until 1862, the year in which he was appointed full professor because of Moleschott's transfer to the University of Turin.

For sixteen years Fick worked successfully in Zürich. Among his achievements was the development of the law of diffusions in 1855.[94] The suggestions for this investigation had come primarily from Ludwig's research concerning the mechanism of secretion. In 1865, together with Wislicenus, he climbed one of the nearby mountains in order to test Liebig's assertion that albumin was the source of muscular energy.

Fick was also the first to apply portions of the second law of thermodynamics to a physiological problem: the muscular contraction. In 1869, probably aided by his friend Rudolf Clausius, Fick calculated the efficiency of muscular work through an experimental determination of the relationship between useful free energy and the total convertible power. He found the values to be excessive, thwarting any plausible thermodynamic explanation of muscular activity.

In Zürich, Fick also carried out important investigations on the necessary conditions for electrical stimulation. He clearly recognized the need for a certain duration of the electrical current in addition to a given intensity in order to achieve the desired effect. Moreover, medical physics, kinetics, muscular work, dioptrics and related fields prompted his attention.

In 1868, Fick went to the University of Würzburg where he worked for another fruitful thirty-one years both as a teacher and investigator, retiring in 1899 at the age of seventy. Many single discoveries and methodological improvements, which cannot all be mentioned here, were the result of Fick's indefatigable labours.[95] Among them was the description of the so-called "Fick Principle", a calculation of the one-minute volume of circulating blood from the actual consumption of oxygen and the arteriovenous difference of oxygen concentration (1870). Finally, he should also be remembered for his method of plethysmography at the level of the arms, a technique developed in 1869.

Beyond the strict scientific activities, Fick was concerned with a variety of issues such as education, alcoholic consumption, dress and other topics which revealed his great social regard. In this connection, he wrote several articles on political and philosophical questions. He died suddenly on August 21, 1901 while vacationing in Blankenberghe.

A series of talented disciples received valuable suggestions from Fick. Among them were *Johannes Gad, Magnus G. Blix* (1849-1904), *Jacques Loeb* (1859-1924), *August Gürber* (1864-1937), *Charles Ernest Overton* (1865-1933) and especially *Friedrich Schenk* (1862-1914) and the pharmacologist *Rudolf Böhm* (1844-1926). Loeb (see Fig. No. 88) was Fick's assistant for some time at Würzburg before going to study with Götz at the University of Strassburg, and later, in 1891, to the United States.

Fig. 88. Jacques Loeb (1859-1924).

There he became professor of physiology at the University of Chicago in 1900, and two years later, the University of California. Next to Verworn, Loeb was one of the most successful early workers in the field of general physiology. His work on heliotropism, the mechanics of growth, artificial parthenogenesis, the effects of salts and colloidal interpretation of certain vital functions were epoch-making in physiology. These discoveries were published in numerous articles and well-read monographs such as *Einleitung in die vergleichende Hirnphysiologie,* (introduction to comparative cerebral physiology) (Leipzig, 1899) and *Die Eiweisskörper* (nitrogenous substances) (Berlin, 1924).

In 1890, *Friedrich Schenk* transferred from Bonn, where he had been Pflüger's assistant, to the University of Würzburg.[96] For eleven years he worked together with Fick and the two men became life-long personal friends as well as scientific collaborators. Schenk had spent most of his early years studying at the University of Bonn where he worked with Pflüger and qualified for a university teaching career with a dissertation concerned with the urinary synthesis in the animal organism. This work was followed by some investigations on the variations of sugar in the blood.

In search of a more independent development of his scientific personality, a goal which was impossible under Pflüger's intellectual dictatorship,[97] Schenck went to Würzburg in 1890. Following some of the suggestions and concepts of Pflüger and Fick, Schenck carried out a series of experiments concerned with muscular physiology. In 1901, he became the successor of Kossel at the University of Marburg where he remained until his death in 1914 as a rather popular teacher and researcher.

Finally, from the circle of Fick's students, *Ernest Overton* should also be mentioned. He became famous for his investigations concerning the action of salts on the muscle and the theory of lipid involvement in narcosis.

Another collaborator of Ludwig at Marburg was *Conrad Eckhard.* Ludwig's influence on Eckhard's career was especially marked. Among the other young people who came together with both Ludwig and Brücke in Vienna were *Willy Kühne* and *Ivan M. Sechenov,* the first Russian to receive his training in physiology in Germany and Austria. Sechenov also studied under Rollet at Graz, and his primary field of interest was the physiology of reflexes and the central nervous system, as evidenced in the work *Physiologische Studien über die Hemmungsmechanismen füdie Reflextätigkeit des Rückenmarks im Gehirn des Frosches* (Berlin, 1863). Together with *Victor Pachutin* (1845-1901), also a Russian physiologist, Sechenov published *Neue Versuche am Hirn und Rückenmark des Frosches.* Later he wrote a textbook of physiology in Russian. The most famous Russian disciple of Ludwig, Pavlov, will be mentioned in more detail later.

After Ludwig had been in charge of the Physiological Institute in Leipzig for two years, another student, who later achieved prominence in the field, arrived at the laboratory: *Hugo Kronecker* (see Fig. No. 89). Kronecker had already displayed excellent talents in college for both mathematical and scientific subjects, and after completing his studies in Berlin, Heidelberg and Pisa, graduated in 1863. Moreover, he had also worked at the University of Heidelberg with Helmholtz and *Wilhelm Wundt* (1832-1920), the latter a prominent physiologist who transferred a few years later to the University of Zürich.

Fig. 89. Hugo Kronecker (1839-1914).

In 1868, Kronecker went to Leipzig where in 1872 he completed his "Habilitationsschrift" or thesis for an academic teaching career entitled "Über die Ermüdung und Erholung quergestreifter Muskeln" (concerning the fatigue and recovery of striated muscles). He remained at Leipzig for ten years, and in 1878 was named section chief at Du Bois-Reymond's Institute at Berlin. Later, in 1885, Kronecker accepted the chair of physiology at the University of Bern, a position which had been previously held by Gabriel G. Valentin and Paul Grützner. He was able to have a new physiological institute erected there which, upon his suggestion, received the name of "Hallerianum" in honor of Albrecht von Haller, the great eighteenth-century physiologist from Bern. Kronecker directed the institution until his death on June 6, 1914.

The years which Kronecker spent in Leipzig with Ludwig were not only highly productive but a period of fruitful and stimulating intellectual exchanges with the many foreign visitors who came to work with Ludwig for several months or years. Kronecker, who spoke fluent French, English and Italian, was greatly instrumental in forging an international community of scholars interested in physiology (see Fig. No. 90). His kindness, helpfulness and scientific suggestions were appreciated, and he became a lifelong friend of such prominent men as Schwalbe, Hüffner, Gaule, von Kries, Bowditch, Heger, Holmgren, Gaskell, Tigerstedt and others. Later, in Bern, Kronecker's international connections were successfully maintained and instrumental in calling the first *International Congress of Physiologists* which met in Bern under the presidency of Friethjof Holmgren. With 124 participating scholars, the meeting was held in Miescher's Institute in September, 1889.

In 1906, Kronecker was elected to the presidency of the Marey Institute in Paris, which was dedicated to the control and development of physiological methods. Its founder, the skilled French investigator *Etienne J. Marey* (1830-1904), was a longtime friend of Kronecker.

Kronecker was also personally involved in the establishment of an international station on Mt. Rosa to study the living conditions prevailing at high altitudes. He himself carried out several observations regarding the causes of mountain sickness. In truth, Kronecker explored and worked with his students on problems concerned with virtually every area of physiology such as vascular innervation, cardiac poisons, irritability and functional capacity of the heart, muscular movements, reflexes, etc. His talent for devising the proper methods and experiments provided physiology with many valuable directions and auxiliary methods of inquiry. Finally, Kronecker originated the widely employed calibration of inductors according to Kronecker units as well as the cardiac manometer still in use.

Among Kronecker's students in Berlin was *Samuel J. Meltzer* (1851-1920) who later emigrated to the United States where he worked with Welch in New York and Curtis at the physiological institute of Columbia University. In 1907, Meltzer was appointed section chief for physiology

Fig. 90. Ludwig's Physiological Institute at Leipzig around 1880.

and pharmacology at the newly founded Rockefeller Institute. The well-known "Kronecker-Meltzer theory" of the swallowing process was a reflection of the early collaboration between teacher and student.

Somewhat later, in Bern, *Leon Asher* (1865-1943) became one of Kronecker's disciples. Originally from Leipzig where he had studied and been one of the last assistants of Ludwig, Asher pursued further studies at Heidelberg with Willy Kühne. In 1894, he came to Bern where he completed his "Habilitationsschrift" under Kronecker the following year. After Kronecker's death in 1914, Asher was unanimously appointed to succeed his teacher, a post which he held until 1936. He belonged to a generation of scientists who could still survey the entire field of physiology and contribute with original works to virtually all of the various branches of the discipline.

Asher was one of the founders of the publication *Ergebnisse der Physiologie* which he edited for forty-one years together with Karl A. Spiro until his death during World War II. Asher was a versatile and original author. His first works concerning the formation of lymphatic fluid brought him international fame. Moreover, he was also interested in glandular secretion, sensory physiology and especially physiological optics. His successor, Alexander von Muralt, declared in an obituary note that "science had lost in Leon Asher one of the last representatives of the great classical tradition in physiology, and a true promoter of the physiological development in all its fields."[98]

Ludwig's interest in chemical questions prompted him to have a person trained in chemistry at the institute who could carry out chemico-physiological investigations. Prominent among those scientists were *Gustav Hüfner* (1840-1908) and *Edmund Drechsel* (1843-1897). Hüfner had received special training in chemical methodology and physical chemistry from the renowned German chemist *Robert Bunsen.* Soon after his graduation, in 1866, Hüfner came to Leipzig and was appointed chief of the chemico-physiological section of the Institute. Later, in 1872, he succeeded *Felix Hoppe-Seyler* (1825-1895) at the University of Tübingen and in 1886, went on to become director there of an independent physiological and chemical laboratory. Hüfner's lifelong work was dedicated to the elucidation of the role played by hemoglobin and its relationships with oxygen and carbonic acid, a field in which he made important discoveries. One of his students, *Karl Bürker* (1872-1957), also studied intensively the physiology of the blood.[99]

Fig. 91. Edmund Drechsel (1843-1897).

Edmund Drechsel (see Fig. No. 91) became Hüfner's successor at Ludwig's institute in 1872. Born at Leipzig, Drechsel received his chemical education with Otto L. Erdmann and Adolph W. Kolbe. He remained at Leipzig for twenty years and influenced decisively the development of chemical physiology there. Later, he was appointed to the chair of medical and physiological chemistry at the University of Bern, receiving at the same time a position as professor of pharmacology. Endowed with an excellent and analytical mind, Drechsel clarified the formation of urea in the organism, and worked on subjects of electrolysis and electrosynthesis. He also studied the lipid catabolism, the extraction of crystallized combinations of proteins and discovered lysin. Drechsel died of a stroke in September, 1897.

The list of German disciples of Ludwig who went on to occupy professorships in physiology cannot be complete without mentioning *Max von Frey* (1852-1932), *Max Rubner* (1854-1932) and *Otto Frank* (1865-1944).

Both Rubner and Frank will be cited in a subsequent section. Max von Frey (see Fig. No. 92) was born on November 16, 1852 in Salzburg, and came into contact with Ludwig's laboratory and physiology during his early studies in Leipzig. Still a student, he solved an important physiological problem which had been given to him by Ludwig himself. Thus, in 1874, he determined that vasodilation and vasoconstriction were two independent and antagonistic actions, a discovery which launched his career in physiology. Frey returned to Ludwig's laboratory in 1880 after receiving intensive training in physics, mathematics and medicine. He remained there for seventeen years until the death of Ludwig in 1895, agreeing to stay another two years thereafter under Ludwig's successor, *Ewald Hering* (1834-1918).

Fig. 92 Max von Frey (1852-1932).

Frey's temperament and mentality did not harmonize with Ludwig's character, and the two persons were not capable of a congenial relationship[100]. Frey was more cautious and careful than his more impulsive master. In 1880, Frey studied the chemical phenomena which occurred during the activities of an isolated mammalian muscle, an investigation which required a highly sophisticated methodology. This research led to the rather remarkable discovery that the muscle used most of its oxygen after the muscular work had already been performed. Frey related these results to the formation of lactic acid, and thus created new horizons for the research of muscular chemistry.

In 1897, Frey became professor at the University of Zürich for two years before accepting the directorship of the physiological institute at the University of Würzburg as a successor of Adolf Fick. For several decades he continued to work and teach until his death on January 25, 1932.

Frey was primarily concerned with three major areas of physiological activity. The first one, muscular physiology, has already been alluded to. Secondly, Frey was instrumental in registering the pulse wave, dealing with various registration systems and their specifications before the work of Otto Frank, as reflected in the work *Untersuchung des Pulses* (1891). Finally, Frey's third field of research was the sense of touch. He tried to establish precise and quantitative stimulation procedures in order to distinguish sharply the various sensory qualities. Frey searched for the specific receptors and their distribution, eventually achieving a great mastery in this area of physiological investigation. In the process, he invented a fair number of ingenious devices which were useful for these studies. Frey's tall figure, his personal warmth and kindness, will always be remembered by his many students.[101] Among those who still occupy chairs of physiology in Germany are Paul Hoffmann, Hermann Rein, Hans Schriever and Hubertus Strughold, while Edgar Wőhlisch died in 1960. Other well known scientists who worked at some time or other with Frey were Rudolf Metzner, Theodor Fahr, Ernest Overton and Fritz Schellong.

Following this review of Ludwig's German disciples, attention should be drawn to some of Ludwig's prominent foreign guests who worked at the Leipzig Institute. As already stated, Ludwig's laboratory between 1865 and 1895 was a busy international meeting place for students interested in physiology. There were *Alarik F. Holmgren, Olof Hammarsten* and *Christian Lovén* (1835-1904) from Sweden and *Christian Bohr* from Denmark, a scientist who received the initial suggestions for his successful analysis of blood gases in Leipzig. In addition, there were *Robert A. Tigerstedt* of Finland, and *Luigi Luciani* and *Angelo Mosso* from Italy. Ludwig's old Russian friend, the physiologist *Ivan M. Sechenov,* has already been mentioned together with *Victor M. Pachutin.* Above all was the mention of *Ivan P. Pavlov* (1849-1936) who worked with Ludwig in 1885-86. Recently, more than twenty Russian collaborators of Ludwig have been compiled, but the development of Russian physiology will be traced in a subsequent section.

English and American students were also active in Ludwig's laboratory. Among those who later rose to prominence were *Walter H. Gaskell, Sir Victor Horsley* (1857-1916), *Sir Thomas Lauder-Brunton, William Stirling, Henry P. Bowditch, Edward Sharpey-Schaefer* and *Leonard C. Wooldridge* (*1889). There was virtually no contact between Ludwig and the representatives of French physiology. Claude Bernard rejected Ludwig's quantitative and physical approach, and only through the Russian physiologist Elie de Cyon was there any personal contact between Leipzig and Paris.

Italian physiology was also greatly influenced and revitalized by Ludwig's group.[102] The excellent tradition of Italian physiological and medical research, connected with men of the stature of Galvani, Spallanzani, Fontana, Rolando and Panizza, had been temporarily interrupted with the exception of Carlo Matteucci. However, the resurgence of Italian physiology only occurred near the middle of the nineteenth century and its main representatives were Jacob Moleschott, Moritz Schiff, Angelo Mosso, Carlo Matteucci and especially Luigi Luciani.

Luciani and Mosso were both students of Carl Ludwig. Luciani (see Fig.

No. 93) was born on November 23, 1840 in Ascoli Piceno and studied at Bologna and Naples between 1862-1868. In the latter year he became an assistant in the physiological institute of the University of Bologna, and from March to November 1873 he visited Ludwig's laboratory at Leipzig.[103] In his unpublished autobiography Luciani wrote: "This sojourn (at Leipzig) was really the most important period in my scientific career which left me with deep and indelible impressions. With a feeling of gratitude and proper justice, I recognize that Carl Ludwig was my real teacher."

Fig. 93. Luigi Luciani (1840-1919).

During his earlier work at Bologna in 1871, Luciani had already established the novel and challenging concept of an "active diastole". Later in Leipzig he discovered and analysed the so-called "Luciani periods" in the frog's heart (1873) which he then related to the periodical Cheyne-Stokes respirations.

After his return from Germany, Luciani was appointed professor of general pathology at the University of Parma in 1875. Five years later he began to teach physiology at Siena, transferring thereafter to Florence where he lectured in 1882-1893. From 1893 until 1914, Luciani worked in the Roman "Sapienza" as the successor of Moleschott. He died, already emeritus, in June 1919.

Luciani greatly advanced two major areas in physiology with his brilliant investigations: the physiology of hunger, carried out in 1889 with a fasting expert named Succi, and cerebellar physiology (1891). His surgical interventions on the cerebellum of dogs and monkeys were among the first successful extirpations carried out on higher animals.

Another of Luciani's great contributions was his five volume textbook on human physiology which was translated into German, English and Spanish. He seems to have been guided, both in his teachings and research, by the idea that physicians must be directed toward thinking in physiological terms, since he was convinced of the unity of medical science and especially the close association between physiology and pathology.

Angelo Mosso (see Fig. No. 94) was known beyond the small group of physiologists because of his series of general and popular works concern-

Fig. 94. Angelo Mosso (1846-1910).

ing actual physiological subjects. Among them was the monograph entitled *Ermüdung* (fatigue), (Leipzig, 1892), and *Die Furcht* (fear). (Leipzig, 1899). The Italian editions of these publications were translated into many languages. Mosso was born on May 31, 1846, in Turin, and after studying under Moleschott and Schiff, he left his native country in order to continue his training with Bernard, Marey, Chauveau and Ludwig, in whose laboratory at Leipzig he received the decisive influences which later shaped his entire career.[104] After 1879, Mosso worked as the successor of Moleschott and became director of the physiological institute in Turin.

A series of successful methodological innovations had their origin in Turin, such as Mosso's ergograph, plethysmograph and myotonometer. Based on one of his suggestions, a laboratory was erected on the south side of Monte Rosa and dedicated to the investigation of high-altitude effects on man, one of Mosso's special fields of endeavor as reflected in his work *Man in the Alps*.

For the first time, Mosso studied systematically the physiology of physical exercise. His research concerning cerebral circulation and the toxicity of parenterally administered eel's blood as a foreign protein in mammals were other fields in which Mosso excelled. In his laboratory he surrounded himself with a great number of capable collaborators.

In this contex, two other prominent Italian physiologists deserve mention: *Filippo Botazzi* (1867-1941) and *Frédéric Batelli* (1867-1941). Botazzi's research was concerned with splenic functions, osmosis, the release of potassium with vagal stimulation and muscular physiology.[105] He also examined with special empathy the significance of Leonardo da Vinci in physiology. On the other hand, Batelli's interests were in chemico-physiological problems. A student of Schiff and Prevost, he became the latter's successor at Geneva in 1913.

Thanks to Carl Ludwig, there were also close scientific ties between Germany and the German-speaking portion of Switzerland during the second half of the nineteenth century. Carl Ludwig himself, like Adolf Fick, had begun his own career in Zürich. Moreover, *Wilhelm His Sr.* (1831-

1904) succeeded Georg Meissner at Basel and was followed by Ludwig's student *Johannes F. Miescher* (1844-1895). Miescher, who had also worked under Wöhler and Hoppe-Seyler, should be considered one of the founders of the new chemical physiology, a scientist who attempted to relate histology and biochemistry.[106]

Wilhelm His Jr. (1863-1934) graduated at the University of Leipzig in 1889. Four years later he discovered the bundle of fibers which still carry his name in the embryonic heart, describing the effects which an interruption of this pathway produced on the coordinated function between the auricles and ventricles (1894). Later, His became professor of internal medicine in Berlin.

Georg J. Gaule (1849-1939), originally from Darmstadt, had trained with both Ludwig and Virchow. Between 1878-1885 he was head of the histological section in Ludwig's Institute. A year later Gaule succeeded to the chair of physiology at the University of Zürich after Luchsinger's tenure, a post which he held until 1916. He had a rather ambivalent relationship with Ludwig. In truth, Ludwig was no friend of morphology while Gaule tried to achieve a concept of cellular physiology and elucidate the role of its microscopical structures. Gaule viewed the cell as the lowest unity of life and was more concerned with the unity of the vital process than Claude Bernard, and less impressed with Ludwig's stress on the work of single organic parts.

REFERENCES

[1]For biographical details see:
W. Haberling, *Johannes Müller, das Leben des rheinischen Naturforschers,* Leipzig, 1924 and U. Ebbecke, *Johannes Müller der grosse rheinische Physiologe,* Bonn, 1951.

[2]J. Müller, Gedächtnisrede auf Carl Asmund Rudolphi 1835," *Abh. Preuss. Akad. Wiss Berlin,* 1837, p. 23.

[3]M. Müller, *Über die philosophischen Anschauungen des Naturforschers Johannes Müller,* Leipzig, 1927.

[4]K.E. Rothschuh, "Johannes Müller and Carl Ludwig," *Dtsch. Med. Wschr.* 78 (1953), pp. 71-74.

[5]E. Du Bois-Reymond, "Gedächtnisrede auf Johannes Müller," *Abh. d. Bln. Akad. d. Wiss.,* Berlin, 1860.

[6]M. Müller, "Über die philosophischen Anschauugen des Naturforschers Johannes Müller, über Erkenntnistheorie und Methodologie," *Arch. Gesch. Med.* 18 (1926), pp. 130, 209, 328.

[7]W. Riese - G.E. Arrington, Jr., "The history of Johannes Müller's doctrine of the specific energies of the senses: original and later versions," *Bull. Hist. Med.* 37 (1963), pp. 179-183.

[8]J. Steudel, "Wissenschaftslehre und Forschungsmethodik Joh. Müllers," *Dtsch. Med. Wsch.* (1952), p. 115.

[9]F. Bidder, "Vor hundert Jahren im Laboratorium Johannes Müllers," *Münch. Med. Wschr.* (1934), p. 50. Also
Gottfried Koller, *Das Leben des Biologen Johannes Müller (1801-1858)*, Stuttgart, 1958.

[10]Rudolph Virchow, "Johann Müller, the physiologist: a eulogy pronounced in the hall of the University of Berlin," transl. and ed. by A.M. Adam, *Edin. Med. J.* 4 (1858), pp. 452-463 & 527-544.

[11]Johannes Müller, *Elements of Physiology*, transl. from German by W. Baley, London, 1837.

[12]Estelle Du Bois-Reymond, *Jugendbriefe von Emil Du Bois-Reymond an Eduard Hallmann*, Berlin, 1918.

[13]W. Haberling, "Alexander von Humboldt and Johannes Müller. With hitherto unpublished letters of Humboldt," *Med. Life* 35 (1928), pp. 529-536.

[14]R.N.S. Rustan, "How a physiologist anticipated a physical chemist," *Bull. Hist. Med.* 14 (1943), pp. 352-365.

[15]E. Bauereisen, "Carl Ludwig as the founder of modern physiology," *Physiologist* 5 (1962), pp. 293-299.

[16]H. Rein, "Dem dtsch. Physiologen Carl Ludwig z. Gedächtnis (Rede 1936)," *Klin. Wschr.* (1936), p. 1887.

[17]H.E. Hoff - L.A. Geddes, "Graphic registration before Ludwig; the antecedents of the cymograph," *Isis* 50 (1959), pp. 1-21.

[18]Erna Lesky, *Die Wiener Medizinische Schule im 19. Jahrhundert*, Graz-Köln, 1965.

[19]H.E. Hoff - L.A. Geddes, "Graphic recording before Carl Ludwig: an historical summary," *Arch. Int. Hist. Sci.* 12 (1959), pp. 3-25.

[20]G. Rosen, "Carl Ludwig and his American students," *Bull. Hist. Med.* 4 (1936), pp. 609-650.

[21]W.P. Lombard, "The life and work of Carl Ludwig," *Science* 44 (1916), pp. 363-375.

[22]H. Schröer, *Carl Ludwig. Begründer der messenden Experimentalphysiologie*, Stuttgart, 1967.

[23]K.E. Rothschuh, "Carl Ludwig (1816-1895) Gestalt und Bildnis des grössten deutschen Physiologen," *Z. Kreislaufforsch.* 49 (1960), pp. 2-19.

[24]P.F. Cranefield, "The organic physics of 1847 and the biophysics of today," *J. Hist. Med.* 12 (1957), pp. 407-423 and
E. Mendelsohn, "Physical models and physiological concepts: explanation in 19th century biology," in *Boston Studies in Philosophy of Science*, ed. by R.S. Cohen & M.W. Wartofsky, New York, 1965, pp. 127-150.

[25]P.F. Cranefield, "The philosophical and cultural interests of the biophysics movement of 1847," *J. Hist. Med.* 21 (1966), pp. 1-7.

[26]O. Temkin, "Materialism in French and German physiology of the early 19th century," *Bull. Hist. Med.* 20 (1946), pp. 322-327.

[27]K.E. Rothschuh, "Johannes Müller und Carl Ludwig," *Dtsch. Med. Wschr.* 78 (1953), pp. 71-74.

[28]H. Schröer, *Carl Ludwig*, Stuttgart, 1967.

[29]Leo Königsberger, *Hermann von Helmholtz*, transl. by F.A. Welby, New York, 1965.

[30]J. Reiner, *H. v. Helmholtz* (=Klass. d. Naturwiss. Band 6) Leipzig, 1905.

[31]E. Du Bois-Reymond, "Hermann von Helmholtz. Gedächtnisrede," *Abh. d. Bln. Akad. d. Wiss.*, Berlin, 1896.

[32]H.E. Hoff - L.A. Geddes, "Ballistics and the instrumentation of physiology: the velocity of the projectile and of the nerve impulse," *J. Hist. Med.* 15 (1960), pp. 133-146.

[33]J. Bernstein, "H. v. Helmholtz," *Naturwiss. Rundschau* 10 (1895), p. 73.

[34]Hermann v. Helmholtz, "On thought in medicine," (Das Denken in der Medizin), introd, by A.B. Luckhardt, *Bull. Hist. Med.* 6 (1938), pp. 117-143.

[35]Hermann L.F. von Helmholtz, *On the Sensations of Tone as a Physiological Basis for the Theory of Music,* New York, 1954. See also
F.W. Sunderman, "Medicine, music and academia," *Trans. Stud. Coll. Phys. Phila.* 37 (1969), pp. 140-148.

[36]Hermann L.F. von Helmholtz, *Treatise on Physiological Optics,* ed. & transl. by J.P.C. Southhall, New York, 1962.

[37]R.M. Warren - R.P. Warren, eds., *Helmholtz on Perception; its Physiology and Development,* New York, 1968.

[38]Leo Köningsberger, *Hermann von Helmholtz,* p. 40.

[39]A.E. Woodruff, "The contributions of Hermann von Helmholtz to electrodynamics," *Isis* 59 (1968), pp. 300-311.

[40]See biographical sketches by E. Abderhalden in *Med. Klinik* 9 (1917) and A. v. Tschermak in *Pflügers Arch.* 174 (1919), p. 1.

[41]Johannes von Kries, *Die Medizin der Gegenwart in Selbstdarstellungen,* Leipzig, 1925, IV, p. 125.

[42]J. v. Kries, "Helmholtz als Physiologe," *Naturwiss* (1921), p. 673.

[43]See biographical sketch by E. Schütz in *Erg. Physiol.* 46 (1950), p. 6.

[44]See biography by H. Ebbinghaus in *Z. Psychol. u. Physiol. d. Sinnesorg.* 27 (1902), p. 145.

[45]See biography by K.E. Rothschuh in *Dict. Scientif. Biogr.* 4 (1971), pp. 200-205.

[46]Heinrich Boruttau, *Emil Du Bois-Reymond,* München, 1922.

[47]Estelle Du Bois-Reymond, *Jugendbriefe von Emil Du Bois-Reymond an Eduard Hallmann,* Berlin, 1918.

[48]E. Metze, *Emil Du Bois-Reymond. Lebensbild eines Naturforschers,* Brackwede, 1912. and E.Metze, *Emil Du Bois-Reymond. Sein Wirken und seine Weltanschauung.* Bielefeld, 1918.

[49]G. Moruzzi, "The electrophysiological work of Carlo Matteucci," *La Storia della Neurologia Italiana,* ed. L. Belloni, Milano, 1963, pp. 139-147, and
G. Moruzzi, "L'opera elettrofisiologica di Carlo Matteucci," *Physis* 6 (1964), pp. 101-140.

[50]K.E. Rothschuh, "Emil Du Bois-Reymond und die Elektrophysiologie der Nerven," in *Von Boerhaave bis Berger,* ed. by K.E. Rothschuh, Stuttgart, 1964, pp. 85-105.

[51]E. Du Bois-Reymond, *Reden in 2 Bänden,* Leipzig, 1912.

[52]Estelle Du Bois-Reymond und P. Diepgen, *Zwei grosse Naturforscher des 19. Jahrhunderts. Ein Briefwechsel zwischen Emil Du Bois-Reymond und Carl Ludwig,* Leipzig, 1927.

[53]J. Marseille, "Das physiologische Lebenswerk von Emil Du Bois-Reymond mit besonderer Berücksichtigung seiner Schüler," (unpubl. med. diss., Münster, 1968).

[54]S.A. Cesnakova und M. Lindemann, "Russische Physiologen in den Laboratorien von Carl Ludwig," *Z. d. Karl Marx Univ. Leipzig. Math. Naturwiss.* 19 (1970), pp. 315-322.

[55]R. Herrlinger - J. Krupp, *Albert von Bezold. 1836-1868. Ein Pionier der Kardiologie,* Stuttgart, 1964.

[56]For a biographical sketch see E. Cyon in *Pflügers Arch.* 132 (1910), p. 1.

[57]C.A. Culotta, "Tissue oxidation and theoretical physiology: Bernard, Ludwig, Pflüger," *Bull. Hist. Med.* 44 (1970), pp. 109-140.

[58]See obituary by R. Du Bois-Reyond in *Berl. Klin. Wschr.* (1910), p. 658 and the article by E. Heischkel-Artelt in *Hessische Lebensbilder* 4 (1950), p. 253.

[59]M. Nussbaum, *E.F.W. Pflüger als Naturforscher,* Bonn, 1909.

[60]E. Schütz, "Rudolph Rosemann," *Med. Klink* (1943), p. 389.

[61]Ludimar Hermann, *Erinnerungen,* Berlin, 1915.

[62]See biographical sketch by R. Tigerstedt in *Erg. Physiol.* 8 (1909), p. 797.

[63]See obituary by P. Grützner in *Pflügers Arch.* 72 (1898), p. 221.

[64]Other biographical data is in J. Bernstein's article in *Naturwiss. Rundschau* 12 (1897), p. 606.

[65]See obituary by R. Du Bois-Reymond in *Berl. Klin. Wschr.* (1912), p. 2063.

[66]See biography by E. Lesky in *Dict. Scientif. Biogr.* 2 (1970), pp. 530-532.

[67]E. Th. v. Brücke, *Ernst Brücke.* Wien, 1928.

[68]See biography by O. Zoth in *Pflügers Arch.* 101 (1904), p. 103.

[69]See obituary by L. Löhner in *Erg. Physiol.* 36 (1934), p. 1.

[70]See account by C. Voit in *Z. Biol* 40 (1900), p. 1. and
Sitz Math. Physik Classe 32 (1902), pp. 249-262.

[71]Willy Kühne, *On the Photochemistry of the Retina and on Visual Purple,* ed. by M. Foster, London, 1878.

[72]See biographical sketch by A. Kreidl in *Z. Sinnesphysiol.* 57 (1926), p. 281.

[73]R.H. Major, "Karl Vierordt," *Ann. Med. Hist.* 10 (1938), pp. 463-473.

[74]C. Vierordt, *Die Schall- u. Tonstärke und das Schalleitungsvermögen der Körper.,* Tübingen, 1885.

[75]G. Rath, "Georg Meissners Tagebuch seiner Triestreise, (1851),", *Sudhoffs Archiv.* 38 (1954), pp. 129-164.

[76]G. Müller, "Georg Meissner, sein Leben und Werk", (unpubl. diss., Düsseldorf, 1936).

[77]See obituary by H. Boruttau in *Pflügers Arch.* 100 (1905), p. 351.

[78]Fr. Hr. Bidder, "Vor hundert Jahren im Laboratorium Johannes Müllers," *Münch. Med. Wschr.* (1934), p. 50.

[79]F. Merkel, *Jakob Henle. Ein Dtsch. Gelehrtenleben,* Braunschweig, 1891.

[80]Rembert Watermann, *Theodor Schwann, Leben und Werk,* Düsseldorf, 1960.

[81]M. Florkin, *Naissance et déviation de la théorie cellulaire dans l'oeuvre de Théodore Schwann,* Paris, 1960. See also
Gilbert Causey, *The Cell of Schwann,* Baltimore, 1960 and E. Mendelsohn, "Cell theory and the development of general physiology," *Arch. Int. Hist. Sci.* 16 (1963), pp. 419-429.

[82]M. Florkin, "Une autobiographie inédite de Théod. Schwann," *Bull. Acad. Roy. Méd. Belg.* 16 (1951), p. 445. See also
J. Oppenheimer, "Lives and letters of Theodor Schwann, a review," *Bull. Hist. Med.* 37 (1963), pp. 78-83.

[83]R. Foregger, "Respiratory apparatus of Theodore Schwann," *Anesthesiology* 27 (1966), pp. 187-194.

[84]K. Sudhoff, "Jakob Henle," *Münch. Med. Wschr.* (1909), p. 1696.

[85]Hermann Hoepke, *Briefwechsel zwischen Jakob Henle und Karl Pfeufer 1843-69,* Wiesbaden, 1969.
A translation of Henle's "On miasmata and contagia", by G. Rosen appeared in *Bull. Inst. Hist. Med.* 6 (1938), pp. 907-983.

[86]B. Kisch, "Forgotten leaders in modern medicine. Valentin, Gruber, Remak, Auerbach," *Trans. Amer. Philos. Soc. N.S.* 44 (1954) Part 2, pp. 227-296.

[87]Erwin H. Ackerknecht, *Rudolf Virchow, Doctor, Statesman, Anthropologist,* Madison, 1953.

[88]R. Virchow, "Standpoints in scientific medicine," transl. by L.J. Rather, *Bull. Hist. Med.* 30 (1956), pp. 537-543.
See also: R. Virchow, "Die naturwiss. Methode und die Standpunkte in der Therapie," *Virchows Arch.* 2 (1849), p. 3.

[89]J.W. Wilson, "Virchow's contribution to cell theory," *J. Hist. Med.* 2 (1947), pp. 163-178.

[90]E. Holländer, *Anekdoten aus der med. Weltgeschichte,* Stuttgart, 1925.

[91]John R. Baker, *The Controversy on Freedom in Science in the 19th Century,* Oxford, 1962.

[92]See obiturary by M. Verworn in *Z. allg. Physiol.* 19 (1921), p. 1.

[93]See biography by K.E. Rothschuh in *Dict. Scientif. Biogr.* 4 (1971), pp. 614-617.

[94]H.J.V. Tyrrell, "The origin and present status of Fick's diffusion law," *J. Chem. Ed.* 41 (1964), pp. 397-400.

[95]See E. Wöhlisch, "Adolf Fick," *Naturwissenschaften* 26 (1938), p. 585.

[96]See biographical sketch by A. Fick in *Pflügers Arch.* 90 (1902), p. 313.

[97]A. Gürber, "F. Schenk," *Erg. Physiol.* 19 (1921), p. VI.

[98]A.v. Muralt, "Leon Asher," *Erg. Physiol.* 46 (1950), p. I.

[99]E. Koch, "K. Bürker," *Münch. Med. Wschr.* (1942), p. 703.

[100]See biographical sketch by P. Hoffmann in *Z. Biol.* 92 (1932), p. I.

[101]See accounts by H. Rein in *Erg. Physiol.* 35 (1933) p. I and
P. Hoffmann in *Sitz ber. Physik. Med. Ges.* Würzburg 57 (1932), p. 56.

[102]S. Baglioni, "L. Luciani," *Erg. Physiol.* 18 (1920), p. XIV.

[103]G. Moruzzi, "L. Luciani," *Scientia Medica Italia* 3 (1955), pp. 403-412.

[104]N. Zuntz, "A. Mosso," *Deutsch. Med. Wschr.* (1911), p. 31.

[105]See biography by L. Belloni in *Dict. Scientif. Biogr.* 2 (1970), pp. 339-340.

[106]H. Buess, "John Friedr. Miescher and the contributions of Basle physicians to the biology of the nineteenth century," *Yale J. Biol. Med.* 25 (1953), pp. 250-261.

CHAPTER VII

Nineteenth and Twentieth Century Physiology: Western Europe-America-Russia

1. CLAUDE BERNARD AND FRENCH PHYSIOLOGY BETWEEN 1848 — 1914

Between 1816 and 1850 Magendie's empirical physiology and experimental vivisectionist methodology found recognition throughout Europe. Two important physiologists emerged from his Parisian laboratory in those years: *Moritz Schiff* (1823-1896) and *Claude Bernard* (1813-1878). Bernard later occupied a position in France akin to that of Johannes Müller or, even better, Carl Ludwig in Germany.

Moritz Schiff, who was born in Frankfurt, graduated in 1844 from the University of Göttingen, and then went to work with Magendie and Longet at Paris.[1] After some years in Bern and Florence, he was active as professor of physiology in Geneva for twenty years (1876-1896). His main field of interest was physiology of the nervous system, and he designated the N. vagus as a motor nerve of the heart. For Schiff, the decreased cardiac rate which followed vagal stimulation was actually a sign of fatigue.[2] As in other works, he went exceedingly far in his conclusions, and thus his results did not achieve the recognition desired from other contemporary scientists.

Schiff was also the first to carry out successful extirpations of the thyroid gland in 1856, and he should be included among the pioneers who described the role of the liver in carbohydrate metabolism. According to Pflüger's monograph on glycogen, Claude Bernard ought to be credited with the priority of this discovery. Finally, Schiff made a series of useful observations in the field of digestive physiology. One of his students, the Italian physiologist Frédéric Batelli, has already been mentioned.

Much more significant for the development of French and European physiology in general was the career of *Claude Bernard*.[3] However, physiology developed rather slowly in France, and there were several reasons for such a delay. First of all, with few exceptions, there was no clear separation between the theory and practice of medicine in the French schools. Moreover, France did not have the well-equipped physiological institutes and laboratories which had mushroomed in Germany. Thus, Bernard complained repeatedly about the deficiencies inherent in the French laboratories, admiring the German counterparts both for their organization and their staffing.[4] By contrast, Magendie and Bernard both worked in the cold, damp and dark rooms at their disposal in the Collège de France, which may well have been the real cause for Bernard's fatal renal disease.[5]

In addition to these external obstacles to the carrying out of proper research, there were also other important internal or ideological factors which delayed the development of French experimental physiology. One of these impediments was the vitalism espoused by Bichat which assigned a clear determinism to the inorganic realm while proclaiming the indeterminate nature of living beings. This idea delayed attempts to apply physics and chemistry to the phenomena occurring in the organism. By the same token, it inhibited those attempts to measure biological variables and discouraged the use of various devices and instruments for such a purpose.[6]

Another important factor which handicapped French physiology was the widespread hesitancy to use the microscope, another attitude which can be traced back to Bichat and led France, with the exception of Strassburg, to ignore the new cellular theory for many years. In fact, for a long time, the concept of a blastema and the existence of so-called organized elements were maintained.[7] Only in 1869 did the French anatomist *Louis-Antoine Ranvier* (1835-1922) establish a pathological histology according to the German model (Cornil and Ranvier). Likewise, histophysiology, an important element in the progress of physiology since the times of Purkinje and Müller, was also introduced by Ranvier.[8]

Fig. 95. Claude Bernard (1813-1878).

Moreover, Cuvier's comparative anatomy exerted an unusually strong and prolonged influence in France, particularly his creed that anatomical comparisons would lead directly to an understanding of the purpose of any given organ. Finally, one cannot underestimate the influence exerted by positivism on the ideas of French physiology, since this philosophical current tended to support vitalism on the one hand and the comparative method of study on the other.[9] Therefore, even Bernard was only gradually able to overcome these ideological prejudices, and his many accomplishments should hence be held in high esteem.

Claude Bernard (see Fig. No. 95) was born on July 12, 1813, in the village of St. Julien near Villefranche, the child of vintners. His biography has already been adequately written several times and presented along with his leading ideas and discoveries.[10] During his early school years, he did not distinguish himself among his classmates, perhaps because physics, chemistry and the natural sciences were then only rarely included in the curriculum. During 1833-34, Bernard worked as an apprentice in an apothecary shop in Lyon, a period of his life in which he began his career as a playwright. Two productions, "La Rose du Rhône" and "Arthur de Bretagne" emerged from Bernard's pen and, although they had some merits, his critics recommended that he pursue another occupation for a living.[11]

Hence, Bernard enrolled as a medical student at the University of Paris in 1835. Shortly thereafter he met Magendie who was doing his experiments in a cellar of the Collège de France. Between 1839-1844, Bernard worked there as an assistant, at the same time carrying out his medical duties at the "Hôtel Dieu" hospital in Paris.

During his daily contact with Magendie, both men learned to respect and appreciate each other, although their personalities and ideas regarding the importance of experiments differed widely. Commenting on Magendie's great stress on laboratory work in medicine, Bernard replied: "I consider the hospital to be a vestibule for scientific medicine; it is the first field of observation to which a physician is exposed. However, the laboratory is the temple of medical science."[12] In 1844, both scientists broke off their association, and Bernard left Magendie's laboratory. Afflicted by financial difficulties, but supported by the chemist Pelouze, Bernard married. His marriage provided him with a financial base but eventually it could not survive the strain connected with the irregular life of a young investigator.

In 1848, Bernard's work received some recognition, and he was employed as a successor to Magendie at the Collège de France, an appointment which led to his reconciliation with the old teacher. A year later, he received his first honor, the Cross of the Legion of Honor, and was appointed in 1854 to the newly created chair of physiology at the Sorbonne. Following Magendie's death in 1855, Bernard also inherited the former's position at the Collège de France. There he developed a fruitful career both as a teacher and researcher, and his growing prestige attracted a series of students from many countries to Paris. Among those who studied at his institute were physicians and scientists, interested laymen such as writers, priests, philosophers and renowned personalities. A similar phenomenon also occurred in Munich with the lectures of the German chemist Justus Liebig.

Due to a persistent malady, Bernard ceased his work in Paris during 1865-68, and retired temporarily to his native village of St. Julien. After his triumphal return, awards and honors began to accumulate once more, a reflection of the growing recognition which he received for his work. Wrote Pasteur in 1866: "When I search for a weak point in Bernard, I cannot find it. His personal determination, the serene beauty of his facial expression, the great gentleness, his kindness, enchant you at first sight. Bernard does not exhibit any pedantry, no one-sided erudition. Rather, he displays a great simplicity and his conversation is devoid of any affectation but is constantly stimulating because of his deep ideas. These are some of Bernard's qualities."

Shortly before the siege of Paris by the Prussian army in 1870, Bernard escaped to St. Julien where he remained for about nine months before returning to Paris. The political defeat suffered by his country left a lasting impression on this patriotic scholar. In addition, his health gradually began to deteriorate again, curtailing his capacity to work, although he continued to explore strenuously the problem of animal heat.

In those eventful years, Bernard often expressed his conceptions regarding the essence of vital phenomena in lectures. These discourses,written down, were eventually edited by Bernard himself and published, thus allowing much of his work to appear in the form of "leçons" or "lessons" about topics which had attracted his special attention. Once, Bernard characterized metaphorically his lecture-room as follows: "In front of me was Aristotle, who held his head with an expression of absolute boredom. I have always thought that there was a hidden irony towards me involved in that gesture. Above my head, Hippocrates gave the impression that he provided me with his blessing. Meanwhile, Buffon was writing on my right and Lavoisier distilling water on my left." Bernard became ill in January, 1878, and died February 10, deeply mourned by his friends and the entire scientific community.

In sheer volume, Bernard's life work was impressive. Between 1834 and 1878 three hundred scientific writings of Bernard were published, among them fifty-three of his own works, including translations.[13] A large number of manuscripts and notes remained unpublished, although some of them have since come into print, such as *La science expérimentale,* the *Cahier de notes* (1850-1860), *Leçons sur les phénomènes de la vie communs aux animaux et aux végétaux,* and *Leçons de physiologie opératoire.* His extraordinary scientific productivity led one author to state that Bernard made discoveries at the rate others would breathe.

Often, Bernard would edit the content of his lectures, such as those of 1855-56 concerning experimental physiology. Those of 1857 were about the effect of tonic and medicinal substances, and those in 1858 on the nervous system. The lectures of 1859 dealt with organic fluids, those of 1866 explained the properties of living tissue while the subject for the 1876 classes was animal heat. With these skillfully composed and comprehensive writings dealing with broad areas of physiology, Bernard actually became, to a great extent, a spokesman for contemporary physiology. However, he presented his subjects in a didactic fashion rather than in a developmental format, thus rarely mentioning the names of earlier workers in the various fields and their accomplishments. Therefore, for those who were unacquainted with a historical perspective, it was difficult to keep Bernard's discoveries apart from those of his predecessors. Moreover, Bernard adhered to the principle of "Peu lire" or "the scientist who discovers something new should not get lost in that which had been done before his time."

Like Magendie, Bernard made most of his discoveries during his vivisections and demonstrations. According to his Russian collaborator, Cyon, "in spite of this he (Bernard) became one of the most brilliant investigators in physiology, a fact which only proves Bernard's truly extraordinary capacities both as an observer and researcher."[14]

Bernard made his discoveries especially in the fields of digestion, carbohydrate metabolism, nervous system and the effects of poisons. He first examined the action of the gastric juice on cane sugar (1843) and then observed the lack of lipid digestion after the ligature of the pancreatic duct. Therefore, Bernard determined the existence of a fat splitting action in the pancreatic juice. In addition, he discerned the digestive effect of the pancreatic secretion mixed with bile on the pancreatic tissue itself, and the reabsorption of fat particles in the lymphatic fluid.

Concerning his other area of research, Bernard was the first to discover the existence of a chemical synthesis in the organism: the formation of sugar in the liver of a dog which had only been fed with proteins (1848). He also found that glucose was a constant constituent of the liver and the blood, and described the so-called "diabetic puncture" in 1850, demonstrating the formation of glycogen in the liver seven years later.[15]

Bernard's studies on the vasoconstrictor and dilator nerves were also significant. The vasodilation was always reflected by an increased temperature of the affected part. This problem prompted Bernard to study the distribution of animal heat throughout the body. He found, as did Georg Liebig before him in 1853, that the blood actually was not warmer in the left ventricle of the heart, concluding that the processes of oxidation and combustion did not take place in the lungs.

The question regarding the actual site of oxidation in the body occupied Bernard throughout his life. In 1850, Georg Liebig had already demonstrated the consumption of oxygen and production of carbon dioxide in a bloodless muscle. Bernard, however, did not believe in Ludwig's ideas of a physical transport mechanism of these gases from the blood to the tissues, and thus distrusted the in vitro experiments regarding respiration at the tissue level.[16] Four years after Pflüger had carried out his decisive experiments of 1872, Bernard finally accepted the idea that the tissues were the actual sites of oxygen consumption.

Bernard's studies concerning curare began in 1844, and in 1857 he finally was able to elucidate through a series of ingenious experiments, the action which took place at the endings of the motor nerves. In 1867, at the same time as Hoppe-Seyler, Bernard recognized the toxic combination between carbon monoxide and hemoglobin. Earlier, during 1858-59, he was concerned with the responses of stimulation and section of the secretory nerves — chorda tympani — after Ludwig had determined the existence of such nerves in 1850.

Next to Bernard's great physiological discoveries stand his efforts regarding a theory of the organism, the establishment of a physiological methodology and a series of physiological principles. Hence, a summary of his biological ideas is in order. According to Bernard, the organism was a unity, an integrated whole. Although its vital manifestations were of a physico-chemical nature, they occurred in units whose organization and structure were dependent upon life and development. The capacity to grow was a property originally given to the living being itself, "a vital force which cannot be denied and is as clear as daylight." To be sure, not every organ possessed an intelligence, but the whole organism was endowed with one.

Bernard declared that in spite of this situation, all the organic functions were strictly determined, since the vital mechanisms worked in a constant milieu or environment. Such a stability was the very precondition for the autonomy of an organism. For Bernard, the vital manifestations were merely expressions of organization. Their primary causes could not be elucidated, but a search of the immediate or secondary causes was in order.[17]

Moreover, Bernard felt that all living phenomena were more qualitative than quantitative in nature, and that measurement would provide exact but often worthless values. Hence, the need in biology was to describe the essential and qualitative nature of something. It follows that Bernard was against a complete reduction of the biological phenomena to physico-chemical processes. On the other hand, he never insisted on the existence of a vital principle which could compete and even act against physical forces.

Therefore, Bernard declared that the formation of an organism was based on more than just physical factors. Life had to be viewed both as organization and coordination, and physiology was to be an independent science. To some extent, it is perhaps plausible and proper to call Bernard's position a "physico-chemical vitalism." He certainly wanted to determine experimentally the "immediate" causes for the vital processes, since this was, for him, the only way in which to predict the living occurrences.

Bernard was one of the very few investigators who concerned himself intensively with questions of methodology. In 1865, his famous work appeared entitled *An Introduction to the Study of Experimental Medicine.* (See Fig. No. 96.) The first part of the publication dealt with observations and experi-

INTRODUCTION

A L'ÉTUDE DE LA

MÉDECINE EXPÉRIMENTALE

PAR

M. CLAUDE BERNARD

Membre de l'Institut de France (Académie des sciences),
et de l'Académie impériale de médecine,
Professeur de médecine au Collége de France,
Professeur de physiologie générale à la Faculté des sciences,
Membre de la Société royale de Londres,
de l'Académie des sciences de Saint-Pétersbourg
et de l'Académie des sciences de Berlin.

PARIS
J. B. BAILLIÈRE ET FILS,
LIBRAIRES DE L'ACADÉMIE IMPÉRIALE DE MÉDECINE,
Rue Hautefeuille, 19.

Londres	Madrid	New-York
Hippolyte Baillière	C. Bailly-Baillière	Baillière-Brothers

LEIPZIG, E. JUNG-TREUTTEL, QUERSTRASSE, 10
1865
Tous droits réservés.

Fig. 96. Title page of Bernard's *Introduction to the Study of Experimental Medicine.*

ments and presented the investigator with his doubts, proofs and counter evidence. By contrast, the second section was concerned with the concrete problem of justification and the possibility of carrying out experiments on living beings in view of the deterministic character of the organic functions. In this context, Bernard urged that the conditions surrounding the animal experiment were to be duly noted, and that the use of complicated instruments had actually a deleterious effect on the research.

In a third portion of the book, Bernard considered that the starting point for any experiment was either a fortuitous observation or an established theory. He was against any blind empirical search which had no definite ideas nor plans on how to proceed. Procedures, observations and their results should be continuously reevaluated with a critical eye, but empiricism and pure statistics were to be rejected since they detracted from the aim to investigate only the most immediate causes. All Bernard's recommendations were broadly and simply described, although there was a certain amount of repetition with a rather distracting interjection of afterthoughts in the book.[18] However, Bernard's *Introduction* was the first clear exposition of the theory and practice surrounding physiological and medical experiments.

Bernard's research approach, like that of Magendie, was predominantly practical and did not require many instruments. As a rule, he carried out vivisection experiments, using ligatures, sections and stimulations on living animals, rarely achieving quantitative conclusions. Thus, Bernard's technique was mostly surgical and he worked on certain organs of the entire animal. For example, he dissected the submaxillary gland of a dog and observed the light or dark color of the outflowing blood due to increased or decreased glandular circulation following the stimulation of certain nerves (1858).[19]

The important concept of a "milieu intérieur" or "internal environment" was only gradually elaborated.[20] Although the idea had already been suggested in 1851, it only achieved a mature formulation in 1854-57. Since the actions of the organism were clearly determined, reasoned Bernard, there had to exist within the body certain unchanging conditions, or an even, constantly regulated environment (1867). Those factors which ensured such a constancy were the blood temperature, water, oxygen and others that were gradually discovered. Therefore, Bernard's internal environment has become in our days a purely physico-chemical concept.[21] It should be pointed out that the term "internal secretion" which appears in connection with the concept of an inner environment, had no relation to our modern understanding of the words. Bernard merely meant by internal secretion the formation of matter in an organ and its passage into the blood, without considering such a phenomenon to be the result or cause of other hormonal secretions.

Bernard's success, his methodological skills, and his growing international reputation attracted a large number of students to Paris from throughout the world. Many of those who worked in his laboratory became famous physiologists in their own right, following the suggestions and impressions gathered with Bernard. Among those already cited were Elie de Cyon, Willy Kühne, Isidor Rosenthal and Angelo Mosso. Others included the Italian physiologist *Luigi Vella* (1825-1886), a student of Bernard's in 1849 who later worked in Modena and Bologna, and the Danish author *Peter L. Panum* (1820-1885), who became professor of physiology at Kiel after studying with Bernard in 1853. In addition, the renowned French histologist *Louis A. Ranvier* (1835-1922) considered himself a disciple of Bernard, and even Louis Pasteur (1822-1895) gratefully acknowledged the physiologist's support and inspiration. From Russia came *Ivan Romanovitch Tarchanoff* (1848-1908), a native of Tiflis, who later occupied the chair of physiology at the University of St. Petersburg.

The United States was gradually beginning its scientific ascendancy during Bernard's tenure at Paris. Among the American students who went to study there with Bernard were *John C. Dalton, Jr., Austin Flint* and *Silas Weir Mitchell,* who have already been mentioned earlier.[22]

A series of French disciples of Bernard greatly contributed to the development of physiology in the second half of the nineteenth century. *Paul Bert* (1833-1886) was one of the earliest collaborators of Bernard.[23] A native of Auxerre, he first became a lawyer and then a scientist, graduating with a dissertation concerning the double innervation of the rat's tail: *De la greffe animale* (Paris, 1863). The work was based on an observation that the sensibility of a severed rat's tail returned if the distal end was implanted on the back of the animal. Bert was an assistant to Bernard in 1868, and became professor at the Sorbonne in 1872. However, he later abandoned this position in order to become a politician, minister of education and finally a governor of Tonkin where he died in 1886.

Bert was one of the first authors to study the effects of barometric pressure on humans, animals and plants: *La pression barométrique* (Paris, 1878). He became well-known for his controversial but valuable observations regarding this topic, and especially for his associates' unfortunate ascent in a balloon with insufficient oxygen containers.[24]

Two other students of Bernard, *Albert J. Dastre* (1844-1917) and *Jean-Pierre Morat* (1846-1920) were associated with the discovery of the law regulating vascular diameters (1884). Dastre was an assistant to Bernard between 1872-76 and was later appointed professor of physiology at the Sorbonne in 1887. Along with his studies on vasomotor problems, he successfully worked on the problem of narcosis. Morat, in turn, studied at Lyon and graduated from the University of Paris. He was appointed professor of physiology first at Lille in 1878, then Lyon in 1882, and his interests lay in the elucidation of the cardiac rhythm.

Other disciples of Bernard included *Nestor Gréhant* (1838-1910) who studied the blood vessels in detail, and *Jacques A. D'Arsonval* (1851-1914), who already worked regularly as a student in Bernard's laboratory. In 1874, Gréhant became an instructor at the Collège de France and director of its laboratory for medical physics in 1882. D'Arsonval, who showed great methodological skills, published a series of works on calorimetry, animal heat, electrical measuring instruments, as well as alternating currents — "D'Arsonvalization." Finally, another student of Bernard and Ranvier should be cited: *Louis C. Malassez* (1842-1909). He contributed a series of excellent works on the morphology and physiology of the blood, studying the development of the red cells and their hemoglobin content.

Outside of Bernard's school of physiology, France had a large number of important physiologists during the second half of the nineteenth century. One of them was *Charles E. Brown-Séquard* (1818-1894) (see Fig. No. 97) who was born on the island of Mauritius.[25] Brown-Séquard studied and graduated from the University of Paris in 1846, and after staying for some years in the United States, he was appointed as Bernard's successor at the Collège de France in 1878.[26] Some of his contributions were made in the field of neurophysiology where he described the intensified reflex excitability in those segments of the spinal cord located below a transverse section. Moreover, Brown-Séquard related, for the first time, the exact results of a unilateral lesion and the method to induce artificially an epileptic seizure.

Fig. 97. Charles E. Brown-Séquard (1818-1894).

In his old age, Brown-Séquard created quite a stir with his experiments which he carried out on himself, employing testicular extracts for rejuvenation purposes (1889). Through this work, he became one of the founders of the new organotherapy in the field of internal secretions, an area in which Moritz Schiff and other French authors made significant contributions[27]

In this connection, the physician *E. Felix A. Vulpian* (1826-1887), professor of pathological anatomy in Paris and a temporary successor of Flourens, should be cited. Vulpian, who was accused of being a materialist, considerably enriched our knowledge of the adrenal glands. Together with Charcot and Brown-Séquard in 1868, he founded the journal *Archives de physiologie normale et pathologique.* Another author who was active in the field of internal secretions was *Eugène Gley* (1857-1930). In 1883, he was appointed a preceptor at the Collège de France and became a member of the educational board for biology in 1908. Gley was a versatile and fertile writer who contributed to the knowledge of the thyroid and parathyroid glands. He was the author of the *Traité élémentaire de physiologie* (Paris, 1906-08) and a historical book: *Essais de philosophie et d'histoire de la biologie* (Paris, 1900).

Fig. 98. Etienne T. Marey (1830-1904).

A group of capable students and collaborators were attracted to the laboratory of the successful investigator *Étienne J. Marey* (see Fig. No. 98). Marey was originally from Beaune and later studied in Paris. In 1864, he established his own physiological laboratory in the French capital and became a professor of natural history at the Collège de France in 1867, succeeding Flourens. Marey was famous for his skillful investigations regarding problems of cardiac and circulatory physiology. He was a master of graphic methods used to depict physiological variables, such as the sphygmograph, cardiograph and myograph. (See Fig. No. 99). Marey also devised the famous Marey-tambour, capsules for the transmission of air, as well as catheters to measure intracardiac pressure. In 1878, Marey was admitted to the Academy of Sciences, and continued to publish several works concerned with questions of animal heat, electro-physiology, motion and especially cardiac physiology. His efforts to introduce photographic registrations in physiology were also very important.[28]

One of Marey's older friends and his collaborator was *J.B. Auguste Chauveau* (1837-1917). Chauveau participated in the experiments to measure intracardiac pressures, invented a hemodromograph and measured the velocity of nerve conduction. He became professor of comparative anatomy in Paris. Marey's other successful student was *Charles E. François-Franck* (1849-1921), a Parisian who worked with Marey at the Collège de France. He contributed valuable studies towards a better understanding of circulatory physiology.

Charles Richet (1850-1935) (see Fig. No. 100) was a student of Bernard, Vulpian and Marey, a successful investigator and fertile author. Also a Pa-

Fig. 99. Marey's double sphygmograph (1878).

Fig. 100. Charles Richet (1850-1935).

risian, Richet worked on questions of neurophysiology and animal heat, founding, in addition, the new field of serotherapy. In 1888, he had already observed that the blood of vaccinated animals possessed a certain immunity against specific diseases. Thus, in 1890, Richet gave the first injection of serum to a human being and two years later he discovered a phenomenon which he called "Anaphylaxis." For this he received the Nobel Prize in physiology and medicine for the year 1913.

In addition to his scientific work, Richet composed a series of philosophical and literary works publishing poems, dramas and novels. Other books dealt with historical subjects and the problem of parapsychology. With the citation of Richet, the modern era of French physiology seems at hand. Among those belonging to the previous generation of successful physiologists were Louis Lapique, J. Lefèvre, L. Tissot, E. Hédon, Michel Pachon, H. Piéron, A. Mayer, Justin Jolly, A. Fessard, Léon Binet, H. Cardot and especially *Alexis Carrel* (1873-1944), who won the Nobel Prize in 1912 for his investigations "on vascular suture and the transplantation of blood vessels and organs."

2.
THE CHEMICAL CURRENT IN PHYSIOLOGY, ESPECIALLY IN GERMANY

> The most sublime task of the human spirit is the investigation of the laws governing life which cannot be considered without a precise knowledge of the chemical forces involved.
>
> *Justus von Liebig*

The application of physical and chemical principles to the explanation of the vital processes was attempted in all historical periods. However, such an approach did not become a generally applied principle until the nineteenth century, when it deeply influenced the development of physiology. Until then, two circumstances had greatly hindered the use of chemical and physical principles for the analysis of physiological phenomena.

In the first place, a strong vitalistic current insisting on the complete autonomy of life existed among physiologists. Secondly, the knowledge and methods available in physics and chemistry until 1840, precluded their employment in the proper elucidation of most physiological problems. Thus, the physiology of the early nineteenth century remained virtually an "animated anatomy" with a morphologically oriented methodology. Such an approach was linked to an experimental vivisection seeking systematically, some of the physiological relationships in the living animal.

From about 1840, the application of quantitative physico-chemical methods in physiology rapidly expanded. Spearheaded by authors such as Ernst H. Weber, Alfred W. Volkmann, Carl Ludwig, Hermann von Helmholtz, Emil Du Bois-Reymond and Ernst Brücke, the current which employed a physical approach has already been described in detail. In turn, the chemical direction had begun with Theodor Schwann, Theodor L.W. von Bischoff and Justus Liebig. The latter author, especially, urged the application of chemical ideas and methods for the solution of physiological problems in two important works, entitled *Die Chemie in ihrer Anwendung auf Agrikulturchemie und Physiologie* (Braunschweig, 1840) and *Die Tierchemie oder die organische Chemie in ihrer Anwendung auf Physiologie und Pathologie* (Braunschweig, 1842).[29]

In order to reach a level of relative sophistication, chemistry went through a long period of evolution which can only be briefly sketched here. The process was characterized by a gradual growth of knowledge concerning the structure and reactivity of chemical substances in both the organic and inorganic realm, as well as a systematization of such information and the establishment of chemical theories. Antoine Lavoisier brought such an order into the field by recognizing the relationships between the absorption of oxygen and the production of water and carbon dioxide, as well as the nature of animal heat during respiration and combustion. Using the discoveries of such eminent chemists as Scheele, Priestley and Black, Lavoisier was able to describe in more detail the properties of elements like carbon, hydrogen, oxygen and their most important combinations.

Lavoisier also introduced the quantitative approach into chemistry, and the progress achieved by organic-chemical analyses in France increasingly facilitated the discovery of new substances while clarifying the complex structure of other compounds. Among those working in this field were *Antoine F. de Fourcroy* and *Michel E. Chevreul* (1786-1889) in the chemistry of fatty substances, and *Hilary M. de Rouell* (1718-1779) on urea. Besides, *Joseph L. Proust* (1754-1826) studied albumin, *Jean L. Prévost* (1790-1850) and *Jean-Baptiste A. Dumas* (1800-1884) urine, blood and milk. Also of great importance were the chemical investigations of *Joseph L. Gay-Lussac* (1778-1850), *Théophile J. Pelouze* (1807-1867), *Auguste Laurent* (1807-1853) and *Louis Pasteur* in France.

In turn, chemical theory was advanced by *Jeremias B. Richter* with his idea of equivalent weights. *John Dalton* (1766-1844), *Amedeo Avogadro* (1776-1856) and Gay-Lussac with their concepts of molecular and atomic weights and multiple proportions. Finally, *Michael Faraday* discovered benzol, established the foundations of electro-chemistry and coined the terms ion and electrolyte. Moreover, the Swedish chemist *Jöns J. Berzelius* (1779-1848) corrected the methods of chemical analysis, created a chemical nomenclature and symbolism still in use today, and established the concepts of isomerism and polymerism.

Berzelius, Wöhler and Liebig were the three figures who successfully contributed to the establishment of a chemical approach to physiology. Berzelius[30], who studied medicine and became professor of medicine and pharmacology at the University of Stockholm, attracted a fair number of students, like *B. Rose, Karl Gustav Mitscherlich* (1805-1871), *Leopold Gmelin* (1788-1853) and *Friedrich Wöhler,* who in turn expanded and greatly helped the growth of physiological chemistry. Gmelin was professor of medicine and chemistry at the University of Heidelberg after 1817, and together with another chemist, *Friedrich Tiedemann,* published the work entitled *Verdauung nach Versuchen* (1826-27).[31] The publication contained a large number of results obtained through experiments as well as chemical and microscopial observations regarding digestion, resorption and assimilation. In truth, the problems of digestion, metabolism, nutrition and blood were the first physiological fields subjected to chemical analysis.

In 1836, a work appeared written by the physiologist *Hermann Nasse,* the son of the well-known clinician Christian F. Nasse of Bonn. Entitled *Das Blut in mehrfacher Hinsicht...,* the publication contained the contemporary results of the physico-chemical and microscopical research concerning the blood in health and disease states. Hermann Nasse's son, *Otto Nasse* (1839-1902), became professor of pharmacology and physiological chemistry at the University of Rostock. He was renowned for his work on the chemistry of carbohydrates and proteins and the actions of ferments.

In those early decades of the nineteenth century, efforts were made to discover the chemical composition of animal and plant structures, their organs as well as the nutritional substances which they required. The situation which existed in the period 1839-1841 with regard to these endeavors was reflected in *J. Franz Simon's* (1807-1843) work *Handbuch der angewandten medizinischen Chemie* (Berlin, 1842).

The contemporary views about oxygen, metabolism and heat formation were reflected in a passage concerning the blood, which appeared in the prologue to Simon's second volume. Here the author stated: "The blood is subjected to a continuous metamorphosis which is the expression of its vitality. During the nutrition in the peripheral system, the blood corpuscles and the entire blood serum are involved in the distribution of the nutrients. Thus, the cells and organs are able to feed from these materials by the power of an indwelling force which helps attract the necessary elements, converts alien products into proper compounds and excretes the residue. The nutritional elements in the serum are mainly albumin, fibrin and fat."

Simon then went on to say that "the products of the metabolism are, for the most part, extractable materials and lactic acid which can be found in the excreted matter, especially in the urine. Urea, bile and carbon dioxide are not to a significant degree products of the metamorphosis of the blood during peripheral nutrition. Rather, they are formed as a result of the vital activity inherent in the blood cells. These elements display the same ability to attract nourishment and dispose of residue as the rest of the bodily cells. They feed on oxygen, albumin and perhaps fat which these blood cells take from the serum. In turn, they produce carbon dioxide, urea, fibrin and other extractable materials, and perhaps parts of bile."

Simon concluded: "The immediate and most important function of the blood cells is to produce animal heat, since its loss would immediately threaten all organic functions and life itself. The formation of such an animal heat consists, first, in a combination between the oxygen and carbon of the globulin which produces carbon dioxide and urea or uric acid in animals with elliptical globules. Thus, the secreted urea is directly proportional to the animal heat produced by the organism. In fact, the formation of blood cells and blood in general are intimately connected and related to nutrition. Deficient diets will produce less blood, while abundant and excessive food intake causes the opposite effect."

Fig. 101. Justus von Liebig (1803-1873).

These efforts to apply chemical principles to vital questions reached a turning point in 1840 with the publication of Liebig's already cited book about the chemical applications to agriculture and physiology. *Justus von Liebig* (see Fig. No. 101) was a native of Darmstadt who began to carry out chemical experiments in his father's laboratory in his early school years. His distaste for the normal school curriculum led to unsatisfactory performances and an inability to progress from one grade to the next. Instead, Liebig devoted all his available time to the study of chemistry and alchemy. Barely seventeen, he went to the University of Bonn, where he studied chemistry, transferring later to Erlangen and studying under Kastner in 1820-21.

Thanks to a stipend from the Grand Duke, Liebig was able to study in Paris for some time under the direction of Guy-Lussac. He clearly recognized the wide lead taken by French chemistry as contrasted with the still highly speculative work being carried out in Germany. Writing from Paris to his friend August Platen in 1823, Liebig commented that German chemistry lacked the necessary rules to bring order into the rapidly swelling body of knowledge, and was still experimenting and formulating hypotheses without any clear ideas on directives. Supported by a recommendation from Alexander von Humboldt and known for his work with fulminic acid, Liebig was appointed professor of chemistry at the University of Giessen in 1824 upon his return from Paris. He was then barely twenty-one years old, lacking a school certificate and any formal degree.

Liebig remained at Giessen for twenty-eight years where he spent the most fruitful years of his career. He erected the most modern German chemical laboratory (see Fig. No. 102), introduced countless students to the intricate chemical methods and carried out some really pioneer work in both analytical and theoretical chemistry. In particular, Liebig studied organic matter and greatly stimulated a chemical interpretation of the vital phenomena.

Fig. 102. Liebig's laboratory at Giessen.

In those years, *Friedrich Wöhler,* a student of Berzelius and Gmelin, collaborated closely with Liebig. Originally from Eschersheim near Frankfurt, Wöhler had studied medicine and analytical chemistry, and, after spending some years in Berlin, became professor of chemistry at Göttingen in 1836. During his stay in Berlin, Wöhler carried out his famous work on the synthesis of urea through the action of cyanogen on liquid ammonia. Wöhler called the experiment "an example of the artificial production of an organic or animal substance from inorganic elements" (1828). However, since urea was only an end-product of the metabolism, and thus not involved in the vital process itself, the synthesis of urea did not threaten the concept of a vital force of "Lebenskraft."[32] Nevertheless, Wöhler's synthesis became the prelude to many other similar efforts.

The friendship between Liebig and Wöhler, two very different personalities, remained untroubled until Liebig's death, while Liebig's relations with Berzelius and other chemists were often strained because of Liebig's impulsive and rather abrasive personality.[33] Some of his contemporaries even dubbed him the "chemical executioner".

Liebig's book about agricultural chemistry was concerned with the general cycle of elements in nature, the assimilation of organic compounds by the plants, their breakdown in animals and man, as well as the consequences derived from these phenomena for agriculture. By contrast, Liebig's second work, entitled *Die Tierchemie oder die organische Chemie in ihrer Anwendung auf Physiologie und Pathologie* (1842), was specifically geared to answer the questions concerning the chemistry of living beings, their metabolism, nutrition, respiration and animal heat.

Liebig's work provided an abundance of new insights into the relationships between chemical phenomena and organic functions. Writing about Liebig's significance for the knowledge of animal and plant metabolism, Carl Voit declared in 1865: "Hitherto, the observations were lying around without connections, like isolated bricks, but thanks to the genius of this man (Liebig) they have been joined together into an orderly whole. This is truly his accomplishment, which in our time has only been recognized by a few physiologists. To appreciate fully his contributions, one has merely to read physiological publications predating Liebig's own works, and then those which followed the famous chemist. In this fashion it becomes clear how much his ideas molded the subsequent concepts regarding bodily functions. Although the chemical knowledge which Liebig employed had already become commonplace, he was the one who applied it to the occurrences in the organism".

Voit continued: "Liebig first recognized the importance of chemical transformation in the body. Thanks to the interaction between the organic components, foodstuffs and oxygen, a series of chemical permutations took place with the formation of metabolic products. Thus, as a result of these activities and motions, there was life. Liebig clearly recognized the relationship between chemical decomposition and its effects, ascribing all animal motion and heat to such chemically induced breakdowns". Regarding Liebig's views on metabolism, Voit wrote: "According to Liebig, the products of such transformations must be contained in the excreted matter, the urine containing the nitrogenous material. Since he examined carefully the entire chemical process of organic transformation, he followed it closely, step by step, to the excreted products. Concerning the value of the various nutrients, all that was known then was the fact that certain substances could not support life, but the reasons for this remained obscure until Liebig studied the topic. He recognized that all parts of the organism which had a defined form, motion and vitality, were composed of a nitrogenous substance derived from the protein contained in the food. In a meat-eating individual, the carbon contained in the ingested protein was sufficient for the formation of the carbon dioxide eliminated during respiration, but the vegetarian had to use carbohydrates for such an excretion."

Voit concluded: "Thus, Liebig came to the momentous separation between the nitrogenous and non-nitrogenous nutrients and their role in organic functions. Nitrogen-containing products replaced the used bodily matter and allowed motions, while the materials without nitrogen served for the production of animal heat. These were ideas which opened-up new directions in our research".

To a large extent, Liebig's ideas are still accepted today. He attempted, as did *Jean Baptiste Boussingault* (1802-1887) before him, to establish a balance between the ingestion and excretion of various elements. He called "respiratory foodstuffs or producers of heat" a group of substances not needed in the material composition of the organs. Thus, the carbohydrates and fats would combine, in Liebig's view, with the inhaled oxygen to produce animal heat. Conversely, Liebig called the protein substances "plastic nutrients" since they were destined to replace the structural elements of the body.[34] Later, Liebig added the mineral salts to the category of necessary nutritive matter.

Liebig's book was a superb synthesis and interpretation of the hitherto gathered chemical information in spite of errors and highly hypothetical notions which seemed the result of a scientifically oriented divination rather than asserted facts. Thus, Berzelius severely criticized Liebig's work referring to it as "arm-chair physiology" and a "delusion of hypotheses". Even Du Bois-Reymond wrote to Ludwig in 1848: "I consider his (Liebig's) physiological fantasies worthless and harmful, since he totally lacks basic knowledge and critical acumen". Although there was undoubtedly some justification for such harsh judgments, the core of Liebig's train of thought was essentially correct. Viewed in a historical context, his ideas prompted a great number of experimental investigations and analyses which, while disproving many details of his work, confirmed Liebig's fundamental concepts.

Liebig's contributions in other areas such as plant nutrition, artificial fertilizers and meat extracts have to be omitted here. In 1852, he transferred to Munich where his popular evening lectures attracted large audiences and constituted a fashionable social event. His work *Chemische Briefe* (1844), a popularization of his scientific discoveries, went through numerous editions. Liebig died in 1873.[35] After his death there was no question about the significance of chemical phenomena for the maintenance of life and the pursuit of vital actions.[36]

A series of violent differences of opinion regarding questions of physiological chemistry occurred between Liebig, Berzelius, Mitscherlich, Mulder and Dumas. Liebig did not avoid these controversies, wielding a skillful and sharp pen. He condemned Berzelius' idea of catalysis as well as the explanations brought forth by Schwann and Pasteur regarding fermentation and putrefaction by the action of microorganisms. Moreover, Liebig was against Henle's concept of contagion and miasmas, the quantitative analysis introduced by Fourcroy, and the total metabolic balances between oxygen and carbon dioxide. In a speech given in 1844 and entitled "Bemerkungen über das Verhältnis der Thierchemie zur Thierphysiologie," Liebig criticized other contemporary chemical views. In addition, he broadly indicted the scientific educational process of his day with its preference for classical languages, history and literature, and the entire humanistic direction prevalent in Prussia, while at the same time praising the humility of the scientist (1840).

On another subject, Liebig steadfastly believed that protein was the source of muscular energy, and he dismissed the classical experiment of Fick and Wislicenus, who could not discover more nitrogenous substances in their urine after an exhausting mountain climb in 1865 than when at rest. Finally, Liebig's activities had great practical consequences for agriculture as well as the chemical industry and technology.

Like Berzelius and Müller, Liebig was a vitalist in the sense that he believed in the existence of a vital force which would create and maintain the organic substances.[37] As already mentioned, Wöhler's synthesis of urea did not constitute a plausible argument against such a vitalism, since urea was merely the end-product of a living process. Only in 1860 did the French chemist *Marcellin Berthelot* (1827-1907) dismiss vitalism using arguments from organic chemistry.[38] Liebig, like Claude Bernard, tended to believe that oxidation and the production of heat in the body did not occur at the level of the lungs. Even before Bernard's experiments, Liebig's son, Georg, discovered that the temperature in the left heart was no higher that in the right chambers. Even the physiologist *Heinrich G. Magnus* (1802-1870), who carried out the first blood analysis in 1837, transferred the place of combustion from the lungs to the capillary system, a concept adopted by Bernard in

1844.

Following Liebig, Germany had a series of famous chemists, such as *Friedrich A. Kékulé von Stradonitz* (1829-1896), *Lothar Meyer* (1830-1895), *Friedrich Wöhler, Felix Hoppe-Seyler* and *Emil Fischer* (1852-1919). Among those physiologists who worked with chemical methods were Carl Voit and Max Pettenkoffer at Munich, Max Rubner, Eduard Pflüger and Nathan Zuntz at Berlin and Bonn. Rubner even combined energetics with chemical metabolism. In Holland, there were *Gerardus J. Mulder* (1802-1880) and *Jacobus Henricus van Hoff* (1852-1911), while Sweden had *Svante Arrhenius* (1858-1927) and France *Louis Pasteur.*

Pasteur[39] had taught at Dijon, Strasbourg and Lille until 1857, when he was called to head the newly established institute in Paris. From the strictly chemical problems of optically active tartaric acid, Pasteur went on to study biological questions, notably fermentation. In contrast to Liebig, Pasteur recognized that the process was actually caused by living organisms without the need of oxygen. He also recognized the causes for the formation of acetic acid in wine, and butyric acid as well as other so-called "diseases" of French agricultural products. Pasteur's experiments definitely refuted the ideas of spontaneous generation.[40] Thus, fermentation, putrefaction and infection were actually processes related to microorganisms, and organic substances which were prevented from exposure to the ordinary air with its bacteria did not experience such phenomena. Pasteur's great discoveries in bacteriology such as gangrene, rabies and immunity, initially stemmed from his chemical works.[41]

The movement in physiology which attempted to clarify the chemical phenomena in the body came gradually to the fore after Liebig. Initially, some physicians and physiologists with chemical knowledge devoted their time to investigate these chemical aspects of the vital processes in addition to their histological and experimental activities. Parallel to this development, the research concerning the chemical structure of organic substances continued to be carried out, especially during the last quarter of the nineteenth century. Such studies were performed by organic chemists and biochemists rather than chemically oriented physiologists.

It would be impossible to consider all those investigators who concerned themselves with questions of organic chemistry related to physiology. Since such an account has already appeared,[42] and some of the authors who contributed to the chemistry of the organism have been considered in preceding sections — Schwann, Kühne, Mulder, Donders, Hammarsten, Drechsel, Bernard, Schiff, Pflüger, Voit, Rubner, Hermann — the subject can be quickly summarized. The following remarks are, therefore, concerned solely with the true founders of physiological chemistry in Germany as an independent medical discipline. They will deal primarily with *Felix Hoppe-Seyler, Franz Hofmeister* (1850-1922) and their most distinguished students.

At the German universities, a physiologically oriented chemistry emerged from those subjects which the medical school called "zoochemistry", "animal chemistry" or simply "medical chemistry", often as an appendix to pathology.[43] Thus, Rudolf Virchow was able to erect a relatively independent chemico-pathological laboratory in Berlin, as early as 1856 where Hoppe-Seyler and Kühne had a chance to work. Later, several physiological institutes inaugurated chemical sections in their laboratories, and only more recently were some chairs in physiological chemistry inaugurated at some universities, independent of physiology.

Such a development was witnessed and influenced greatly by *Felix Hoppe* (see Fig. No. 103), known after 1864 as Hoppe-Seyler because of his adoption by a brother-in-law. A native of Freiburg, he was already busy with chemical experiments and botanical questions during his early school years. Following the contemporary rules, and in spite of his chemical interests, Hoppe-Seyler went to study medicine at the University of Halle where he simultaneously became active in Steinberg's chemical laboratory. A casual encounter with the Weber brothers prompted him to leave for Leipzig, where he worked especially with Ernst H. Weber.

Fig. 103. Felix Hoppe-Seyler (1825-1895).

During the winter semester of 1855-56, Hoppe-Seyler taught as "Privatdozent" or unsalaried lecturer at the University of Greifswald. His subject matter was physiological chemistry and pathology[44] and he worked in a small laboratory of his own. In 1856, Hoppe-Seyler went to Berlin and studied with Virchow, Kühne, Alexander Schmidt, Leyden, Botkin and Wilson Fox, until his appointment at the University of Tübingen in 1861.

At Tübingen, Hoppe-Seyler occupied the chair of applied chemistry and carried out fruitful and versatile research on the pigments of the blood: hemin, hematin, hematoporphyrin, methemoglobin and crystallized hemoglobin.[45] In 1865, he employed, for the first time, Bunsen's method of spectral analysis in order to check the spectrum of the blood. Moreover, Hoppe-Seyler examined the content and type of chemical bond between the oxygen and hemoglobin. Together with Georg Liebig, Heinrich Magnus and Eduard Pflüger, he searched for the location of the oxidation phenomenon in the tissues rather than in the blood, a problem which Pflüger finally resolved in 1874. Many later famous physiologists worked with Hoppe-Seyler in his laboratory at Tübingen, especially *Friedrich Miescher Jr., Ernst Salkowski* (1844-1923), *Eugen Baumann* (1864-1896) and *Zalesky.*

In 1872, Hoppe-Seyler was appointed professor of physiological chemistry at the University of Strasbourg. Until the completion of the new building in 1884,[46] the physiological institute was located in the old medical school. Hoppe-Seyler had numerous Russian students there, like Pachutin, Popoff, Sokoloff, Tårchanoff, Rajevsky, Frédéricq, together with Mering, Mauthner, Zweifel, Ledderhose and others. Thus, he contributed decisively to the establishment and development of a current physiological chemistry.[47]

In addition to his work on the blood, Hoppe-Seyler also studied the role of glycogen, lecithin and cholesterol in the body. He clarified the structure of lecithin, analyzed and classified proteins, researched the action of biliary acids and chlorophyll, and together with Miescher, examined the nuclear proteins in pus cells. Therefore, Hoppe-Seyler contributed to the knowledge of almost all substances which played significant roles in the organism. In 1877, he founded the still existing journal *Zeitschrift für physiologische Chemie.* He died at the age of seventy.[48]

Among Hoppe-Seyler's early students was the already mentioned physiologist Willy Kühne, who wrote the textbook entitled *Lehrbuch der physiologischen Chemie.* Kühne was especially interested in the breakdown of proteins by proteolytic enzymes, the formation of hippuric acid and bile, as well as the chemical aspects of twilight vision through the action of visual purpura.

In 1866, *Friedrich Miescher Jr.* (see Fig. No. 104) came to Hoppe-Seyler's laboratory at Tübingen. A son of Müller's student, the anatomist Fritz Miescher, he studied in Basel during 1860-61 with Wilhelm His, and then in Göttingen, especially with Friedrich Wöhler. During his stay with Hoppe-Seyler, Miescher began his investigations concerning nuclein, the start of a series of works within the class of nucleoproteins. He spent some time with Carl Ludwig in Leipzig, but tended to be disposed more towards a synthetic approach to physiological problems rather

Fig. 104. Friedrich Miescher, Jr. (1844-1895).

than the analytical method prevailing at the Leipzig Institute.[49]

In 1871, Miescher went to Basel as an instructor and was named to the chair of physiology the following year, succeeding Wilhelm His. Miescher worked on so-called "Histochemical" subjects there, examining spermatocytes and the change in musculature which took place in the ovaries of spawning salmon in the Rhine River.

In addition to his research, Miescher was an enthusiastic teacher. He wrote that "a professor without young students is like a mutilated trunk of a willow-tree". On the other hand, he resented the need to study extensively the contemporary literature in order to garnish his lectures with learned references. Writing to Wilhelm His, Miescher declared: "Yesterday I was finally able to leave the treadmill. It was about time to quit this massive reading and pre-digesting of foreign material which for didactic reasons has to be polished and somewhat falsified. It has a strange but stultifying effect on one's own thinking. Thus, one can understand why Bunsen has urgently recommended his students, who are not quite prepared for it, to avoid as many lectures as possible, so that the thrust of their activities would not be spoiled by shabby pedantry".

If Hoppe-Seyler had derived his studies from medical problems and physiological considerations, his student *Eugen Baumann* was more devoted to the structural chemistry of organic bodies. Born in Cannstadt and destined to become an apothecary, Baumann (see Fig. No. 105) studied in Lűbeck, Gothenburg and Tűbingen. In the latter city he came in contact with Hoppe-Seyler while he took courses in pharmacology. The physiological institute was housed in the former kitchen of the local palace, where "the large fireplaces which had served to roast whole oxen had now been adapted to the more delicate demands of physiological chemistry".[50]

Fig. 105. Eugen Baumann (1846-1896).

Baumann made his first great discovery during 1875-76 as a young assistant: the paired sulphuric acid in the urine. One year later, he became head of the chemico-physiological section of the Physiological Institute in Berlin, under the direction of Du Bois-Reymond, and in 1883, professor of medical chemistry at Freiburg. In addition to his findings on the cyanogen combinations, benzol derivatives and the structure of indoxyl, Baumann discovered an organic compound containing iodine in the thyroid gland the "iodothyrin" (1895-96) a substance which was later used to correct thyroid insufficiency. He died shortly after Hoppe-Seyler, in 1896.

Baumann's successor in Hoppe-Seyler's laboratory at the University of Strassburg was *Albrecht Kossel* (1853-1927). Originally from Rostock, Kossel came to Strassburg in 1877 and completed his "Habilitationsschrift" for physiological chemistry and hygiene in 1881. Three years later, he was appointed to succeed Baumann at Du Bois-Reymond's institute in Berlin, a post which he left to go to Marburg in 1895. There, Kossel assumed the chair vacated by the death of the physiologist and chemist *Eduard Külz* (1845-1895), renowned for his work on glycogen.

Finally, Kossel succeeded Willy Kühne at the University of Heidelberg in 1901. His main field of interest was the chemistry of the cell nucleus, nuclear acids, protamines and histones. Kossel became the recipient of the 1910 Nobel Prize in physiology and medicine for "contributions to the chemistry of the cell which he had made through his work on proteins, including the nucleic substances". Moreover, with H.D. Dakin he discovered histidine and arginase, making fundamental contributions to the chemistry of purines.[51] Among his students were D. Ackermann, H. Steudel, F.A. Kutscher and S. Edelbacher.

Hoppe-Seyler's successor at the Pathological Institute in Berlin was *Ernst Salkowski* whose primary endeavor lay in the chemistry of the protein metabolism, especially the formation of urinary substances, sulphuric acid and the fate of phenol compounds in the body. Lastly, *Alexander Schmidt* (1831-1914) carried out basic studies on bloodclotting with Hoppe-Seyler and studied the gas exchange in the muscle with Carl Ludwig.

Before considering the younger German chemists interested in physiology and led by *Franz Hofmeister,* a few more notable authors of the older generation should be mentioned. One of them, *Carl Schmidt* (1822-1894) has already been cited, since along with Friedrich H. Bidder at Dorpat, he contributed a series of fundamental works on the physiology of digestion and the protein metabolism. One of Carl Schmidt's students, *Gustav von Bunge* (1844-1920) was interested in the physiology of mineral salts in the organism. A native of Dorpat, von Bunge became professor of physiology at the University of Bern in 1885. He studied the chemistry of milk and iron in the blood and his textbook achieved a wide circulation.

Franz Hofmeister (see Fig. No. 106) became the center of a large group of students and followers, both at Prague and later Strassburg. The son of a physician in Prague, Hofmeister was greatly influenced by Ewald Hering and Ernst Mach during a sojourn at Leipzig. Later, he studied analytical chemistry in Prague under the direction of *Karl Hugo Huppert* (1832-1904), a disciple of the German physician August Wunderlich. In 1883, Hofmeister was appointed to the newly created chair of experimental pharmacology at the University of Prague, a position which he held until his transfer to Strassburg in 1896. There, he became Hoppe-Seyler's successor in physiological chemistry, working together with Friedrich Kraus, A. Czerny and Friedrich Czapek.

Fig. 106. Franz Hofmeister (1850-1922).

Hofmeister concerned himself with the protein and peptone chemistry, the resorption and assimilation of foodstuffs, the effect of salts on organic colloids, the function of leucocytes and other subjects. In Strassburg, he made an arrangement with Friedrich L. Goltz and taught the physiology of the vegetative system. Between 1896 and 1918, Hofmeister had a varied and extensive career at Strassburg, studying the chemistry and biology of proteins such as casein, egg-albumin, peptone, cystine, asparagine, as well as questions of intermediate metabolism and urine formation. Moreover, Hofmeister also carried out research on blood plasma and clotting, hormones, ferments, vitamins and other important topics.

Among Hofmeister's collaborators were O. Fürth, A. Ellinger, J. Pohl, A. Magnus-Levy, O. Porges, F. Alexander, E.P. Pick, H.S. Raper, S. Loewe, B. Embden, C. Neuberg, J. Parnas, W. Stepp, K. Spiro, M. Pfaundler, F. Knoop, P. Morawitz, O. Loewi and J.L. Henderson. All these figures constituted a group of highly talented scientists who surrounded Hofmeister, occasionally ten or more of them at a time. The master was not a friend of appearances, avoided meetings and lived in virtual seclusion devoting all of his time to research. Hofmeister was a stimulating teacher as well as a fanatic partisan of precise and exacting investigations.[52] After the first World War, he had to leave Strassburg, now in French hands, and went to Würzburg, where he worked continuously until his death in 1922 at the age of seventy-two.[53]

One of Hofmeister's most distinguished students, *Otto Loewi* (1873-1961), should be briefly mentioned here. In 1900, Loewi earned his teaching degree in pharmacology under H.H. Meyer at Marburg and went to the University of Graz in 1909. He discovered the cholinergic vagal effect on the heart, a feat which earned him and Henry H. Dale in 1936 the Nobel Prize in physiology and medicine "for their discoveries relating to the chemical transmission of nerve impulses." In 1938, Loewi was forced to leave Austria following the German annexation, and he went to New York City where he became a research professor.

Since the end of the nineteenth century, new positions in physiological chemistry have been erected almost everywhere. A great number of investigators have continued to concern themselves with the chemistry of the organic structures and contributed to the gradual elucidation of problems associated with nutrition, intermediary metabolism, hormonal action and the effects of ferments and vitamins. A few outstanding representatives of this special physiological discipline will be summarily mentioned. One of Hofmeister's well-known disciples was *Alexander Ellinger* (1870-1923), who studied chemistry and medicine, working for some time in Carl Voit's metabolic laboratory at Munich. Ellinger, however, considered himself primarily a student of Hofmeister since he worked with his friend Karl Spiro in Strassburg until his graduation from medical school there in 1897. Later, he went to Königsberg, where he succeeded *Max Jaffé* (1841-1911) until his appointment in pharmacology at the newly created University of Frankfurt in 1914.

Ellinger worked predominantly with problems of structural chemistry: the aminoacid ornithine, and topics of the intermediate metabolism: tryptophane, indol, indican and the processes of acetylization. Other more physiological subjects studied by Ellinger included the action of peptones on blood coagulation. He was one of the editors of Albrecht Bethe's *Handbuch der normalen und pathologischen Physiologie.*[54]

One of Ellinger's students was *Otto Riesser* (1882-1949), one of the most famous experts in muscular physiology. The same topic was also studied by *Gustav Embden* (1874-1933), a native of Hamburg who worked with Hofmeister at Strassburg between 1899-1903. Embden also spent some time with Justus G. Gaule at Zürich and Richard J. Ewald in Strassburg before going to C. von Noorden's laboratory in Frankfurt and becoming its director in 1907. Embden especially studied the carbohydrate metabolism employing pharmacological methods and animal experiments, and combining physical and chemical viewpoints in the formulation of his research. Among Embden's disciples were E. Gräfe, F. Laquer, E. Adler, E. Lehnartz, H.J. Deuticke and H. Jost.

Two important investigators connected with muscular chemistry deserve to be mentioned in this summary: *Otto Meyerhof* (1884-1951) and *Otto Warburg* (1883-1970). Meyerhof was from Hannover and began his teaching career at Kiel in 1913. Later, in 1924, he became a member of the Kaiser-Wilhelm Institut in Berlin and five years thereafter director of the physiological research institute at the University of Heidelberg. Meyerhof was one of the first to demonstrate that the apparently stationary situation of living systems was actually due to a series of cyclical chemical and energetic processes of construction and destruction of substances such as the lactic acid cycle in the muscle. Meyerhof's quotient expressing numerically anaerobic decomposition and aerobic reconstitution has become famous in physiology.[55]

Together with the English physiologist Archibald V. Hill, Meyerhof received the 1922 Nobel Prize in physiology and medicine "for his discovery of the fixed relationship between the consumption of oxygen and the metabolism of lactic acid in muscle."[56] His comprehensive text, entitled *Die chemischen Vorgănge im Muskel,* appeared in Berlin (1930). Between 1928 and 1938, Meyerhof discovered the cycle of guanidine phosphate and adenosine triphosphoric acid. In the latter year, he was forced to leave Germany and died in Philadelphia on October 6, 1951.[57] His student, Hans H. Weber of Tűbingen, has provided us with a valuable obituary,[58] while another disciple, *Karl Lohmann* (*1898) was the discoverer of the adenosine triphosphoric acid.

Otto Warburg, a native of Freiburg, studied chemistry and medicine, beginning his teaching career at the University of Heidelberg under the direction of *Rudolf von Krehl* (1861-1937). Later, Warburg became director of the Kaiser-Wilhelm Institut for cellular physiology in Berlin. For his contributions in this field, and especially "for the discovery of the nature and mode of action of the respiratory enzyme," Warburg received the Nobel Prize in physiology and medicine for the year 1931. In addition to this important work, he carried out important research concerning energy conversion and photosynthesis.

Among the rather large number of biochemists and physiological chemists who made substantial contributions, only a few will be singled out. One was *Hans Thierfelder* (1858-1930), who worked on the elucidation of the chemical structure of a number of important biological substances, especially the cerebrosides and phospholipids.[59] Another author who carried out a series of classical investigations in pure chemistry was *Emil Fischer.* Fischer studied the chemistry of sugars and their optical qualities, and received the Nobel Prize in chemistry for the year 1902. Although his work on the structure of proteins was already in progress, Fischer's main discoveries in this field came somewhat later when he was able to recognize the formation of polypeptides and synthetize these proteins artificially in the laboratory.

One of Fischer's students was the Swiss *Emil Abderhalden* (1877-1950) (see Fig. No. 107) who published more than a thousand papers on protein chemistry, metabolism, ferments, hormones and vitamins.[60] In addition, Abderhalden completed textbooks in physiology and chemical physiology, and was the editor of works on biological methodology entitled *Handbuch der biologischen Arbeitsmethoden* and *Handbuch der biochemischen Ar-*

beitsmethoden. Between 1911 and 1945, Abderhalden held the chair of physiology and physiological chemistry at the University of Halle, and became well-known for his efforts in ameliorating the miseries resulting from both World Wars. He was an honorary and corresponding member of over sixty scientific societies throughout the world.

Fig. 107. Emil Abderhalden (1877-1950).

In the field of pure biochemistry one should also mention *Adolph Windaus* (1876-1959) of Göttingen, and his student, *Adolf Butenandt* (*1903) of Tübingen and Munich. Both Windaus and Butenandt carried out a series of successful biochemical investigations into the structure of sterols and their connections with vitamins and sexual hormones, a work which brought Windaus the 1928 Nobel Prize in chemistry while Butenandt was awarded his Nobel Prize, together with Leopold Ruzicka, in 1939.

Another Nobel Prize winner in chemistry, for the year 1907, was *Eduard Buchner* (1860-1917)[61] who discovered the ferment called "zymase" in 1897. Much had been added since Pasteur's fundamental investigations regarding the nature and the causes of fermentation. However, while Pasteur insisted that such a process was intimately connected to the vital activity of yeast fungi, Liebig believed that the phenomenon was the effect of unknown but purely chemical substances with catalytic activity which were actually elaborated by the microorganisms. In turn, Buchner was able to separate the fermentation-producing agent "zymase" from the yeast, and then proceeded to demonstrate that such a substance could achieve the chemical reduction of the sugars without the presence of living organisms. Thus, fermentation appeared to be a truly chemical process, catalyzed by enzymes.

Toward the end of the nineteenth century, the structural chemistry of most biologically important categories of substances achieved its greatest development. In this regard, the names of *Emil Fischer* and *Hans Fischer* (1881-1945) should be remembered. Then, in the more recent decades of our century, there has been a shift in objectives within biochemical research. Structural biochemistry has gradually given way to another direction concerned more with the dynamics of biochemical matter, its formation and transformation in the intermediate metabolism.

Thus, the chemistry of enzymes, vitamins and hormones has been gradually making inroads into the central problems of matter. After *James B. Sumner* (1887-1955) first demonstrated the possibility of creating a ferment twenty-five years ago — he crystallized the ureolytic enzyme urease and demonstrated its protein character — the structure and action of many enzymes has been elucidated.

Just as important have been Otto Warburg's discoveries concerning the oxidation mechanism with the help of cytochrome. Finally, the Hungarian physiologist *Albert v. Szent-Györgyi* (*1893) discovered the central role of adenosinetriphosphoric acid in the mechanism of muscular contraction. Of great significance for future research is the present employment of tagged radioactive isotopes which can be traced through the body.

3.
OTHER GERMAN SCHOOLS OF PHYSIOLOGY: CARL VOIT, OTTO FRANK, FRIEDRICH L. GOLTZ AND EWALD HERING

The second half of the nineteenth century was one of the golden ages of German physiology. In addition to the great schools of physiology of Johannes Müller and Carl Ludwig, there were a number of smaller ones associated with such prestigious figures as *Carl von Voit, Otto Frank, Friedrich L. Goltz* (1834-1902) and *Ewald Hering.*

The oldest of these men, *Carl (von) Voit* (see Fig. No. 108), was born on October 31, 1831, in the Bavarian city of Amberg. In 1848, he began his medical studies in Munich, transferring for a year to Würzburg where he attended the lectures of Kölliker, Virchow, Scherer and others. After his return to Munich, Voit studied under Justin Jolly, the anatomist Theodor Bischoff and above all, the chemist Justus Liebig. Thereafter, he went to the laboratory of *Max von Pettenkofer* (1818-1901), and the two men became life-long friends.

Fig. 108. Carl Voit (1831-1909).

After another year of training in analytical chemistry with Friedrich Wöhler at the University of Göttingen, Voit returned to work as an assistant in Bischoff's laboratory. Bischoff's chemical interests were based on his friendship with Liebig, and Voit came to share in the respect for the great chemist whose works he used as guidelines through the still confusing field of metabolism. In 1863, following the death of the physiologist *Emil Harless* (1820-1861), Voit went on to occupy the chair of physiology at the University of Munich where he remained until his death in 1908.[62]

Voit's small school of physiology in Munich aimed at a better understanding of the metabolic processes through the use of more precise techniques of analysis. Thus, Voit himself devoted many years of his research to establishing reliable methods of investigation whereby the material intake and output of the organism could be calculated. A result of these efforts was the establishment of metabolic balances which were much more sophisticated than the previous estimates made by Friedrich H. Bidder and Carl Schmidt in their work *Die Verdauungssäfte und der Stoffwechsel* (1852). To be successful, these new studies had to determine accurately the composition of both the foodstuffs and the excreted substances. Such analyses had already begun in Voit's early years at Munich in collaboration with Bischoff, and they were diligently continued by their many talented students.[63]

In the course of time, these investigations yielded important results. Among other things, it was discovered that the total amount of nitrogen present in the food was recovered in the urine and stool without a deficit. Moreover, Liebig's theory of protein as the primary source for muscular energy was refuted experimentally. In collaboration with Pettenkofer (see Fig. No. 109), Voit constructed the famous open-circuit metabolic apparatus for the analysis of the gas-exchange in man, together with a calorimeter to measure heat loss.[64] These instruments enabled the researchers to explore the metabolic turnover in the organism and trace the pathways of individual types of substances in the same fashion as one would be able to surmise indirectly the work of a factory by examining the entering raw materials, end-products and emerging waste.

Fig. 109. Pettenkofer's metabolic apparatus.

Such an approach had its obvious limitations, but the Munich school of metabolic physiology pursued this knowledge with great dedication and success. Hence, the combined efforts of these scientists led to the establishment of a dietary regime, which became known all over the world under the name "Voitsches Kostmass." Voit's students continued to pursue these metabolic studies, and among them were Max Rubner, Hans von Hoesslin, Erwin Voit, Graham Lusk, Wilbur O. Atwater and Francis Benedict.

Max Rubner (1854-1932) was one of Voit's most distinguished students (see Fig. No. 110). His name is forever connected with the development of the quantitative calorimetric method and the energetic aspect of metabolism. Rubner studied medicine in Ziemsen's clinic before coming to Voit's laboratory. Initially, he participated in the investigations concerned with respiration and metabolic balance with certain selective diets. However, Rubner soon got involved in questions of energy consumption in the body, the caloric value of food and its various components. Unfortunately, Voit was not in agreement with Rubner's conclusions, and delayed their publication for quite some time.[65]

In 1880, Rubner went to Leipzig for one year in order to work with Carl Ludwig, but returned to Voit's laboratory in order to complete his calorimetric investigations. Eventually, Rubner was able to convince Voit of the correctness of his arguments. He was called to Marburg in 1885 as a hygienist, occupying a similar position in Berlin after 1891 until his appointment to the chair of physiology there in 1908 as the successor of Theodor W. Engelmann.

Fig. 110. Max Rubner (1854-1932).

Rubner's discoveries included the law of isodynamics, the law of surfaces as well as the proof that energetic principles could be applied to metabolism through the use of combined direct and indirect calorimetry and metabolic balances. These achievements, including the questions of heat formation and regulation, were the first important points of departure for many questions related to nutrition. Rubner continued to be active in this field for several decades in addition to other research regarding climatology, the hygiene of apparel and the doctrine of disinfection.

Finally, Rubner founded and directed an institute concerned with the physiology and hygiene of labor, the Kaiser-Wilhelm Society in Berlin. There, he worked with *Edgar Atzler* (1889-1939), the true founder of this new branch of physiology.[66] Later, during Atzler's directorship, the institute was transferred from Berlin to Dortmund, where it has been under the direction of Gunther Lehmann since 1940. Other students of Rubner included Karl Thomas, Arndt Kohlrausch and Otto Krummacher. Rubner received the highest scientific as well as political and public honors in recognition for his achievements. He died on April 27, 1932.

Carl Voit's stepbrother, *Erwin Voit* (1852-1932), was also a physiologist interested in metabolism. In 1885, he completed his "Habilitation" under Carl Ludwig in Leipzig, and then went to Munich where he became professor at the Veterinary School. Voit's work was concerned with questions of protein catabolism, calcium deficient diets, lipid formation from carbohydrates, and especially the problem of the biological atomicity of proteins.

Another student, the American physiologist *Graham Lusk* (1866-1932), often mentioned his training years spent in Voit's laboratory. Reminiscing about his investigations there concerning diabetes, Lusk wrote: "At Columbia my instructors had been of two types: gentlemen who were good teachers but who did not know very much, and rather rough people who knew a great deal. In Munich, for the first time in my life, I had found a teacher who represented a highly developed form of culture which was both intellectual and personal." In 1891, Lusk returned to the United States and became professor of physiology in New York. He continued to study metabolic problems, especially diabetes.[67]

Voit's other American students were *Wilbur O. Atwater* (1844-1907)[68] and *Francis G. Benedict* (1870-1957),[69] who, endowed with generous resources and excellent instruments, carried out a great number of experiments on energy and food balances, confirming the validity of the energetic principle for human metabolism.[70]

Next to *Heinrich von Hősslin* (1878-1955) and *Friedrich von Műller* (1858-1941) who applied the experiences of Voit's school in clinical medicine, *Max Cremer* (1865-1935) should be mentioned. Cremer worked in Voit's laboratory from 1890 until 1908, and was successful in a series of physiological fields, many of which he studied from a mathematical and physical standpoint. Cremer investigated glycogen, lipid formation from proteins, phlorhizin-induced diabetes, and later developed electrocardiography (esophageal and intracardiac derivations) as well as establishing the foundations for bioelectrical appearances: excitability and the conduction of stimulations. Cremer's personality has been aptly described and he was succeeded by *John Seemann* (1874-1912) at Cologne.[71]

Otto Frank (see Fig. No. 111) was Carl Voit's most talented student. Like his teacher, who was the founder of the metabolic area in physiology, Frank pioneered another field, the exact mathematical and physical analysis of the circulation. Born on June 21, 1865, in Gross-Umstadt, Frank studied medicine and, after his graduation in 1889, chemistry, physics, mathematics, anatomy and zoology. In 1891, Frank went to Ludwig's laboratory in Leipzig where he soon displayed his special talents in the investigation of lipid resorption. He concluded that all fatty substances were broken down in the gut before they could be absorbed.

After 1894, Frank became Voit's assistant and his extraordinary thoroughness and exactitude found a more congenial milieu than Ludwig's more temperamental and imaginative geniality. If Ostwald's types are indeed applicable, then Ludwig and Liebig belonged to the romantic group while Voit and Frank were in the classic division. Frank remained in Voit's institute until 1905, perfecting his methodology and achieving his first successes. In his "Habilitationsschrift" or thesis for an academic teaching career, Frank transferred his experiences concerning isometric and isotonic contraction in skeletal muscles to the cardiac muscle. This work, entitled *Zur Dynamik des Herzmuskels* (The dynamics of the cardiac muscle), was followed by the well-known investigations of the heart's isometric, isotonic and distention curves.

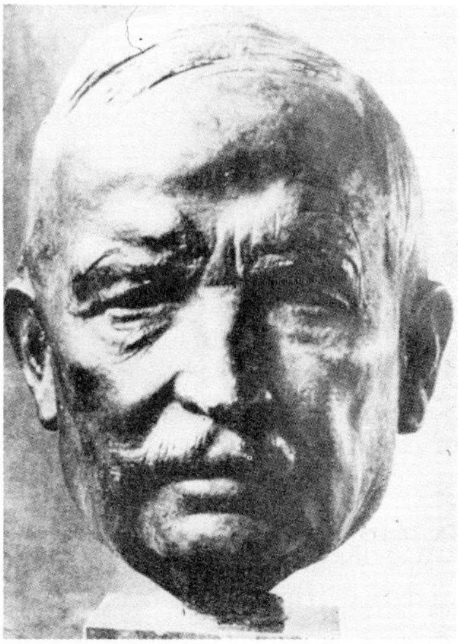

Fig. 111. Otto Frank (1865-1944).

In a later work, Frank established the foundations for an exact calculation of cardiac work and precise criteria for the construction of manometers to register the pulse. In 1904, he was able to obtain the first perfect pulse curves with his impeccable manometers — the so-called Frank capsules — as well as with some optical devices. A year later, Frank responded to a call from the University of Giessen, occupying the chair of physiology until 1908, the date of his transfer to Munich, when he succeeded his former teacher Carl Voit.

In those years, Frank attempted to calculate the physical coefficient for the vascular pulse, establishing the doctrine of vascular elasticity and a theory of pulse wave. Finally, Frank enunciated a method for the determination of the cardiac stroke volume in man and animals on the basis of the wave and air chamber theories. These calculations were related to the theory of arterial blood pressure and its dependence upon pressure amplitude, stroke volume and the total elastic resistance in the vascular system.

In addition to this fundamental work in vascular physiology, Frank also analyzed mathematically and physically the oscillatory properties of sound transmission in the ear. His book *Thermodynamik des Muskels* (1904) was also far advanced for his time. Frank's physiological accomplishments were impressive for their conciseness and have only been supplemented and developed further since his time. Frank's meticulous approach, his exactitude, punctuality and tidiness reached the limits of pedantry and was combined with an incorruptible and feared criticism exercised upon his own activities as well as the work of others. Thus, in spite of his great success, and the recognition he received from the international scientific community, Frank remained full of inner tensions which considerably hampered his relationships with students and colleagues.[72]

A series of Frank's disciples continued his work. *Hermann Straub* (1882-1938) and *Arthur E. Weber* (*1879) applied Frank's results and methods to clinical medicine, while *Carl J. Wiggers* (*1883) propagated the new knowledge in the United States. Other students included the already deceased *Philip Brőmser* (1886-1940), *Emil von Skramlik* in Berlin, *Richard Wagner* in Munich, *Karl Wezler* (*1900) in Frankfurt and *Otto F. Ranke* (1899-1959) in Erlangen.

While Voit's school concentrated on metabolic studies and Frank's circle of students worked on the physico-mathematical aspects of circulation, *Friedrich L. Goltz* and his group investigated the physiology of the nervous system. Goltz (see Fig. No. 112), a native of Posen, was born on August 14, 1834, and attended the University of Kőnigsberg. After his graduation, he became a surgical assistant and later a prosector in anatomy at the university. His free time was devoted almost exclusively to physiological experiments with frogs; investigations which Goltz carried out under very simple and rather primitive conditions.[73]

Fig. 112. Friedrich L. Goltz (1834-1902).

Nevertheless, in purely autodidactic fashion, Goltz made a series of brilliant observations concerning the functions of the nervous system during his years at Kőnigsberg from 1858 to 1870. Among them were the so-called "Klopfversuch" on the mechanism of shock (1862) and the "Quakversuch" or croak experiment on the decerebrated frog (1869). Some of these experiments on frogs were first demonstrated during the 1865 meeting of the German scientists in Hannover. On that occasion, Goltz demonstrated to his audience, in a rather dramatic as well as humorous fashion, the effects of cerebral excision in frogs. These experiments were followed by the description of the clinching reflex and its actual explanation.[74]

Another area of investigation for Goltz was the role played by the semicircular canals in the maintenance of equilibrium (1870). In that year, Goltz was appointed to the chair of physiology at the University of Halle and, in 1872 he went to the newly created University of Strassburg. In the beginning, three institutes were housed in the building of the former Faculté de Médecine: those of Goltz, Hoppe-Seyler and Schmiedeberg. Not until 1884 was Goltz able to move into better quarters at the new physiological institute.

Until 1876, Goltz's investigations were the result of rather fortuitous observations and a series of brilliant ideas which allowed him to penetrate with great sagacity and ingenuity the complex functions of the central nervous system. Thereafter, however, his field of endeavor narrowed progressively and his work became much more systematic, especially the investigations concerning decerebrated dogs which convinced Goltz of the fallacy concerning strict cerebral localizations.

This question remained foremost in Goltz's mind until retirement in 1900 at the Strassburg Institute. Among the highlights of that final period were three dogs which survived the total excision of the brain for three full years. Such experiments required a great deal of surgical skill as well as careful subsequent care of the operated and handicapped animals.[75]

Goltz had an exceedingly attractive personality. His modesty, sense of humor and spirit of cooperation with the students made him a very popular figure at the university. Moreover, Goltz was a good lecturer and carried out his public duties as a dean with disarming open-mindedness and honesty. As stated by one of his students, the physiologist *Richard Ewald* (1855-1921), "Goltz remained a biologist at a time when physiology was being degraded to applied physics and chemistry."[76] After a prolonged illness, he died on May 5, 1902, at the age of sixty-seven.

One of Goltz's most talented students was *Max E. Schrader* who studied the inhibitory center of the heart, and *Justus Gaule* (1849-1939) who became Ludimar Hermann's successor in Zurich. The well-known physiologists *Rudolf Wlassak* (1865-1930), *Rudolf O.A. Höber* (1873-1953), *Wilhelm Brünings* (1876-1958) and Otto Meyerhof also worked in Gaule's institute.

Both German physiologists, *Richard Ewald* and *Albrecht Bethe,* considered themselves members of Goltz's school of physiology. Ewald, a native of Berlin, was an assistant of Goltz after 1880 and both men developed a harmonic and friendly relationship. Ewald was a skillful investigator who excelled both in the design and construction of physiological devices and loved to carry out difficult experiments where he displayed his uncanny surgical abilities. However, he was less enthusiastic about the prospects of writing down his findings and preparing publishable manuscripts. Both Goltz and Ewald allowed their assistants a great deal of freedom in the choice, range and execution of their research activities.

Ewald's early scientific work was concerned with questions of hemodynamics and the thyroid function, and he was one of the first individuals to study the latter subject in a true systematic fashion. Following some suggestions of Goltz, Ewald examined the functions of the labyrinth, clarifying the function of the eighth cranial nerve (Wiesbaden, 1892). Especially interesting were his observations of the recovery which various types of experimental animals experienced after the establishment of lesions in the labyrinth.

Ewald was also interested in the field of physiological acoustics, and his theory of sound impressions became well-known in contrast to Helmholtz's theory of resonance. In spite of his excellent scientific qualifications, Ewald was only appointed professor of physiology and successor to Goltz at Strassburg in 1900. The activities of the Strassburg Institute began to languish after the outbreak of the first World War, and Ewald's long ailment severely curtailed his teaching and research during the subsequent years until his death in 1921.

Another student of Goltz and Ewald, *Albrecht Bethe* was until recently the dean of German physiology. In spite of his years, he was very active and full of ideas to the very end.[77] Bethe was a steadfast representative of a biological direction in physiology, a man who viewed physico-mathematical analysis merely as an auxiliary method which could not be routinely employed in a useful fashion. Among Bethe's students were W. Steinhausen (Greifswald), E. Fischer, R. Thauer (Bad Nauheim) and E. von Holst (Wilhelmshaven).

Ewald Hering's contributions to physiology were largely made in another rather specific area of study, the physiology of vision. This sensory physiology suited well Hering's own capacities which allowed him to become prominent in the field and an equal to Helmholtz after the publication of his first articles. Hering (see Fig. No. 113) was born in Altgersdorf[78] and

Fig. 113. Ewald Hering (1834-1918).

studied medicine at the University of Leipzig between 1853 and 1858, where he attended the lectures of Ernst H. Weber, Gustav T. Fichner, Otto Funke and Carl G. Carus. He especially admired Johannes Müller who was at the end of his career and actually died during Hering's senior year in school. Between 1860 and 1865, he became an assistant to Ernst Wagner in the polyclinic of Leipzig, and was authorized to teach physiology after 1862.[79]

Hering's first important publication, *Beiträge zur Physiologie* (Leipzig, 1861-64), demonstrated his superb talents as an investigator. The work contained five sections which dealt exhaustively with the sense of locality in the retina, the concept of horopter and single vision, as well as the idea of binocular perception of depth on the basis of a great number of observations and new conclusions. In 1865, Hering received a call to become Carl Ludwig's successor at the Academy for Military Medicine in Vienna, and there he developed the principle of the two eyes as one single organ. Moreover, Hering formulated the concept of identical retinal locations, and congenital sensory correspondence of retinas and ocular muscular movements, both expressed in the work *Die Lehre vom binokularen Sehen* (Leipzig, 1868).

Hering's contributions in other areas of physiological research were likewise noteworthy. He studied the histology of the liver (a tubular gland with netlike anastomoses), the autonomous, reflex-controlled process of respiration (in collaboration with Josef Breuer), and the centrally dominated vascular dilatations and blood-pressure variations — the so-called "Hering waves." In 1870, Hering was appointed the successor of Purkinje at the University of Prague, and in contrast to Purkinje's Czech nationalism, he stressed his German background and the significance of German science. In 1882, he was deeply involved in the creation of a German university in Prague, rejecting an invitation to transfer to the University of Strassburg.

During his tenure at Prague, Hering summarized his previous investigations in the visual field in a publication entitled *Die Lehre vom Gesichtssinn,* a work which sought to oppose the growing psychological tendencies which were penetrating into sensory physiology. Wrote Hering: "I have tried to see the phenomena of consciousness as dependent and conditioned by organic processes. I have, moreover, investigated the developments and relationships of consciousness from a biological point of view inasmuch as such an approach is presently feasible."

Hering also expressed his opinions about fundamental contemporary problems in physiology in a series of speeches, excellent both in style and content. These utterances reflected his superb qualities of synthesis as well as philosophical reflection. In 1870, for example, Hering spoke about "memory as a general function of matter," dealing with the concept of a functional dependency between both the material and psychic phenomena. Another discourse in 1884 was concerned with "the specific energies in the nervous system."

Hering gave another especially interesting talk in 1888 about the general role of assimilation and catabolism in organic matter entitled "Zur Theorie der Vorgänge in der lebendigen Substanz." He based these processes not only on his interpretation of color vision, but also included the function of the sensory organs concerned with temperature variations and an explanation of electromotor manifestations. Those phenomena which prompted Hering to adopt such a theory were the result of sensory and electrophysiological observations carried out on nerves and muscles at the University of Prague.

After Ludwig's death in 1895, Hering was appointed to succeed him at the famous Physiological Institute in Leipzig. Already sixty years old at the time, Hering continued his rather busy schedule there and was able to attract many students who admired him as the world's foremost investigator of physiological optics. He received for his accomplishments the Graefe-Medal and the "Pour le Mérite" order, dying in Leipzig in 1918.

Hering was a good teacher who supplemented his lectures with numerous physiological experiments, especially in the area of optics. He spent a great deal of time with his students and collaborators, and his criticisms and polemics were feared just as much as those of Otto Frank. Barely twenty-nine years old, Hering had already accused Georg Meissner, then professor at the University of Göttingen, of gross inaccuracies and erroneous conclusions in his work on visual physiology, since it had not been formulated in accordance with the laws of geometry and trigonometry. In turn, Hering was severely criticized by the veteran physiologist Helmholtz regarding some of the newer approaches and interpretations.

Like Johannes Müller before him, Hering began his analysis of color and space from sensations in order to achieve a clear relationship between these sensory impressions and the occurrences on the retina. Hering felt that the stimulus only acted as a trigger on an independently existing retina. Helmholz, quite to the contrary, started from measurable physical stimuli and then searched for relationships between the variable factors of stimulation and sensation instead.

During Hering's tenure at the University of Prague, numerous students from many countries came to study in his laboratory. Among them were the English neurologist *Henry Head* (1861-1940), the German opthalmologist *Carl von Hess* (1863-1923), and especially the physiologist *Wilhelm Biedermann* (1852-1929). A Bohemian by birth, Biedermann studied in Prague and remained there working with Hering until 1888, when he was appointed professor of physiology at the University of Jena to succeed the physiologist *Wilhelm Preyer* (1841-1897).

Biedermann remained at Jena for almost forty years, a man of great knowledge and an infallible memory. He became well-known for his experimental work in electrophysiology and the analysis of amylolytic enzymes. Biedermann also wrote three extensive volumes in Hans Winterstein's handbook of comparative physiology. A bachelor and an eccentric, his daily schedule was highly unusual.[80]

Biedermann's most renowned students were *Hermann J. Jordan* (1877-1943), *Ernst Mangold* (1879-1961) and above all *Max Verworn*. Verworn, who was born in Berlin on November 4, 1863 (see Fig. No. 114), claimed to be actually a disciple of Ernst Häckel, from whom he received, to a large degree, his interest in unicellular organisms. In any event, Verworn had already completed his education when he came to work in Biedermann's laboratory. In 1901, he became heir to Meissner's chair of physiology at the University of Göttingen, a post in which he experienced the most productive years of his career. Later, in 1910, he went as Pflüger's successor to the University of Bonn, dying shortly after the first World War at the age of fifty-nine.

Verworn can be considered the true founder of general cellular physiology. In addition, he was especially interested in the physiology of the nervous system: excitability, fatigue, exhaustion, narcosis and hypnosis. Ver-

Fig. 114. Max Verworn (1863-1923) and his circle of collaborators at the Physiological Institute in Bonn. From left to right: A. Lipschütz, E. Küppers, K. Grube, P. Junkersdorf, G. Elrington, B. Schöndorff; seated: J.Vészi, H. Frédéricq, L. Löhner, M. Verworn, A. Pütter.

worn also wrote a series of monographs which reflected his keen intellect and tendency to discern the general and theoretical issues behind the phenomena. The titles of some of these publications speak for themselves: "Allgemeine Physiologie" (general physiology), "Vitalismus und Neovitalismus" (vitalism and neo-vitalism), "Kausale und konditionale Weltanschauung" (causal and conditional world-view), "Biogenhypothese" (biogenic hypothesis) and others. They all were ample proof of Verworn's broad philosophical concerns.[81]

In addition to his physiological work, Verworn was especially interested in documents pertaining to the historical development of man. Hence, he acquired a large collection of stone-age tools, prehistoric paintings and amulets. A large number of students and guests came to his laboratory in Jena and later Bonn. Their names form part of Hering's "genealogical" tree, and include Hans Winterstein, August Pütter, Friedrich W. Fröhlich, Leopold Löhner, Kurt Wacholder and Rupprecht Matthaei.

In this connection it is appropriate to discuss the Swiss physiologist *Walter R. Hess* (*1881), one of the most important physiologists in the first half of the twentieth century. After studying in Switzerland, Berlin and Kiel, Hess first devoted his time to ophthalmology. In 1911, he became an assistant to Verworn at the physiological institute in Bonn, and a year later worked with Justus G. Gaule in Zürich. Hess' "Habilitationsschrift" or thesis for an academic teaching career was entitled "Das Prinzip des kleinsten Kraftverbrauches im Dienste haemodynamischer Forschung" (the principle of minimum energy consumption in the service of hemodynamic research), and it appeared in 1913. Later, in 1917 he was appointed Gaule's successor to the chair of physiology at the University of Zürich.[82]

Hess' idea was to establish an economical factor as the regulative principle for the interpretation of physiological functions. Such an approach was applicable to the general questions studied by Verworn as well as the goal-oriented view towards living matter proposed by Gaule. Indeed, such an economical principle as enunciated by Hess was the main thread which pervaded his entire and highly successful work in all areas of physiology. In the publication *Die Regulierung des Kreislaufs* (1930), Hess interpreted the cardiac and general circulation as a sensitively regulated unit where ergotropic nutritional reflexes were placed in opposition to trophotropic discharge reflexes.

Especially important were Hess' analyses of the performance of the autonomic nervous system and its central regulatory centers, viewed from the standpoint of a meaningful alternative steering by actions and rest. To this effect, he found subcortical diencephalic activation and inhibition centers for functions such as sleep, food absorption, aggression and flight with concomitant changes in circulation, respiration, tone, etc. Hess' results were published in the work *Vegetative Funktionen und Zwischenhirn* (1947).

In 1949, Hess shared the Nobel Prize in physiology and medicine with A. C. de Egas Moniz, receiving his award "for his discovery of the functional organization of the interbrain as a co-ordinator of the activities of the internal organs." Hess' contributions have generally stimulated new research, especially in the United States. His improved methodology — use of the viscosimeter, techniques of localized stimulation — have been widely accepted. Hess shared with Claude Bernard and Charles S. Sherrington a tendency towards establishing syntheses on the basis of an "organizational" principle.

After this digression, it is necessary to return to Hering's collaborators and disciples. Among them was *Franz B. Hofmann* (1869-1926) who accompanied Hering from Prague to Leipzig in 1896 and became an assistant and later instructor there.[83] After occupying several other positions, Hofmann became heir to Max Rubner's chair at the University of Berlin in 1923.

Hering's influence on Hofmann's work was rather marked, and thus the student shared with his teacher an interest in problems of sensory physiology as well as concern for the muscular aspects of tetanus. Finally, Hofmann also contributed important new knowledge to an understanding of the myogenic origin of the heart beat, and its transmission within the heart. A Japanese assistant, H. Yoshida, also participated in these electrophysiological investigations. Hofmann's work on the effect of various ions on cardiac activity was continued by the late *Bruno Kisch* (1890-1966) in the United States.[84]

In 1896, another disciple, *Siegfried Garten* (1871-1923), returned to Hering's institute after a temporary professorship at the University of Giessen.[85] Garten's emphasis lay on experimental methodology, a field in which he made extraordinary contributions: photokymograph, phonograph and electrical sphygmograph. His investigations regarding the bleaching of visual purpura, the work concerning the pre-existence of bioelectrical tensions and many others still have contemporary significance. Among those who were students and guests at Garten's laboratory in Leipzig were B.J. Daniel Achelis (Heidelberg), Rudolf Dittler (Marburg), H.W. Sulze (Leipzig), and Y. Renqvist-Reenpää and Dusser du Barenne.

An Austrian physiologist, *Armin von Tschermak-Seysenegg* (1870-1952), spent three years in Hering's laboratory at Leipzig between 1896 and 1899, and then went to study with Julius Bernstein at Halle (1900-1906). After some work at the Viennese Veterinary School, Tschermak-Seysenegg went to Prague in 1913, and resumed his physiological career in Regensburg after the second World War. His primary fields of interest were general and sensory physiology, especially physiological optics. The latter topic was also studied by his students *Max H. Fischer* (West Berlin) and *Gustav Schubert* (Vienna).

Another of Hering's disciples was *Theodor von Brücke* (1880-1941), a grandson of the famous physiologist Ernst Brücke. He started his teaching career in 1907 at the University of Leipzig, and came to Innsbruck in 1916 as the successor of Wilhelm Trendelenburg. There he remained until his retirement in 1938.[86] Brücke's fields of interest were physiological optics, muscular- and neurophysiology, and vegetative functions. His works on the theory of nervous inhibition were particularly significant.

Together with Friedrich Plattner, Brücke studied the humoral transmission of the inhibitory action on the heart. Regarding the functions of the central nervous system, Brücke was close to Albrecht Bethe's position when he declared that "all our efforts to interpret the adaptability of the nervous function as the partial expression of general tendency towards organic conformity have to be derived from other sources than the mere structural knowledge of a machine." In this vein Brücke attempted to achieve an experimental vegetative "conversion" in the nervous system. One of his students, *Ludwig Haberlandt,* became known for his studies on a "cardiac hormone."

Finally, there was Hering's son, *Heinrich E. Hering* (1886-1948), who also became a physiologist. He taught general and experimental pathology in Prague and was called in 1913 to head the institute for pathological physiology in Cologne. Hering, Jr., was the discoverer of the carotid sinus reflex, and contributed also to knowledge of the formation and transmission of cardiac excitability. Among his students were Bruno Kisch, Eberhard Koch (Giessen) and Heinz Mies (Cologne).

In addition to the already mentioned true disciples of Ewald Hering, there was another group of renowned physiologists who worked with him for short periods of time. Prominent among them were *Boris P. Babkin* (1877-1950), for many years an assistant of Pavlov, then a professor in Odessa and since 1924 active in the United States, and *Eugen Steinach* (1861-1944), Hering's long time assistant who became famous for his rejuvenation experiments. Moreover, there were also *Jacob Singer* (1853-1926), *Egmont Münzer* (1865-1924), both interested in neurology, and *Richard Burian*, who in 1920 occupied the newly founded chair of physiology at the University of Belgrad and had a large number of Japanese students.

4.
BRITISH PHYSIOLOGY SINCE 1848

"In the middle part of the nineteenth century Great Britain was far behind France and Germany in the development of physiology. We had no pure physiologists and it was considered that any surgeon or physician was competent to teach the science. Hence, during a period of time when other experimental sciences were progressing, physiology in this country could show no names worthy to be mentioned with those of Magendie, Bernard, Müller, Helmholtz or Ludwig, to mention but a few of the brilliant physiologists of France and Germany."

This characterization of the British scene came from the pen of the physiologist Sir Edward *Sharpey-Schafer* in his history of the Physiological Society in England.[87] The situation was clearly a break with the brilliant past which included such names as Willis, Lower, Boyle, Harvey, Hooke and Mayow in the seventeenth century, and Whytt, Hewson, Hunter and Walker in the eighteenth. Even Charles Bell and Marshall Hall in the early nineteenth century formed part of that seemingly lost tradition.

Towards the middle of the last century, some British anatomists and clinicians dealt with physiology as a side-line, but they had no useful laboratories in which to pursue their studies.[88] The leading scholar who attempted to improve the physiological teaching and research at that time was *William Sharpey* (1802-1880), an anatomist and physiologist at the University College in London since 1836.[89] Much of the subsequent development of British physiology was intimately associated with his name. In 1866, Sharpey made the rather crucial decision to bring the young thirty year old *Michael Foster* to teach physiology at University College. (See Fig. No. 115).

Fig. 115. Michael Foster (1836-1907).

With Foster, British physiology experienced a veritable rebirth. Foster not only had great knowledge and organizational talents, but he was an excellent teacher, and well acquainted with Continental physiology. His special interests lay in a histophysiology envisioned by Ernst Brücke, and he created the first physiological laboratory in the British Isles. Moreover, Foster was instrumental, together with Sharpey, Burdon-Sanderson and others, in the founding of the Physiological Society at London in 1876.[90]

The first English journal exclusively dedicated to physiology — the *Journal of Physiology* — was another of Foster's accomplishments in 1878. His influence extended beyond his immediate students such as *John N. Langley* (1852-1925), *Walter H. Gaskell* (1847-1914), *Henry Newell-Martin* and *Sheridan Lea,* and affected medical schools beyond Cambridge. Foster was also instrumental in organizing the first International Congress of Physiologists at Basel in the year 1889, and was, for many years, the secretary of the physiological section existing within the Royal Society of London. He died at the age of seventy-one[91] and his textbook of physiology, published in 1877, was the first authoritative work in the field written in English. Foster was also well-known for his experimental investigations concerning cardiac, circulatory, muscular and neurophysiology.

Also working with Sharpey and Foster was the physiologist *John Burdon-Sanderson.*[92] A native of Jesmond in Northumberland, Burdon-Sanderson studied medicine at the University of Edinburgh and then on the Continent. (See Fig. No. 116) After his return to London and following Foster's transfer to the new lectureship of physiology at Trinity College, Cambridge, he became Foster's successor at University College in 1870. Four years later, Burdon-Sanderson also succeeded Sharpey as the Jodrell Professor of Physiology, occupying the second academic position dedicated exclusively to physiology in the British Isles at that time. Later, between 1882 and 1895, he was professor of physiology at Oxford. His main interests were in the field of electrophysiology which he studied on animals and plants.[93]

Fig. 116. John Burdon-Sanderson (1828-1905).

Almost twenty years behind events in Germany, British physiology seemed to emerge from its temporary slumber and both Trinity College in Cambridge and in London University College[94] became the centers of a rapidly developing science of physiology. The growth during the last decade of the nineteenth and the early years of this century was such that Britain's physiology soon achieved world fame and finally displaced Germany as the leading country in physiological research after the deaths of Brücke (1892), Helmholtz (1896), Ludwig (1895) and Du Bois-Reymond (1896). This successful development was achieved by a group of young British physiologists who by and large had trained in the French and German laboratories.

Burdon-Sanderson also attracted a number of excellent students to London, such as *Edward Sharpey-Schafer, Francis Gotch* (1853-1913), *Victor A. Horsley, William M. Bayliss, John S. Haldane* (1860-1936) and others who became famous for their research. Another physiologist, *Augustus D. Waller* (1856-1922), was also inspired by Burdon-Sanderson's electo-physiological work. In 1888, Waller was able to obtain the first human electrocardiogram using a capillary electrometer which was extended from mouth to foot on the volunteers. He registered the electrical potentials of the heart through optical enlargement, an achievement which became the starting point for further investigations conducted by the Dutch physiologist *Willem Einthoven* (1860-1927). Earlier, in 1878, Waller carried out studies concerning auricular tension in the heart at Ludwig's institute in Leipzig.

Another of Burdon-Sanderson's collaborators was Sir *Edward Sharpey-Schafer,* who in 1918 added the last name of his teacher, William Sharpey, to his name. Sharpey-Schafer was an assistant professor under Burdon-Sanderson until the latter moved to Oxford in 1883. He then succeeded to the chair of physiology at University College where he taught for sixteen years.[95]

Sharpey-Schafer's interests lay in the areas of histology, brain research and internal secretions. His *Textbook of Physiology* was published in 1898-99. Among those who worked with him were *Sidney Ringer* (1835-1910), famous for the parental solution which still carries his name, and *Leonard E. Hill* (1866-1952) who worked with Bayliss and was interested in problems of organ perfusion as well as the environmental influences of atmospheric pressure and temperature on man. Hill later became professor of physiology and director of the section of applied physiology at the National Institute of Medical Research in Mt. Vernon, U.S.A. He developed the cuff method for the determination of blood pressure independently of Riva Rocci.

Fig. 117. Victor A.H. Horsley (1857-1915).

Together with Sharpey-Schafer and Francis Gotch, *Victor A. H. Horsley* (see Fig. No. 117), the renowned pathologist, surgeon and physiologist worked at University College in London. Horsley was the first to induce myxedema artificially by removing the thyroid gland in apes. After 1884, he studied cerebral and spinal localizations with the help of surgery.[96]

Horsley had worked with Carl Ludwig while Gotch spent some time with Hugo Kronecker at the Physiological Institute in Berlin. Both Gotch and Horsley were embroiled in a dispute over priority of discovery in 1888. They had derived an electrical potential from the exposed cerebral surface of an animal unaware that *Richard Caton* (1842-1926), professor of physiology at University College in Liverpool, had already achieved similar results in 1874. The same situation was also repeated on the Continent, where in 1810 the physiologist *Adolf Beck* (1863-1944) in Krakow and *Ernst Fleischl von Marxow* in Vienna arrived at similar findings.[97]

Another student of Foster and Burdon-Sanderson was *John N. Langley* who studied in Cambridge and replaced Newell-Martin in 1875 after Martin's departure for the United States. In 1903, Langley became Foster's successor at Trinity College in Cambridge (see Fig. No. 118). He was especially interested in the effects of drugs and poisons on the nervous system. Among other topics, Langley studied the actions of pilocarpine and recognized the blocking action of nicotinic acid on synapses in 1889. Later, in 1900, he described the axonic reflexes and a year later the role of adrenalin in the stimulation of the cells belonging to the sympathetic system, salivary glands, etc. In 1921, Langley coined the concept "autonomic nervous system," and the term "parasympathetic."

Sir *Henry H. Dale* (1875-1968) also worked in Cambridge, together with Foster, Gaskell and Langley. Later, Dale collaborated with Bayliss and Starling in the department of physiology at University College in London. After a sojourn at Frankfort, where he was active with the immunologist *Paul Ehrlich* (1854-1915), Dale took over the directorship of the Wellcome Physiological Research Laboratory, where he investigated ergotoxin, histamine and a pituitary substance called oxytocin.

In 1919, Dale described the sympathicomimetic action of amines and the role of chemical transmitters in nerve action. He discovered the inhibitory activity of acetylcholine on the heart, clearing the way for the German physiologist Otto Loewi to conclude that acetylcholine was identical with the so-called 'Vagusstoff" produced at the nerve endings of the vagus nerve. Both Dale and Loewi received the Nobel Prize in physiology and medicine for this research. Moreover, Dale coined the terms "cholinergic" and "adrenergic" in order to designate the proper chemical transmission. Dale became director of the National Institute for Medical Research in 1928 and was knighted in 1932.

Other collaborators of Foster and Langley at Trinity College in Cambridge were Sir *Frederick Gowland Hopkins* (1861-1947) (see Fig. No. 119) and *Walter M. Fletcher* (1873-1933). Hopkins, who was born in Eastbourne, studied zoology and chemistry in his early years, followed by the study of medicine in which he received his M.D. degree in 1894. Four years later, he was called by Foster to Cambridge to teach physiological chemistry. At that time, there were no true biochemists with the possible exception of one of Ludwig's students, *Ludwig J.W. Thudicum* (1829-1901) who had

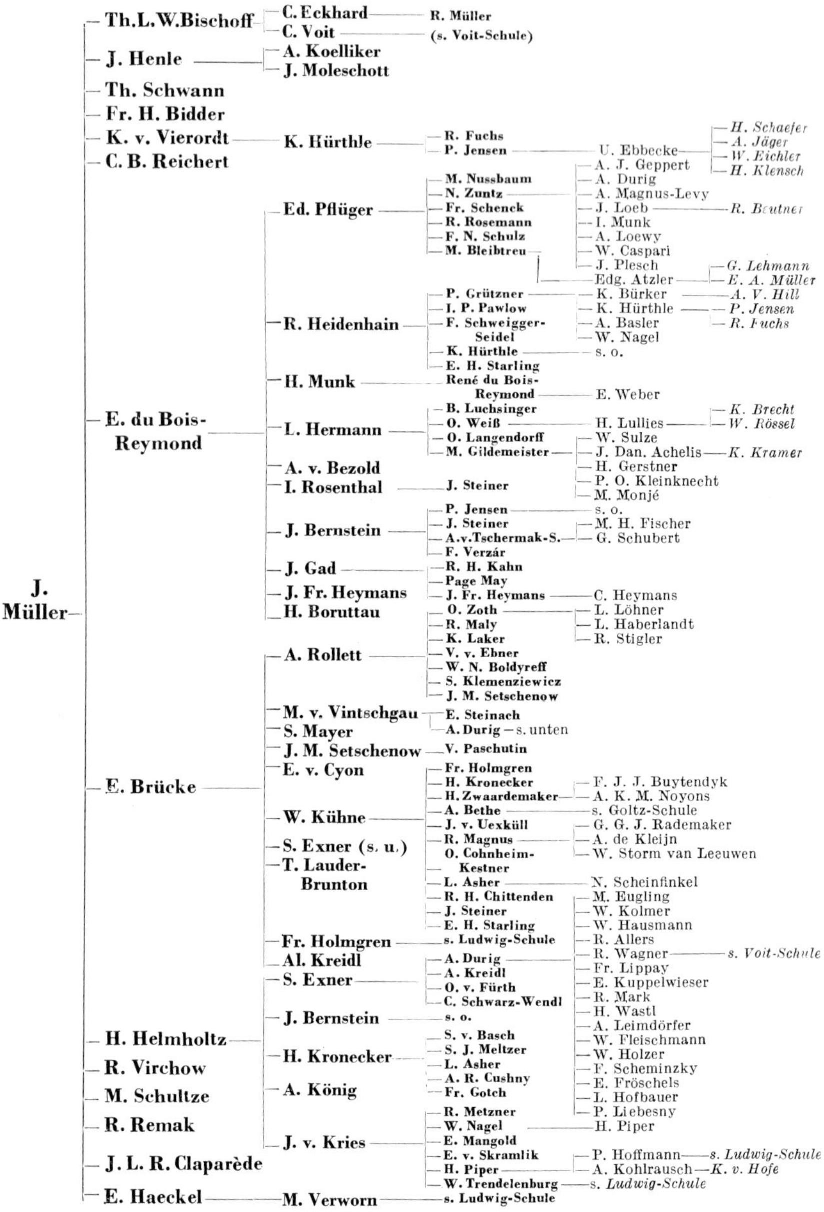

Table of Müller's disciples.

C. Ludwig

- **C. Eckhard** — R. Müller
- **Ad. Fick**
  - R. Boehm — C. G. Santesson
  - J. Gad — H. Boruttau
  - M. G. Blix
  - J. Loeb
  - Fr. Schenk
  - A. Gürber
- **W. Kühne** — s. Müller-Schule
- **I. M. Setschenow** — O. Hammarsten — Iv. Bang
- **Fr. Holmgren**
  - J. E. Johansson
  - Hj. Öhrwall — T. L. Thunberg
- **V. Paschutin**
- **Chr. Lovén** — R. A. Tigerstedt
  - E. Lindhagen
  - C. G. Santesson
  - J. E. Johansson
  - C. Tigerstedt
  - Y. Renqvist-Reenpää
  - E. M. Widmark
- **H. Kronecker** — s. Müller-Schule
- **O. Hammarsten** — s. o.
- **G. Hüfner**
  - K. Hürthle — R. Fuchs
  - K. Bürker
- **E. Drechsel**
  - M. Siegfried
  - Ernst Fischer
- **L. Luciani** — S. Baglioni
- **Ang. Mosso**
- **J. v. Kries**
  - E. Mangold
  - W. Trendelenburg
    - L. Haberlandt
    - R. Wagner
    - E. Holzlöhner — *H. J. Trurnit*
    - E. Schütz
      - *F. Buchthal*
      - *Chr. Maltesos*
      - *K. E. Rothschuh*
      - *Eb. Lerche*
    - B. Lueken
    - I. Schmidt
    - H. O. Meitner
    - U. Luft
    - H. Hellauer
  - V. v. Weizsäcker
  - R. Metzner
  - E. v. Skramlik — W. Eichler
  - W. A. Nagel — H. Piper
    - *P. Hoffmann*
    - *A. Kohlrausch*
  - H. Piper
- **W. H. Welch**
- **V. Horsley**
- **T. Lauder-Brunton**
- **E. A. Schaefer**
- **M. v. Frey**
  - Y. Renqvist-Reenpää — M. Schneider
    - *W. Noell*
    - *W. Rössel*
  - R. Metzner
  - E. Wöhlisch
  - J. Bizzozero
  - H. Rein
    - K. Kramer
    - W. Schoedel
    - Chr. Maltesos
    - G. Kahlson
    - E. Opitz
    - Fr. Valdecasas
    - D. Smyth
    - H. Mercker
    - J. Antal
    - G. Domini
    - H. Loeschke
    - J. Aschoff
    - H. Brüner
    - D. Mertens
  - H. Schriever
  - H. Strughold
    - E. Opitz
    - U. Luft
    - Fr. Palme
  - P. Hoffmann — H. Rein — s. o.
- **Al. Schmidt**
- **Fr. Miescher**
- **A. D. Waller**
- **R. H. Chittenden**
- **W. Pl. Lombard**
- **Chr. Bohr**
  - A. Krogh
  - J. Lindhard
    - F. Buchthal
    - E. Hansen
    - E. H. Christensen
  - K. Hasselbalch
  - V. Henriques
- **Fr. P. Mall**
- **R. A. Tigerstedt** — s. o.
- **J. Gaule** — R. Höber
  - R. Mond — *H. Luckner*
  - H. Netter
  - F. Scheminzky
- **J. Dogiel**
- **P. H. Bowditch**
  - W. B. Cannon
  - H. Cushing
  - W. T. Porter
  - W. H. Howell
  - F. Pratt
- **W. H. Gaskell**
- **El. v. Cyon**
- **I. P. Pawlow**
  - A. D. Speransky
  - G. v. Anrep
  - W. N. Boldyreff
  - L. Popielski
  - B. P. Babkin
  - L. A. Orbeli
  - O. Cohnheim-Kestner
- **W. Stirling**
- **A. Budge**
- **M. Rubner** — s. Voit-Schule
- **O. Frank** — s. Voit-Schule
- **L. Asher** — N. Scheinfinkel
- **C. Voit** — s. Voit-Schule

Carl Ludwig and his disciples

analyzed the chemical components of the brain and, in 1884, described cephalin, sphingomyelin, kerasin and phosphatides among other substances.

Fig. 119. Frederick Gowland Hopkins (1861-1947).

Hopkins became the true founder of biochemistry in Britain, an author who in 1925 received the directorship of the "Institute of Biochemistry" at Cambridge in order to carry out his research. He is considered the principal discoverer of the so-called "accessory food factors," substances which, next to proteins, lipids, carbohydrates and minerals are absolutely necessary in the diet (1912).[98] In 1929, Hopkins shared the Nobel Prize for physiology and medicine with the beri-beri disease researcher *Christiaan Eijkman* (1858-1930), "for his discovery of the growth-stimulating vitamins." He also studied and isolated tryptophane, an essential amino acid widely distributed in animal and some plant proteins as well as arginine, histidine and glutathione.

In 1907, the physiologist *Walter M. Fletcher,* together with Hopkins, discovered lactic acid, an important compound in the glycolysis of the working muscle. Based on this finding, Archibald V. Hill and Otto Meyerhof were able to carry out their brilliant investigations. Fletcher spent some time in Meyer's laboratory at the University of Marburg, and he developed extremely sensitive devices for the measurement of the oxygen-carbon dioxide turnover in the muscle. He concluded that the muscular contractility remained intact in an atmosphere free of oxygen. Fletcher concluded his laboratory work in 1914.

Archibald V. Hill (*1886) also received his education at Trinity College, Cambridge, but after 1923 worked at University College in London. His greatest contribution was the analysis of heat formation during muscular contraction.[99] Hill described the classic four phases of heat production, which he correlated with the mechanical processes occurring in the muscle (1913-1920). Together with Otto Meyerhof, he received the Nobel Prize in physiology and medicine.

In 1897, *Joseph Barcroft* (1872-1947) (see Fig. No. 120) went from Kings College, London, to the physiological laboratory in Cambridge. Under the direction of Langley, Barcroft began his investigations on salivary secretion and the consumption of oxygen in glands. In 1902, he developed with

Haldane a method of analyzing blood gases with the help of ferricyanide. Moreover, he examined the effect of diminished oxygen pressure on the dissociation curve of hemoglobin (1907-1914), investigating also the oxygen consumption in the kidneys, the storage of blood in the spleen and other important topics. In 1925, Barcroft became Langley's successor at Cambridge and late in life he chose a brand new field of endeavor, the physiological problem of fetal respiration (1932-1947).[100] Among Barcroft's British collaborators were K.J. Franklin, G.R. Mines, J.H. Burn, A.V. Hill, F.J. Roughton; while Morawitz, Straub and Feldberg came from Germany and L. Orbeli from Russia.

Fig. 120. Joseph Barcroft (1872-1947).

From Scotland arrived *John S. Haldane,* the son of Mary Burdon-Sanderson. Haldane was primarily concerned with questions of gas analysis. Haldane, graduating from the University of Edinburgh in 1884, was also interested since his early years in philosophical questions related to life.[101] After physiological and chemical studies in Berlin, Haldane returned to England where he went to work in the Cambridge laboratory of his uncle, John Burdon-Sanderson. His special field of interest was the physiology of respiration, and he developed a gravimetric method for the determination of carbon dioxide.[102]

In 1905, Haldane examined the stimulating effect which carbon dioxide had on respiration as well as the combination of carbon monoxide with hemoglobin. He constructed, in 1898, a device for the analysis of blood gases which is still employed today. In 1907, he was named an "Appointed Reader in Physiology" at Oxford, and in 1935 his work on theoretical biology entitled *The Philosophy of a Biologist* was published. In it, Haldane propounded a holistic philosophy based on his studies of the gases existing in the blood.

William M. Bayliss[103] and *Ernest H. Starling* (1866-1927) (see Fig. No. 121) constituted a truly remarkable pair of brilliant investigators who worked closely together and were great personal friends as well as relatives.[104] Bayliss was born in Wednesbury near Birmingham and studied medicine at University College London. In 1885, he went as an undergraduate to Wadham College Oxford and became an assistant to Burdon-Sanderson in his department of physiology at University College, London, in 1888. Since the laboratory facilities, especially the instruments, were deficient there, Bayliss kept a private laboratory in his own house. Unfortunately, there were violent disputes at the time concerning the antivivisectionist law of the year 1876, favored especially by W.B. Coleridge. Bayliss was accused of having operated on a dog without proper anesthesia, but he was eventually acquitted in what was called the "brown dog case." In 1912, Bayliss was appointed professor of general physiology at University College, London.

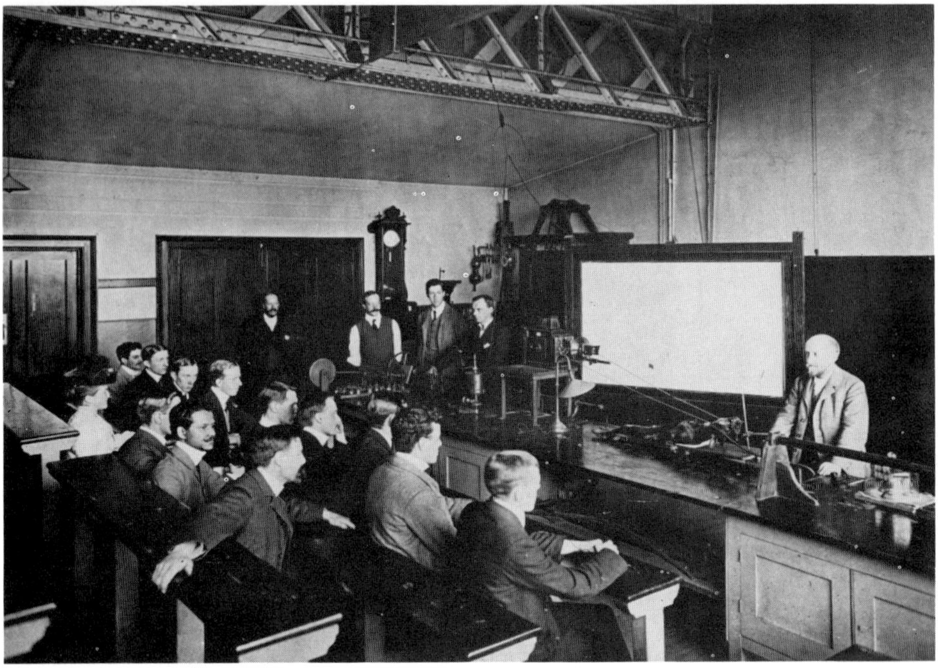

Fig. 121. Bayliss (right) and Starling (left) with Sir Henry Dale, May 1903.

During the same time Ernest H. Starling was also working in the same laboratory. Starling, somewhat younger and better educated than his friend, was born in Bombay, India, and studied medicine at Guy's Hospital Medical School after 1882. Not satisfied with the predominantly practical instruction there, he went to work with Willy Kühne at Heidelberg in 1885, returning to his country with great enthusiasm for physiological research. After a short appointment as physiological demonstrator at Guy's Hospital, Starling went to work at the laboratory of physiology of University College in London during the year 1890. There, he met Bayliss, and the two men extablished a close working relationship.

Starling and Bayliss complemented each other admirably. While Bayliss was rather cautious, slow and careful in his work, Starling was enthusiastic, full of ideas and methodical in his approach. In 1892, Starling returned to Germany in order to work with Rudolf Heidenhain at the University of Breslau investigating the formation of the lymphatic fluid. Later, in 1899, he became Sharpey-Schafer's successor to the Jodrell chair, and worked in the new physiological institute created at University College, London, in 1908. Starling remained in this academic post until 1922.

The fruits of Bayliss' and Starling's research were rich and numerous. Both authors discovered the humorally triggered secretion of pancreatic juice by a substance which they called "secretin," and in 1906 Starling coined the term "hormone" for certain chemical substances which could influence different bodily organs other than those in which they were produced. Previously, Starling had discovered the significance of the osmotic and capillary blood pressures for the equilibrium between the fluids in the tissues and capillaries.[105]

Bayliss participated in the research on colloidal and osmotic pressure, membrane permeability, electrical charges, vascular innervation and other topics. During the years 1915-1918, Starling developed with his collaborators Knowlton, Evans, Patterson and Piper the theory of cardiac regulation — the so-called "law of the heart." In connection with Otto Frank's experiments on the frog's heart carried out in Munich, Starling explained the increased supply of blood to the heart and the augmented resistance to its outflow with the help of an ingeniously constructed "heart-lung preparation."[106]

In addition to the already mentioned assistants, Starling had a number of excellent collaborators who became quite famous: *G.V. Anrep* later a physiologist in Cairo, Egypt, *Boris Babkin* from Russia, later the U.S., Leonard E. Bayliss, Jr., J.B. Daly, L.N. Katz, Sir Thomas Lewis, W.O. Fenn, Herbert S. Gasser and R.W. Gerard from the United States. Germany was represented by the pharmacologists H. Gremels and R. Eichholtz as well as Hans Piper and E.A. Müller.[107]

Ouside of his rather specialized activities, Bayliss wrote a textbook of general physiology in 1914, a book which reflects the spirit of a man who was much more than a research physiologist. It is a rather fascinating book which reflects the spirit of Bayliss the scientist as well as human being, a publication translated into several languages. Likewise, Starling's "law of the heart" elicited great interest and was also printed in other languages.[108]

One of Starling's friends was *Arthur R. Cushny* (1866-1926) of Scotland, an important kidney researcher. After some studies at Aberdeen, Cushny went to work with Otto Schmiedeberg at Strassburg and Hugo Kronecker at Bern. In 1893, he came to the United States to study under John J. Abel at the Johns Hopkins University, returning to his native Scotland twelve years later to become professor of pharmacology at the University of Edinburgh.

Cushny's book, entitled *The Secretion of the Urine* (1917), was dedicated to Starling who was also interested in renal questions.

Another of Starling's students was *Sir Thomas Lewis* (1881-1945) of Cardiff. He was not only an enthusiastic clinician but had a "life-long burning desire to understand how things happen." After his training at University College Hospital in London, Lewis went to Germany for a year in 1906, and returned to work with Starling. He was one of the pioneers in the research of irregular cardiac rhythms which he studied with the sphygmograph and Einthoven's string-galvanometer. Lewis's attempts to explain the cause and transmission of the cardiac excitability were later imitated and perfected elsewhere. Another of his interests was the "white skin" phenomenon, a capillary contraction of the skin. In addition to his clinical works, in 1945 Lewis wrote another book entitled *Exercises in Human Physiology.*

British physiology reached new heights with the achievements of *Charles Scott Sherrington,* who worked primarily in only one area of physiology, the nervous system. (See Fig. No. 122.) When he finally stopped his activities after a long and successful career, neurophysiology had changed completely both in matters of detail as well as general concepts.[109] Born in Islington, Sherrington was attracted not only to the natural sciences but also interested in history, poetry and painting. His interest in the nervous system was awakened by the studies of *Sir David Ferrier* (1843-1928) entitled *The Function of the Brain* (London, 1876). Ferrier was able to produce muscular contractions by stimulating the cerebral surface of a monkey, thus establishing, experimentally, a map of distinct cerebral localizations. As already mentioned, Leopold Goltz held similar views following his own neuro-physiological experiments.

Fig. 122. Charles Scott Sherrington (1857-1952).

Sherrington, who worked in Foster's laboratory at Cambridge since 1880, had occasion to examine histologically the brains of some of Goltz's dogs in 1882-1883. Later, in 1886, he returned to Germany to study cholera, visiting Virchow and Koch in Berlin, Pflüger in Bonn and Goltz in Strassburg. In 1890, Sherrington became a lecturer in physiology at St. Thomas Hospital in London, and a year later at the Thomas Brown Institute of University College. Thereafter, he occupied the chair of physiology at the University of Liverpool.

In Liverpool, Sherrington began his long and intensive analysis of the anatomy and physiology of the nervous system by using surgical methods of ablation and severance as well as degeneration and stimulation techniques. His approach was primarily morphological, but he had an uncanny talent for translating structural data into physiological concepts. In 1913, Sherrington assumed the chair of physiology at Oxford, dying on March 4, 1952 at the age of ninety-five.[110]

Beginning in 1892, Sherrington went rapidly from one discovery to another. In 1892 he analyzed the dissemination of motor fibres in the lumbosacral plexus, and in 1894, the sensory nerves emerging from the muscles. Thereafter, Sherrington examined the role of both the extension and scratch reflex as well as their antagonistic innervation. Later, in 1897, Sherrington showed the coordination of inhibitory and facilitating motor reflexes in the spinal cord and their so-called "reciprocal innervation." Also in 1897, he coined the expression "synapse" to express the point at which an impulse passes from an axon of one neuron to a dendrite or cell-body of another.[111]

Moreover, in 1894, Sherrington discovered the extended rigidity in decerebrated animals, while in 1903 he found the coordinated functions of the spinal cord during his studies of the scratch reflex. Further concepts, such as the "common path," pointed to the true integrative character of the nervous functions. Among other accomplishments, Sherrington distinguished in 1906 the deep dermal receptors (proprioceptors) from those located in the superficial layers of the skin. All these results were combined into a remarkable concept of the total neurofunction, published under the title *The Integrative Action of the Nervous System* (1906). The book was dedicated to David Ferrier.

Sherrington's characterization of an "integrative action" expressed the idea that all the elements, parts, organs, centers and areas of the nervous system actually constituted a homogeneous and cohesive unity. Thus, the system becomes at once truly understandable and can be more easily surveyed since it is basically built on reception, transmission and central response. From a simple neuron through elementary and complex reflexes, the integration of the system is a gradual ascension towards the central organ, the brain. For Sherrington, the cerebrum was the organ of initiative and guidance for the entire living organism and the fundamental elements in the regulation of the body's behavior.

Sherrington's scientific qualities and his kind personality attracted a great number of important collaborators. Among the members of his school of physiology were *Rudolf Magnus* (1873-1927), *John F. Fulton* (1899-1960), *Ragnar Granit* (*1900) and *Sir John Eccles* (*1903). It is said that Sherrington expressed a certain distrust towards complicated laboratory devices, and his research was more morphological in nature instead of tending towards a functional, numerical and qualitative expression of the phenomena. The "how" was more important to him than the "why" and "how much."[112]

In this connection, Sherrington's second great achievement in life ought to be mentioned. At an advanced age and at the conclusion of his physiological work, Sherrington began to study the history and philosophy of research in the life sciences. He was especially interested in a figure from the sixteenth century, the physician who for the first time had coined the term "physiology" instead of talking of "res naturales" as did the older practitioners. The result of his historical research was a book entitled *The Endeavour of John Fernel* (1946), in which Sherrington revealed the thoroughness of his search for Renaissance source material and Fernel's work in particular. In this publication, Sherrington seemed to dominate the historical method like an expert, critically interpreting Fernel's rather difficult Latin with great precision and thus furnishing posterity with a stunning historiographical achievement.

Especially original was Sherrington's last book entitled *Man on his Nature* (Cambridge, 1946). In this publication, Sherrington often followed his historical hero Fernel and his concepts, re-examining the life-soul problem, nature, sensory functions, conciousness, will localization, reciprocal effects, sense of value and many other topics. To all these issues Sherrington brought a deep, as well as informed and historically based, perspective. The entire book is directed to the knowledge and conscience of modern man. It is neither systematic nor does it offer solutions but instead expresses the wide range of the unsolved questions which surround a scholar concerned with research in the biological sciences.[113]

5. AMERICAN PHYSIOLOGY: 1870-1930

In a previous section, American physiology was divided into three phases: a pre-experimental period extending until 1860, followed by a decade of gradual organization without research laboratories and a third era which began in 1870. This section proposes to discuss the latter phase during which the European, especially German, physiological methodology was consciously and extensively adopted and applied in America. However, considerable difficulties had to be surmounted before such an important development could actually take place in the United States.

The creation of independent physiological research in America was largely the result of efforts made by the excellent physician and pathologist *William H. Welch.* Welch was born in Norfolk and studied at Yale as well as at the College of Physicians and Surgeons in New York, whose faculty was composed of practicing physicians from the city. After a tour through

Germany, where he studied in 1876 under Waldeyer, Hoppe-Seyler and Recklinghausen in Strassburg, Ludwig in Leipzig, Cohnheim and Heidenhain in Breslau, Welch returned to the United States. There, he became a key figure in the organization of the newly founded Johns Hopkins University in Baltimore.[114]

Welch's primary field of interest was pathology, and he taught the subject at Johns Hopkins beginning in 1885. In addition, Welch was involved to a great extent in the foundation and staffing of other laboratories at the university. In 1893, he was instrumental in securing the chair of physiology for one of Ludwig's students, *William H. Howell* (1860-1945), while *John J. Abel* (1857-1938) became professor of pharmacology and chemistry, and *Franklin P. Mall* (1862-1917) occupied the chair of anatomy. Between 1896 and 1900, Welch successfully weathered the attacks of the American antivivisection movement, and insisted on the system of full-time basic science faculty members at the universities. He died of cancer at the age of eighty-four.[115]

Fig. 123. Henry Pickering Bowditch (1840-1911).

In spite of some local progress, no physiology in the true sense of the word was being taught nor practiced before 1860.[116] Only with the appointment of *Henry P. Bowditch* at Harvard in the year 1871 and *Henry Newell-Martin* at Johns Hopkins University in 1876 did American laboratory physiology mark its beginnings. Bowditch, a Bostonian, (see Fig. No. 123) studied and graduated from the Harvard Medical School in 1868.[117] After a short sojourn in Claude Bernard's laboratory in Paris and some studies under Ranvier and Kühne, Bowditch went to Leipzig where he worked with Carl Ludwig from 1869 to 1871. At the Leipzig Physiological Institute he carried out a series of investigations which led him to formulate the "all or nothing" principle of muscular contraction in the heart as well as the "staircase" (Treppe) phenomenon. Bowditch met many other international figures in Leipzig, such as Mosso, Cyon, Kronecker and others; a very stimulating period in his life.[118]

After his return from Europe, Bowditch became assistant professor at the Harvard Medical School under the direction of the famous physician *Oliver Wendell Holmes* (1809-1894). He was assigned a laboratory in which he could begin his activities with the aid of those instruments he had brought with him from Europe. In 1876, he became a full professor of physiology, dying in 1911.

Bowditch was not only an excellent teacher but also an able organizer who attracted numerous students. Many of them followed his advice and went on to study with Carl Ludwig in Leipzig for some time. At Harvard, he worked on the problems of cardiac nerves and blood pressure, chloroform and the effects of ether, as well as the influence of breathing on the pulmonary circulation. In 1885, Bowditch proved the untiring quality inherent in nervous transmission. His "clock" to measure blood flow rates also became widely known.[119]

The second physiological research laboratory was established in 1876 under the direction of *Henry Newell-Martin* (see Fig. No. 124) at the Johns Hopkins University. Martin was originally from Ireland and had studied at University College London where he came to the attention of Michael Foster. The famous British physiologist took the young Martin with him to Cambridge as a demonstrator, recommending him for the chair of biology at Johns Hopkins in 1876 despite Martin's lack of publications.[120]

Fig. 124. Henry Newell-Martin (1848-1896).

Once in Baltimore, Martin received space and resources to install and equip his laboratory, and he gathered a galvanometer, recording machines, myographers and other instruments then available in Europe. In addition, Martin prepared some courses in experimental physiology and constructed his own devices, including an isolated heart preparation in 1881, a precursor of Langendorff's heart. After becoming ill in 1893, Martin returned to England where he died, barely forty-eight years old.

Martin's successor was one of his own students, William H. Howell. Other disciples included *Henry Sewall* (1855-1936), *William T. Sedgwick* (1855-1921), *Henry H. Donaldson* (1857-1938), *Archibald B. Macallum* (1858-1934), *Charles S. Minot* (1852-1914), *Warren P. Lombard* and *Frederick S. Lee* (1858-1939). Sewall, Minot, Lombard and Lee also went to study with Carl Ludwig in Leipzig.

The American Physiological Society was founded in New York in 1887. Among its seventeen founding members were Bowditch, Ott, Warren, Howell, Lombard and Hall, as well as Newell-Martin, Minot, Sedgwick, Sewall, Mitchell, Chittenden and others. After 1902, the society also allowed membership to women, and its fiftieth anniversary in 1938 gave rise to great festivities.[121]

The generation of young collaborators of Bowditch at Harvard occupied a crucial position in American physiology. Many of these men were able to occupy, in the 1890's, the newly created chairs in physiology which emerged throughout the country at that time. A real wave of well-trained physiologists spread across the land in those years. By contrast, some of Bowditch's students in Boston did not remain very long in physiology. *Joseph W. Warren* (1849-1916) went into medical practice and *Granville S. Hall* (1844-1924) began a series of studies in philosophy, psychology and physiology with Helmholtz, Ludwig and Wundt. Such a broad training allowed Hall to become professor of psychology and physiological psychology at several American universities upon his return from Germany. Two students of Bowditch, *Frederick W. Ellis* and *Isaac Ott* (1847-1916) went into ophthalmology and pharmacology respectively.

Among Bowditch's contemporaries who made physiology their career, one should first mention *William H. Howell,* a native of Baltimore (see Fig. No. 125). Howell studied first under Newell-Martin and was later appointed professor of physiology and histology at the University of Michigan in 1889, succeeding Sewall. Finally, he returned to Johns Hopkins as the successor of Newell-Martin himself, where he remained active between the years 1893 and 1913. In 1931, Howell became president of the thirteenth International Congress of Physiology which held its meeting in the city of Baltimore. On that occasion, he could proudly boast: "We look upon this meeting of the Congress in Baltimore as a recognition that American physiology has come of age and been admitted to full membership in that old world group to whom we have so long looked for guidance and inspiration." Howell died in 1945 at the age of eighty-five.[122]

Fig. 125. William Henry Howell (1860-1945).

Another disciple of Bowditch, *Warren P. Lombard* (see Fig. No. 126), was born in West Newton, Massachusetts, and also studied at Harvard.

Following Bowditch's advice, he spent three years with Carl Ludwig at Leipzig during 1881-1884, becoming interested there in muscular and neurophysiology. Lombard examined the reciprocal muscular innervation, the pain- and vasomotor reflexes, capillary circulation and other topics. Between 1892 and 1923 he was a professor of physiology at the University of Michigan.[123]

Fig. 126. Warren Plimpton Lombard (1885-1939).

William T. Porter was a graduate of the St. Louis Medical School where he became professor of physiology in 1887. Six years later, he transferred to Harvard, worked with Bowditch and retired in 1928. Porter was instrumental in the foundation and financing of the *American Journal of Physiology* in its difficult first years. In the scientific field, he worked primarily with problems of coronary circulation and the nerves responsible for coronary constriction.

Especially successful was the imaginative physician *Harvey Cushing* (1869-1939),[124] a native of Cleveland, Ohio, who also went to Harvard. Cushing trained in surgery with the famous *William S. Halsted* (1852-1922), studied physiology at Bern with Kronecker and neurology in Liverpool under Sherrington. Upon his return to America, he went to Johns Hopkins in 1903, Harvard in 1912 and finally Yale in 1932. Cushing's accomplishments lay in the integration of neurophysiology with neurosurgery. In 1912, he published his work on the pituitary gland while his investigations of cerebral tumors and trauma made him famous the world over. His love for the art of writing and his bibliographical interests reflected Cushing's rather broad interests.[125]

The last and probably most prominent student of Bowditch was *Walter B. Cannon* (1871-1945) (see Fig. No. 127), who was born in Prairie du Chien, Wisconsin. Cannon also went to study at the Harvard Medical School where he was graduated in 1900. For many years he worked with Bowditch in Boston, first as an instructor, then in 1902 as assistant professor and finally he succeeded Bowditch in the chair of physiology in 1906.[126]

Cannon considerably enriched the field of macroscopic physiology, introducing new techniques and presenting novel concepts. Between 1911 and 1915 he was the first to apply x-rays and barium sulphate as contrast material to study the physiological motility of the stomach and intestines.

During those investigations, Cannon also examined the emotional influences on peristalsis. He discovered the role of adrenalin in the so-called "stress situation" and came close to discovering neurohumoral transmission. One of his students, *Arturo S. Rosenblueth* (*1900), described in 1933 the sympathin I and E. Cannon's more philosophical work, entitled *The Wisdom of the Body* (1932), expressed great admiration for the internal order of the living organisms. His autobiography, *The Way of an Investigator* (1945), represents the valuable testimony of a renowned scientist and great human being.[127]

Fig. 127. Walter Bradford Cannon (1871-1945).

As noted before, Henry Newell-Martin worked for seventeen years (1876-1893) in the physiological laboratory of the Johns Hopkins University. In that period of time he trained a number of scholars who formed the "second generation" of authentic American physiologists. Among the most prominent authors was *Henry Sewall* who came to Johns Hopkins in 1876 and was an assistant and later associate professor there between 1881 and 1889.

In the latter year, Sewall was appointed to the chair of physiology at the University of Michigan, where he succeeded Howell. Later, he went to Denver, becoming professor of medicine at the University of Colorado. Sewall had worked under the direction of Carl Ludwig at the University of Leipzig in 1879, and at Cambridge in 1880. He was very interested in the role of the depressor nerves on blood pressure as well as the summation of stimuli studied by Johann von Kries. As a clinician, Sewall was concerned with clinical physiology, immunology and tuberculosis.[128]

Another of Newell-Martin's collaborators was *William T. Sedgwick,* who worked with Martin for two years between 1879 and 1881. Thereafter, Sedgwick went to the Massachusetts Institute of Technology and devoted his efforts to hygiene and problems of public health.

Other less known figures of Newell-Martin's circle included *Henry H. Donaldson,* who became a physiologist, psychiatrist and neurologist in Chicago, and *Henry G. Beyer* (1850-1918) of Saxony, who graduated from Bellevue Hospital in New York and later became a naval physician and hygienist. To be mentioned is *Charles S. Minot* of West Roxbury, who studied physiology and histology in Paris and Leipzig, becoming professor of embryology and histology at the Harvard Medical School in 1892. There was also *Frederick S. Lee,* of Canton, Ohio. Lee received his Ph.D. from Johns Hopkins University in 1885, and collaborated with Newell-Martin on the problem of arterial tone. In 1886, he went to work with Carl Ludwig and Max von Frey in Leipzig, publishing in 1887 an article on electrophysiology in Du Bois-Reymond's *Archiv.* In 1895, Lee was appointed professor of physiology at Columbia University, and a research professor in 1920. He contributed some valuable studies on the regulation of body weight, muscular fatigue and other topics which made him famous.

Belonging to the same generation was *Hobart A. Hare* (1862-1931), a physiologist who also worked with Ludwig and Kronecker in Leipzig and later entered clinical medicine. Another student of Kronecker was *Samuel J. Meltzer* (see Fig. No. 128), originally from the Kurland. Meltzer worked with Kronecker in Berlin on the physiological aspects of swallowing and became the chief of the section for physiology and pharmacology at the Rockefeller Institute in 1906. Moreover, he founded the *Society for Experimental Biology and Medicine*, the so-called "Meltzer-Verein." Meltzer's later research dealt with problems of nervous inhibition, reciprocal innervation, adrenalin and the effects of magnesium.

Fig. 128. Samuel James Meltzer (1851-1920).

Finally, *Franklin P. Mall* carried out some anatomical work with Wilhelm His, Jr., in Leipzig, and studied with Ludwig the existence of motor nerves in veins as well as subjects related to embryology.

Regarding the development of American chemical physiology, three figures can be justly considered as the founders of this physiological field: *Archibald B. Macallum, Russel H. Chittenden* and *Wilbur O. Atwater* (1844-1907). Macallum studied in Toronto, Canada, where he became a lecturer in biology in the year 1883. After spending some time in Newell-Martin's laboratory at Johns Hopkins University, Macallum returned to Toronto in order to occupy the chair of physiology at the university in 1888. Later, Macallum received an appointment as professor of biochemistry and was active both in Ottawa and Montreal. He made valuable observations regarding the role and distribution of salts in the cells and their existence in the blood, as well as searching for the localization of potassium in the muscular fibre.

After receiving his Ph.d., Wilbur O. Atwater, a native of Johnsburg, New York, went abroad for two years in order to study in Leipzig and Berlin. He returned to the United States in 1871 to become professor of chemistry at Wesleyan University, being also the founder of the first agricultural-chemical institute in the year 1875. During 1882-1883, Atwater went back to Germany where he studied with Carl Voit in Munich.

The result of this sojourn on the Continent was that Atwater continued to develop respiratory calorimetry in the United States between 1892 and 1897. For this purpose, he constructed a large calorimeter, an apparatus which determines the heat production of an individual by measuring the gaseous exchange of the lungs. The device allowed simultaneous determinations of both direct and indirect calorimetry, and Atwater was thus able to gather conclusive proof regarding the validity of the law of energetics in man. Shortly before his death, he installed with his collaborator, Benedict, a nutritional laboratory in Boston.

Another physiological chemist was *Russel H. Chittenden,* a native of New Haven, Connecticut. Chittenden studied at the Sheffield Scientific School of Yale University and then worked for two years in Europe, especially under the direction of Willy Kühne in Heidelberg. Between 1882 and 1922 he was professor of physiological chemistry at Yale. Chittenden worked successfully in the areas of digestion and nutrition. One of his most important collaborators was *Lafayette B. Mendel* (1872-1935).

Another physiologist, *Graham Lusk,* was also originally from Connecticut. Lusk worked with Carl Ludwig and especially Carl Voit regarding the influence of carbohydrates on protein metabolism. As early as 1885, he became a full professor of physiology at the Yale School of Medicine, going later to New York. Lusk worked in the controversial field of lipid and sugar transformation, and studied the specific dynamic effect of food on the gaseous exchange which occurred during respiration. During his lifetime, he was the best American expert in the field of nutritional science as reflected in the publication *The Science of Nutrition* (1906).

Other renowned representatives of physiological chemistry were *Victor C. Vaugham* (1851-1929) and *Walter E. Garrey* (1874-1951). Vaugham was the first physiologist to create a thoroughly modern biochemical laboratory in the U.S. and was especially interested in the chemistry of proteins. Garrey, in turn, was a student of the physiologist *Jacques Loeb,* who had emigrated to the United States after some studies with Adolf Fick in Würzburg and Friedrich L. Goltz in Strassburg. Both Garrey and Loeb worked on related topics involving ionic activity, artificial embryological stimulation, chemistry of colloid substances and problems of cardiac physiology.

6.
THE DEVELOPMENT OF RUSSIAN PHYSIOLOGY IN THE NINETEENTH AND TWENTIETH CENTURIES

Little is known regarding the beginnings of Russian physiology outside of the Soviet Union. This is due primarily to a lack of proper sources, secondary literature and a language barrier. Despite the inherent difficulties, an attempt to depict the Russian physiological development will be made here based principally on a recent doctoral dissertation.[129]

In the *first period* extending between 1800 and 1860, Western physiological concepts clearly predominated in Russia and were taught at the existing universities. Independent research and significant discoveries were rare. The *second period* was characterized by a mass exodus of young Russian students to the German and Austrian laboratories of physiology, especially those of Ludwig in Vienna and Leipzig, Du Bois-Reymond and Kronecker in Berlin, and Brücke in Vienna. This wave of eager students began about 1860 and continued well into the 1890's. The young Russian investigators who returned to their country following such a training, established institutes and research laboratories everywhere in Russia according to the German model. Moreover, they often communicated and printed their results in Western publications.

At the end of the nineteenth century a *third period* of physiology got under way in Russia with the emergence of the famous physiologist *Ivan P. Pavlov* who was able to establish a genuine Russian physiology and was primarily interested in neurophysiological subjects.

A. *First period until 1860.*

Under its rulers Peter I and Catherine II, Russia had repeatedly attempted to erase the scientific backwardness of the country, so obvious when compared with the Western European standards. To this effect, various Western scientists and scholars were invited to come to Russia while many of the young Russian physicians went to Germany, France and England in order to learn about the newer developments in their field.

In the nineteenth century, for example, *Danilo Mikhailovitch Vellansky* (1773-1847) attended the lectures of Schelling and Oken in Germany for several years, becoming an adherent of the theories of polarity and animal magnetism. Paris became, however, a more popular city in which to study, and some Russian students worked successfully under the physicians Corvisart, Bichat and Magendie.

After some studies at the University of Dorpat, the pathologist and physiologist *Alexander P. Sagorsky* (1807-1888) established a physiological laboratory in St. Petersburg in the year 1837. As a result of a royal decree, *Nicolai I. Pirogov* (1810-1881), *A.M. Filomafitsky* (1807-1849) and Sagorsky went to Dorpat for further training in 1829. Filomafitsky subsequently became professor of physiology at the University of Moscow, signaling the fact that physiology gradually was becoming an independent science in Russia. He also published the first Russian textbook of physiology (1836-40) and concerned himself with questions of blood transfusion and narcosis.

After carrying out some studies in Halle, Berlin and Paris, *I.T. Glebov* (1806-1884) received an appointment in physiology and comparative anatomy at the University of Moscow. He translated Magendie's *Précis* and Budge's *Lehrbuch* into Russian. In turn, the physiologists *I.O. Kalenichenko* (1805-1876) and *N.F. Belezky* (1851-1882) taught at the University of Kharkov. Another author who had worked with Johannes Müller and Czermak, *Alexander P. Walter* also lectured in physiology. In Kasan, *Alexej A. Sokolov* (1822-1891), who apparently had studied electrophysiology with Du Bois-Reymond in Berlin, occupied the chair of physiology. His successor, *Ivan G. Navalychin* (1842-1884) published some scientific articles in German journals after studying abroad with several physiologists including Heidenhain.

B. *The assimilation of Western experimental physiology* 1860-1890.

In the second half of the nineteenth century, Russia became even more strongly attached to the European research centers and at the same time also more independent and productive from the scientific standpoint. Among the better known figures one can cite *Nicolai M. Jakubovitch* (1817-1879) who worked in St. Petersburg and carried out histophysiological studies. His assistant, *Filip W. Ovsianikov* (1827-1906) (see Fig. No. 129) had first worked under Bernard in Paris, and then with Ludwig in Leipzig during the years 1860, 1871, and 1874. Ovsianikov concentrated on neurophysiological questions, especially the action of the vasomotor centers, and he later went to work at Kharkov, Kazan and St. Petersburg.

Fig. 129. Filip W. Ovsianikov (1827-1906).

In the latter city, a building was erected in 1863 for the purpose of teaching physiology. The institution had important and qualified members such as *Elie de Cyon, N.N. Baxt* and Pavlov. Cyon studied in Warsaw, Kiev and Berlin, where he graduated in 1864. Thereafter, he worked for three years in Ludwig's laboratory at the University of Leipzig, where he discovered with his teacher the N. depressor. Cyon also studied the effect of heat on the isolated heart of a frog with the help of an ingenious device. In 1872, he was appointed to the chair of anatomy and physiology in St. Petersburg at the age of twenty-nine. Later, he had to leave Russia for political reasons and worked in Paris with Claude Bernard. His accomplishments were recognized throughout Europe.

Another well-known Russian figure was *Vladimir M. Bechterev* (1857-1927) (see Fig. No. 130), who became famous for his work in neurology. In his formative years, Bechterev studied with Wundt and Westphal in Germany and Charcot in France. Another scholar, *Alexander I. Babuchin* (1827-1891), at the University of Moscow, was very active in the field of histophysiology. After studying with both Brücke and Ludwig in Vienna during 1865, another Russian physiologist, *Nicolai O. Kovalevsky* (1840-1891), became the successor of Ovsianikov at Kazan. Kovalevsky was especially interested in the physiology of the blood gases and the circulation.

Fig. 130. Vladimir M. Bechterev (1857-1927).

By contrast, *Nicolai A. Mislavsky*, who had studied with Bechterev and Einthoven, was more attracted to problems of neurophysiology. In turn, the Russian scholar *Vassily J. Tchagovez* became famous for his studies on the theory of bioelectrical appearances. He assumed a chair of physiology at the University of Kiev in 1910. Finally, the physiologist *Vassily J. Danilevsky* (1852-1939) of Kharkov and Kiev should be mentioned. After studies with Adolf Fick in Würzburg on muscular force (1876), Danilevsky concentrated on problems of endocrinology, parasitology and neurophysiology. He was also the founder of the first Russian journal of physiology.[130]

Of great significance, not only for Russia but for physiology in general, was *Ivan M. Sechenov* (see Fig. No. 181). Originally from a small town on the Volga River, Sechenov became the founder of an entire school of physiology which included among others Pavlov, Pachutin, Trachanov, Samojlov, Chaternikov and Verigo. In 1850, he went to the University of Moscow where he received his doctorate. This was followed by a tour abroad which brought him into contact with Ludwig and Brücke in Vienna during 1858-1859. Sechenov also trained with Helmholtz in Heidelberg, Du Bois-Reymond in Berlin and finally Bernard in Paris. Together with Ludwig he investigated the gases in the blood, and at Berlin he was

Fig. 131. Ivan M. Sechenov (1829-1905).

concerned with problems of electrophysiology. His greatest accomplishment was the discovery of a central inhibition mechanism for the reflex processes occurring in the nervous system.[131] Sechenov became professor of physiology in Odessa and St. Petersburg, but fell into disgrace with the political powers in 1880, dying in Moscow in 1905.

Sechenov adopted Claude Bernard's ideas of determinism and "milieu intérieur." His book entitled *Die Reflexe des Grosshirns* (1863) aimed at a closer connection between the mental reactions and functions of the nervous system. He tried to portray these psychological changes as reflexes which originated through the action of external factors. Sechenov received strong criticisms for these ideas and the book was actually confiscated. He also described the summation of subliminal stimuli and dealt with other electrophysiological topics. Pavlov called him "the father of Russian physiology" since he was responsible for the great interest in neurophysiology which characterized most of later Russian physiology.[132]

Fig. 132. Victor W. Pachutin (1845-1901).

Sechenov's student, *Victor W. Pachutin* (see Fig. No. 132), especially advanced physiopathology. He was initially interested in neurophysiological questions concerned with the action of inhibitory nerves, and spent one year with Ludwig and Huppert in Leipzig (1872-73) and another with Hoppe-Seyler in Strassburg. There, he studied physiological chemistry, especially problems of fermentation and putrefaction. Pachutin later went to teach physiology at Kazan and St. Petersburg.

Another disciple of Sechenov was *Ivan R. Tarchanov* who,

after some initial studies in St. Petersburg, also went to work with Hoppe-Seyler. Tarchanov examined the phenomena of summation in the central nervous system, galvanic psycho-reflexes, blood substitutes and other topics. Another of Sechenov's followers, *Nadesha P. Suslova* (+1918) discovered the inhibitory effect of the thalamus on organs of the vegetative system.

The area of electrophysiology was enriched by the work of *Alexander F. Samojlov* (1867-1930) of Odessa. He also worked until 1892 in various foreign laboratories of physiology before joining Pavlov in St. Petersburg and then Sechenov in Moscow. Samojlov was interested in the action potentials in the cardiac as well as skeletal muscle, and also studied the humoral factors in the central inhibition mechanisms of the nervous system. He was the author of many articles which appeared in German journals.

Two other well-known figures of the same school were *Mikhail N. Chaternikov* (+1939) and *Borislav F. Verigo* (+1925). The former studied in Carl Voit's laboratory in Munich and later went to Moscow where he carried out some studies concerning the gaseous exchange and metabolic problems. Verigo, in turn was the discoverer of the depressive effect of the cathodes.

Finally, Sechenov's student *Nicolai E. Vedensky* (1852-1922) should be mentioned. Before 1868, Vedensky (see Fig. No. 133) studied in St. Petersburg except for a three-year prison term imposed on him for political reasons. After 1880, he worked in Sechenov's institute on phenomena of excitability and inhibition. When his former teacher transferred to Moscow, Vedensky became his successor in physiology at the University of St. Petersburg. Between 1881 and 1887, he made several trips to Western Europe, visiting Heidenhain, Du Bois-Reymond and Kronecker in Berlin and Bowman in London.

Fig. 133. Nicolai E. Vedensky (1852-1922).

Many important researchers emerged from Vedensky's laboratory in St. Petersburg, such as *Aleksei A. Uchtomsky* (1875-1942) (see Fig. No. 134) and *D.S. Voronzov* (*1886). Uchtomsky developed the principle of dominant and continuously stimulated nervous centers and eventually succeeded to Vedensky's chair in St. Petersburg. Voronzov, in turn, was a frequent contributor to German journals and later taught in Kazan. His main interest was centered on the electrophysiology of the nerves and the heart. He was the first investigator to employ oscillography based on cathode rays.

Next to the professional physiologists, there were a series of Russian clinicians interested in physiology, just as in Britain and France. Among them one can name *Sergei P. Botkin* (1832-1889), clinical professor of medicine and therapeutics at the University of St. Petersburg and personal physician to the Czar. Pavlov himself was a representative of Botkin's school. Botkin was primarily interested in physiopathology and physiological chemistry.

Fig. 134. Aleksei A. Uchtomsky (1875-1942).

Another important scholar was *Kliment A. Timirjasen* (1843-1920) who greatly contributed to the elucidation of plant photosynthesis. He also worked in Germany and France where he learned methods of spectral analysis. A Russian scientist who acquired international fame was *Ilya I. Metchnikoff* (1845-1916) (see Fig. No. 135). Metchnikoff worked on problems of phagocytosis and resided after 1888 in Paris for political reasons.

Among the Russian physiologists who spent some years in Carl Ludwig's laboratory was *Ivan M. Dogiel* (1830-1916). Dogiel was in Leipzig during 1866-1867 working on the origin of the first heart sound. Later, he became a physiologist and pharmacologist at the University of Kazan. The following Russian physiologists worked with Ludwig during his appointment in Vienna: *E.A. Junge* (1833-1898), *B.W. Thomas* (1831-1895) and *P.P. Einbrodt* (1835-1865). Later, after Ludwig's transfer from the University of Leipzig, the following Russian scientists studied

Fig. 135. Ilya I. Metchnikoff (1845-1916).

with him: Bakst, Savarykin, Ustimovitch, Cheremetevski, Navalichin, Afanasyev, Tchiryev, Morochovez, Tchuyevski and others.[133]

C. *The Pavlov era* (1890-1956)

Ivan P. Pavlov (see Fig. No. 136) exerted the greatest influence on the subsequent development of physiology in Russia. Not only was he personally instrumental in establishing Russian supremacy in certain areas of neurophysiology for many decades, but he also achieved an international status for Soviet-Russian physiology in general. At the same time, however, Pavlov had a very deleterious effect on the general Russian physiological outlook: the thrust of his research stressed the supremacy of the environmental conditions over the endogenous hereditary qualities as determinants of the human existence. Therefore, Pavlov's veneration in Russia has a double origin: his physiological achievements and the ideological implications of his theories.[134]

Pavlov was born on September 14, 1849, in Ryazan, the son of a priest. Instead of following in his father's footsteps, as intended, Pavlov studied natural sciences at the University of St. Petersburg in 1870. Thanks to the influence of Elie de Cyon, he became interested in physiology, and with the help of the clinician Sergei P. Botkin, he investigated topics of circulation, experimental pharmacology and neurophysiology. At the same time, Pavlov studied at the medico-surgical academy of St. Petersburg and obtained his medical degree in 1883. A year later, he launched his career in physiology by visiting Germany for two years. He worked with Rudolph Heidenhain in Breslau during 1884-1885 and Carl Ludwig in Leipzig the following year. This training afforded Pavlov an excellent knowledge of the existing physiological techniques as well as inspiration for future topics of research since Heidenhain, for example, was just beginning to study the problems of glandular secretion.

Fig. 136. Ivan P. Pavlov (1849-1936).

In 1890, Pavlov became professor of pharmacology at St. Petersburg, and between 1896 and 1925 professor of physiology at the Academy for Military Medicine there. His first investigations dealt primarily with the physiology of the circulatory organs and their innervation. Moreover, Pavlov was concerned with the trophic effect of the nerves and the autoregulation within the cardiovascular system based on reflexes. He also developed new techniques for the establishment of a permanent fistula of the salivary gland and a small gastric pouch which had the advantage of allowing the collection of gastric juice without mixing it with the foodstuffs. Thanks to the novel methodology, Pavlov was able to make some significant new contributions to the digestive glands in 1897.[135]

Pavlov possessed an unusual capacity for work and an unflinching determination, carrying out his research without sparing himself or his assistants. According to an eyewitness, he did not care about his personal appearance, always dressing rather modestly (Popovsky). Pavlov was extremely punctual for his appointments and lectures, barely missing one hour due to illness in ten years of activity at the Academy. On the other hand, Pavlov's patience was limitless in the execution and repetition of experiments duplicating some of the investigations over a hundred times with the help of his large staff of collaborators. Nearly fifty assistants were at his disposal in St. Petersburg, and he possessed one of the best equipped physiological laboratories in the world.

In 1904, Pavlov was the first physiologist and Russian scientist to be awarded a Nobel Prize. Further research produced new insights about the influence of the cerebrum on the secretion of the digestive glands. Pavlov established temporal correlations between acoustic or visual signals and the salivary secretion which he called "conditioned reflexes". The necessary precondition for such phenomena was that the conditioning signal had to be repeatedly connected with an offering of food in order to become the necessary stimulus for the secretion.

Thus, this technique allowed Pavlov to create experimental "neuroses" in dogs. For example, he would train a dog to react positively with stimulation and salivary secretion when confronted with a visual image of an ellipse while remaining unimpressed when a circle was drawn. Once the conditioned reflex had been adequately learned by the animal, Pavlov would gradually change the elliptical image into a circle thus plunging the animal into a mixed state of stimulation and inhibition, a restlessness he considered similar to a neurosis.[136]

Pavlov studied the processes of central stimulation and inhibition in the nervous system by the use of many different experimental settings, arriving at conclusions about the so-called "higher nervous activity" represented by such functions as sleeping and awakening. He developed the doctrine of a first and second signal system within the human cerebral function.[137] The Russian government provided Pavlov in 1910 with a specially arranged and staffed institute for his research. Located in Koltuchi, near Leningrad, the so-called "tower of silence" allowed Pavlov to continue his research with the help of many assistants until his death in 1936.

A number of excellent investigators continued Pavlov's work concerning conditioned reflexes and their role in the functional regulation of the internal organs. Among them were the following authors: *Anton A. Walther* (+1902), *Marija K. Petrova* (*1874), *Leon A. Orbeli* (+1958), *K.M. Bykov* (+1959) and *A.D. Speransky* (*1888).[138]

The number of physiological chairs in the medical schools of the Soviet Union has increased to over a hundred. Not only have the Russian physiologists been able to acquaint themselves with the research carried out elsewhere in the world, but they have, in certain areas, achieved international prominence and even supremacy. The physiological investigations are presently carried out in the Academy of Sciences, the Academy of Medical Sciences in Moscow and the institutes which exist in Leningrad.

Next to Pavlov, the Russian physiologists C. Kostojanc, L.G. Voronin, V. N. Chernigovsky, Kupulov, K.M. Bykov and L.A. Orbeli have given modern physiology in the Soviet Union its high interantional standing. The International Congress of Physiology held in Soviet Russia in 1956, constituted a brilliant demonstration of the contemporary Russian efforts and results in the field.[139]

Two countries claim *Boris P. Babkin* as their own. Born in the Russian city of Kourk, Babkin worked initially with Bechterev and, after 1902 with Pavlov with whom he studied pancreatic functions. In 1915, he became a physiologist in Odessa but had to emigrate in 1922 because of the difficulties with the new Soviet regime. After spending two years with Starling in London, Babkin went as a physiologist to Halifax, Canada, and in 1928 to McGill University in Montreal. He was an excellent investigator of the glandular system and also studied the innervation of the salivary glands, the action of histamine on gastric secretion and other related topics.

7. THE DEVELOPMENT OF SCANDINAVIAN, DUTCH AND BELGIAN PHYSIOLOGY IN THE NINETEENTH CENTURY WITH AN APPENDIX ON JAPAN

Throughout the centuries the Scandinavian countries played an important role in scientific evolution and brought forth a number of famous investigators. Among them was the botanist *Carl von Linné* or *Linnaeus* (1707-1778) who studied medicine in Lund and Upsala and later became the founder of botanical systematics and nomenclature. Other famous authors were the chemists *Carl W. Scheele* and *Jöns J. Berzelius,* the latter being the teacher of Friedrich Wöhler, and the anatomist *Anders Retzius* (1796-1860).

Peter L. Panum was the founder of modern Danish physiology. Born on the island of Bornholm, he studied at Kiel, Copenhagen and Berlin, graduating in 1851 with a dissertation on fibrin. Two years later, Panum became one of Claude Bernard's assistants for a short time and was appointed professor of physiology, medical chemistry and pathology at the University of Kiel from 1853 until 1864. In 1864, he went to Copenhagen where he organized the first Danish laboratory of physiology. In addition to his important investigations about blood transfusion, embolisms and blood volume (1857), Panum wrote the first textbook of physiology in Danish (Copenhagen, 1865-1872). He also made valuable contributions to the development of public health and the application of scientific dietetics to the nutrition of the Danish people.[140]

One of the representatives of Panum's school was *Christian Bohr,* a native of Copenhagen. Bohr (see Fig. No. 137) studied medicine there in 1872 and wrote his first scientific work under the direction of Panum himself. In 1881 and again in 1883, Bohr went to Leipzig in order to carry out research with Carl Ludwig. His investigations were primarily concerned with muscular contraction and the gases in the blood. Barely thirty years old, Bohr became in 1886, professor of physiology in Copenhagen. His later scientific efforts were linked to the absorption of gas in fluids varying in temperature and salt content as well as the oxygen and carbon dioxide combination with the blood.

Bohr discovered the effect of carbon dioxide tension on the oxygen dissociation curve of the hemoglobin — the so-called "Bohr effect" — and then searched for the forces which facilitated the absorption of oxygen in the lungs. He believed that such an effect was due to oxygen secretion as opposed to the theory of plain physical pressure gradients. Bohr was an important teacher, very independent in his scientific judgment and strongly inclined towards empiricism, a true pioneer in the field of physiological gas exchange.[141]

Fig. 137. Christian Bohr (1855-1911).

Among Bohr's collaborators and students were notably *Karl A. Hasselbach* (1874-1962), *August Krogh* (1874-1949) and *Valdemar Henriques.* Born in Jutland, Krogh began to work with Bohr in 1897 and received at Copenhagen a position as zoophysiologist, especially created for him in 1908. He was full professor there between 1916 and 1945. Krogh's work was also concerned with the gas exchange in the blood and his investigations were characterized by his clear formulations and imaginative methods. He invented the microtonometer for the measurement of gaseous tensions in the flowing blood and significantly contributed to our present knowledge of the forces present in those exchanges which take place at the level of the lungs.

Together with *Johannes Lindhard* (1870-1947) Krogh developed the nitrous oxide method to determine the minute volume of circulation in 1912. For "his discovery of the capillary motor regulating mechanism," Krogh received the Nobel Prize in physiology and medicine for 1920. Also remarkable were his investigations of the fluid and particle exchange occurring in the capillaries and membranes through osmotic regulation. Among his newly designed instruments were a spirometer and a bicycle-ergometer. Krogh was a stimulating teacher, an objective individual and always a helpful friend.[142] Lindhard, who first worked at the zoophysiological laboratory in Copenhagen, contributed to the physiology of circulation, respiration and muscular contraction (action potentials, histological structure).

Finally, Valdemar Henriques was initially a lecturer for physiology at the Royal Veterinary and Agricultural School, becoming professor of physiology at the University of Copenhagen somewhat later.

Swedish, Norwegian and Finnish physiology were connected with two major figures: *Alarik F. Holmgren* and *Christian Lovén.* Born in Asen, Holmgren (see Fig. No. 138) graduated from Upsala in 1861 and was commissioned a year later to study abroad in order to establish the first physiological laboratory in Sweden. Thus, Holmgren went to work with Brücke and Ludwig at Vienna and Du Bois-Reymond at Berlin during 1861-1864. Somewhat later, in 1869, he studied with Helmholtz at the University of Heidelberg.

Fig. 138. Alarik Frithjof Holmgren (1831-1897).

Armed with his German experiences and the new methodology, Holmgren became professor of physiology at Upsala between 1864, and 1897, and became well-known for his studies regarding retinal electric potentials, color vision and blindness, especially in relation to traffic safety. In 1891, Holmgren founded the *Skandinavisches Archiv für Physiologie,* a journal which was published predominantly in German until the second World War.[143]

Christian Lovén was born in Stockholm and became a student

in Upsala and later Stockholm where he worked under the direction of Retzius. After his graduation in 1863, he was named an instructor in anatomy and physiology at the Karolinska Institute, traveling for further training to Paris. There in 1858, he carried out a series of studies under the direction of Claude Bernard.[144]

Lovén went later to Leipzig where in 1865 he performed investigations under the direction of Carl Ludwig. This work was concerned with the influence of the vascular nerves on the dilatation of the arteries pertaining to the ear, penis, etc. After his return, he carried out some research in anatomy, since the Karolinska Institute had no physiological laboratory. In those circumstances Lovén discovered, together with Gustav Schwalbe, the taste buds.

In the winter semester of the year 1871, Lovén began to deliver his first physiological lectures in Stockholm, and three years later a chair of physiology was created there under his direction. In spite of the restricted working conditions — small laboratory, few instruments and a small budget — Lovén was able to carry out valuable investigations such as the work on a capillary electrometer and the discontinuous nature of the voluntary muscular contraction. His lectures were interesting, the delivery lively, and he attracted, as did Ludwig, a large number of students whom he guided in their research.

Thanks to his command of foreign languages, Lovén became an essential link between the Nordic and Middle European physiologies. A majority of the subsequent Scandinavian physiologists were related to the school of both Holmgren and Lovén as exemplified in the following genealogical tabel (Renqvist-Renpää):

C. Lovén — R.A. Tigerstedt
- J.E. Johansson
  - Liljestrand — U.S. Euler
  - Zotterman — Gernadt
  - Abramson
- C.G. Santesson
- Carl Tigerstedt
- Y. Renqvist-Renpää
  - E.O. Jalavisto
  - A. Wilska

A.E. Holmgren
- Olof Hammersten — I. Bang
- Hj. Oehrwall
- J.E. Johannson
- T. Thunberg
  - E.M. Widmark
  - G. Ahlgren
  - H. Enghoff

One of Holmgren's most prominent students was *Olof Hammarsten* who later became an important physiological chemist. From Norrkőping near Stockholm, Hammarsten studied in Upsala where a chair of medical chemistry existed under the direction of *August T. Almén* (1833-1903) since 1861, thanks to the efforts of Berzelius. Later, Hammarsten also worked with Holmgren on topics of physiology and physiological chemistry. After a short stay with Ludwig in Leipzig during 1871, he began his teaching career at Upsala which culminated in his appointment to the chair of medical and physiological chemistry vacated by Almén in 1883. He remained in the academic position until 1906.

Hammarsten's first scientific activities were concerned with the action of the gastric juice on proteins. His most important achievements, however, came in the fields of milk and blood coagulation, the fermentative transformation of fibrinogen, the role of calcium salts and the chemistry of proteins. Hammarsten's textbook of physiological chemistry was translated from Swedish into many other languages. Several prominent students worked with him in Upsala such as *Svante Arrhenius, Christian Ivar Bang* (1869-1918) and *Otto Folin* (1867-1912) of Boston.[145]

Lovén's student, the physiologist *Robert A.A. Tigerstedt* also achieved international reputation. He was born in Helsingfors, Finland, where he spent his early life and initial career in physiology. Thereafter, Tigerstedt (see Fig. No. 139) went to study with Lovén at the Karolinska Institute in Stockholm, becoming professor of physiology there between 1886 and 1900 after the departure of his teacher, Christian Lovén. This academic appointment was followed by a second similar one in his native Helsingfors, where he was active from 1900 until 1919.

Fig. 139. Robert Armand Tigerstedt (1853-1923).

Tigerstedt went to study under Carl Ludwig at Leipzig in 1884, interested primarily in questions of circulatory physiology. This field became his fundamental area of research together with the consideration of metabolic problems. Several of Tigerstedt's research projects were concerned with measurements of the blood perfusion in certain organs with the help of Ludwig's "Stromuhr." He also studied the pulmonary circulation, pulse and vagal effects on the heart.

Tigerstedt published most of his findings in a four volume work on the physiology of circulation, as well as his large textbook entitled *Handbuch der physiologischen Methodik,* in five volumes (1908-1914). The latter publication was one of the standard works in the field for many years and widely known outside Finland. His extensive metabolic research was carried out with the help of *Klas Sondén* (*1895) in a large "respiration chamber." Tigerstedt's most important discovery was renin, a kidney protein which exhibits a vasopressor activity.

Tigerstedt was a man who possessed an extensive general culture and broad interests, showing a great responsiveness to public questions and exhibiting considerable personal kindness. His most important students were Jöns E. Johansson, C.G. Santesson and E.M.P. Widmark. In 1903, *Jöns E. Johansson* (1862-1935) (see Fig. No. 140) became Tigerstedt's successor at the Karolinska Institute in Stockholm. He had worked with Carl Ludwig in the summer of 1889, investigating blood pressure and the nervous system. Johansson was a gifted student of physics and mathematics and his research dealt with color vision, muscular activity, respiration and metabolism. He was likewise very concerned with the nutritional state of inmates in hospitals and prisons. In that capacity, Johansson altruistically helped the famished masses of German people after the first World War. Johansson was also for many years chairman of the Nobel Prize committee in Stockholm.

Fig. 140. Jöns Erik Johansson (1862-1935).

The successor of Robert Tigerstedt in Helsinki was his son *Carl Tigerstedt* (1882-1930) who represented the third generation of physiologists in his family. He was followed in 1930 by *Yrjö R. Reenpää* (*1894) who is especially interested in sensory physiology. Another author of Finnish descent, *Ragnar Granit,* was appointed research professor at the Neurophysiological Institute in Stockholm in 1953. His investigations related to the retina and the spinal cord have made him widely known in scientific circles. Many young German students went to work with Granit after the second World War.

Also at the Karolinska Institute was *Goran Liljestrand* (*1886). He was an instructor in physiology after 1917 and in 1927 became professor of pharmacodynamics and pharmacognosy. Liljestrand was also very interested in the history of Scandinavian physiology and was one of the editors of the *Skandinavisches Archiv für Physiologie,* a journal which since 1940 has been called *Acta Physiologica Scandinavica.* The change in name was a consequence of the second World War and signaled the shift which the Nordic countries have made towards the Anglo-Saxon cultural circle.

Space does not allow the consideration of other renowned physiologists of these countries. Among them were *Hans K.A. von Euler-Chelpin* (1873-1964)[146] from Augsburg, Germany, and his son, *Ulf S. von Euler-Chelpin* (*1905). Both received the Nobel Prize for their respective research. Euler-Chelpin, Sr., shared the Nobel Prize in chemistry with Arthur Harden for his studies on fermentation, and he was also interested in vitamins and enzymes. Ulf Svante von Euler-Chelpin shared the Nobel Prize in medicine and physiology for the year 1970.

The founder of the physiological institute in Lund and professor of physiology there since 1885 was *Magnus G. Blix.* Blix was widely known for his studies in sensory physiology and psychology as well as for constructing a new ophthalmometer. Today, another physiologist named Georg Kahlson is the director of a new and modern institute erected in Lund during 1952.

Finally a few words should be said about *Allvar Gullstrand* (1862-1930) who was a practicing ophthalmologist but also had mathematical talents. He enriched the knowledge concerning the pathways which the light takes within the eye and developed a theory of astigmatic vision. Gullstrand received the Nobel Prize in 1911 in physiology and medicine "for his work on the dioptics of the eye."[147] Among the latest representatives of physiology are B.Y. Zottermann and E.H. Christensen in Stockholm, Fritz Buchthal, E. Hansen and E. Lundsgaard in Copenhagen and F. Leegard, as well as R. Nicolaysen in Oslo.

In Holland, it was not forgotten that Leyden had been the medical center of Europe in Boerhaave's time. Although the Dutch scientific activities had temporarily declined as in many other countries of Europe, there was a revival in the second half of the nineteenth century at the various universities, especially at Utrecht. These centers of learning resumed their previously excellent teaching and research activities. The physiologist *Jacques L. Schroeder van der Kolk* (1797-1862) had already made some important discoveries. He was one of the first to carry out an analysis of the gases present in the blood, and also published valuable studies about the anatomy and physiology of the nervous system. Schroeder was active from 1827 onwards as a teacher of both anatomy and physiology, and at the same time practiced psychiatry.

Also from Utrecht was *Gerardus J. Mulder*, an excellent investigator in the field of protein chemistry. Mulder became professor of chemistry at Utrecht in 1842 and began his work in physiological chemistry with the help of *Franciscus C. Donders* (1818-1889). Both authors made important contributions to the chemistry of organic tissues. In those same years, the physiologist *Isaac van Deen* (1804-1869), actually called Isaac Abraham, taught at Groningen. Although born on German soil near the border of Holland, van Deen considered himself a Dutchman and from 1851 was professor of physiology in Groningen where he also founded a physiological laboratory. He was especially active researching problems of neurophysiology.

Utrecht's most brilliant period in physiology began with Franciscus C. Donders (see Fig. No. 141) who was born on May 27, 1818 in Tilburg.[148] After studying medicine in Utrecht between the years 1835 and 1840, Donders became a lecturer in anatomy and physiology at the Military Medical School there until 1848. Later, he was appointed professor of ophthalmology at the University of Utrecht and, following the death of Schroeder, professor of physiology. Together with Jakob Moleschott and Isaac van Deen, Donders edited the journal *Holländische Beiträge zu den anatomischen und physiologischen Wissenschaften* between 1846 and 1848. From 1845 on, he was also the publisher of the journal *Het Nederlandsch Lancet.*

Fig. 141. Franciscus Cornelius Donders (1818-1889).

Donders' initial physiological work was devoted to physiological optics, eye movements and rotation of the eyeball. He investigated the phenomenon of the swimming objects before the eyes and the use of prismatic lenses to correct strabismus. For the first time, Donders was able to distinguish clearly between presbyopia and longsightedness, defects of insufficient accommodation and abnormal refraction. He also established the use of cylindric lenses for the correction of astigmatism, founding a hospital for eye disease in Utrecht and practicing his ophthalmological specialty until 1862.[149]

Donders' other physiological research was concerned with the negative pressure in the intrapleural space as well as the velocity of the psychic process.[150] Many capable physiologists trained in his school, including *Herman Snellen* (1834-1908), *Theodor W. Engelmann* (1843-1910), *Hartog J. Hamburger* (1859-1924) and the already mentioned *Willem Einthoven.* Donders was a friend of the surgeon *Albrecht von Graefe* (1828-1870), the founder of modern scientific ophthalmology in Germany. In 1888, he gave up his academic position to Engelmann and died a year later in Utrecht.[151]

Donders' assistant, Theodor W. Engelmann (see Fig. No. 142) was originally from Leipzig and very early had shown some interest in the natural sciences. At the age of sixteen, he published his first paper concerning the infusoria. Engelmann studied in Jena, Leipzig, Heidelberg and Göttingen, finding in Ernst Haeckel at the University of Jena a friend who was similarly interested in those unicellular living creatures. In 1867, he went to work with Donders at Utrecht, where he married his teacher's daughter and finally succeeded him to the chair of physiology in 1888.

Starting with some of the problems exhibited by unicellular organisms, Engelmann investigated predominantly the general physiological questions of excitement, excitability and the transmission of excitement, followed by studies on ciliary movement as well as protoplasmic and muscular contractility. In this connection, the author studied the problem of cardiac automatism and conduction, becoming with Walter H. Gaskell, one of the founders of the myogenic theory of cardiac transmission. In a somewhat related field, Engelmann carried out some investigations on muscular structure and the bases for muscular contraction.[152]

Fig. 142. Theodor Wilhelm Engelmann (1843-1910).

In 1897, Engelmann transferred to Berlin where he became the successor of Du Bois-Reymond for the chair of physiology. He was a kind and modest man, endowed with a good sense of humor and musical talents — he was a friend of Johannes Brahms.[153] He died in 1909 of diabetes.[154]

Another of Donders' students, *Hartog J. Hamburger,* was born in Alkmaar, Holland, and became an assistant to Donders before assuming the chair of physiology at the University of Groningen in 1901. At that time, he was generally considered to be the most important physiological chemist in the Netherlands. Hamburger's physico-chemical investigations covered topics such as permeability, hemolysis, and phagocytosis in which he made some notable contributions. Other work was devoted to the chemical substances secreted by the vagal and sympathetic nerves.

Finally, Donders' disciple *Willem Einthoven* (see Fig. No. 143) was born in Samarang, Dutch Indies, and had a rather unusual career.[155] During his medical studies at the University of Utrecht he clearly excelled his fellow students and was therefore appointed professor of physiology and histology at Leyden even before his own graduation in 1885. Einthoven kept this academic appointment, bestowed upon him at the age of twenty-six, until his death in 1927.

Einthoven should be considered the founder of modern electrocardiography. His corrections of the capillary electrometer also furnished him with useful ideas. Einthoven's construction of the string galvanometer provided the necessary prerequisite for an instrument which could be usefully employed in both physiology and clinical medicine. The theoretical bases for the interpretation of the electrocardiogram were also laid by Einthoven with the introduction of the limb derivations, the so-called "Einthoven triangle" which allowed computations. Moreover, the Dutch scientist used his string galvanometer to register heart tones, and received the Nobel Prize in physiology and medicine for the year 1924.[156]

Fig. 143. Willem Einthoven (1860-1927).

Einthoven was a very careful and cautious researcher who did not publish his completed work immediately but locked it into his desk for a whole year before revising it again. Eventually, he would publish the results provided that they still agreed in all aspects with his current research. Together with Hartog J. Hamburger, Einthoven founded the journal *Archives néerlandaises de physiologie, de l'homme et des animaux* in the year 1918.

A few years after Einthoven received his highest scientific award, another Dutch author, *Christian Eijkman,* shared the Nobel Prize in physiology and medicine "for his discovery of the anti-neuritic vitamin" in 1929. Eijkman had recognized the relationship between the disease called "Beri-beri" — a peripheral type of neuritis — and the polished vitamin-free rice. Stimulated by his findings other scientists began the highly successful modern research to discover other so-called "vitamins."

Engelmann's successor at Utrecht was *Hendrik Zwaardemaker* of Haarlem, Holland.[157] Zwaardemaker studied physiology with Willy Kühne in Heidelberg and otology with the famous *Adam Politzer* (1835-1920) in Vienna. After his return in 1887, he was appointed a teacher of anatomy and physiology at the Veterinary School in Utrecht. At this time, Zwaardemaker made the acquaintance of Donders and contributed an article about the physiology of the sense of smell for Donders' "Festschrift."

Zwaardemaker achieved an international reputation with his studies on olfaction, olfactometry as well as phonetics and presbycusis. The so-called "camera inodorata" (chamber without odors) and "camera silenta" (silent room) were auxiliary methods employed by Zwaardemaker for his sensory analyses. His instrument used to check the patency of the nasal passages is still in use today. In later years, Zwaardemaker investigated some questions of cardiac automatism. He believed that the cause of the heart beat, still being debated today, lay in the radioactive properties of potassium.

Much younger than the above named figures was the physiologist *Rudolf Magnus* (see Fig. No. 144) who also studied under Willy Kűhne in Heidelberg and with the pharmacologist Gottlieb. Magnus then went to Great Britain where he trained under Edward Sharpey-Schafer in 1900, John N. Langley in 1905 and Charles S. Sherrington in 1908. After acquiring the latest physiological knowledge available in Europe, he was called back to Utrecht as a pharmacologist in 1908.[158]

Fig. 144. Rudolf Magnus (1873-1927).

Magnus' interests lay in the field of pure physiology and he established a well-known method for experimental work with the intestines. His most important achievements came, however, in neurophysiology and he published a monograph entitled *Kőrperstellung* (Berlin, 1924) which summarized his long research on positional reflexes as well as labyrinthine and non-labyrinthine standing reflexes. In addition, Magnus was concerned with the cerebellar and thalmic functions. He possessed considerable surgical skills, superb teaching abilities and exercised great caution in reaching experimental conclusions.

Some of Magnus' students should be named here: A. de Kleijn, G.G.J. Rademaker and the German neurologist G. Schaltenbrand. Among those Dutch physiologists of the last generations were S. de Boer, F.J.J. Buytendijk, J.Ten Kate, H.J. Jordan, C.A. Pekelharing, A.K.M. Noyons, G.A. van Rijnberk, A. de Waart and C.A.G. Wiersma. Within this group, *Hermann J. Jordan* was the founder of Dutch comparative physiology and zoophysiology. Born in Paris and the son of German parents, Jordan studied in Würzburg and was an assistant to Pflüger at the University of Bonn during 1887-1898. Later, he was active in the zoological station in Naples under the direction of Dorn. His teaching career culminated in 1919 when he became full professor of zoophysiology at the University of Utrecht. He died a victim of the Jewish persecution during the second World War.

Until the year 1875, physiology was a largely undeveloped field of the biological sciences in *Belgium.*[159] Most of the teaching positions were in the hands of clinicians who used the poorly remunerated chairs in Gand, Liège and Brussels only as temporary stepping stones until a more profitable clinical appointment became available. Even Theodor Schwann, (see Fig. No. 145) the student of Müller, was unable to change the existing conditions after receiving his appointment to the anatomical chair at the University of Louvain in 1839.[160] In fact, Schwann's scientific productivity virtually ceased after his arrival in Belgium, although part of his difficulties were more internal than environmental reflecting his conflict between the rational-scientific and mystico-religious aspects of his personality.[161] In any event, Schwann did not publish any valuable work in Belgium except for the article on biliary fistula in 1844. Schwann became professor of physiology at Louvain in 1849 and transferred to Liège in 1858 where he erected the first "cabinet de physiologie" or miniature physiological laboratory.

Fig. 145. Theodor Schwann (1810-1882).

Schwann's successor at Liège was *Léon Frédéricq* (1851-1935), the real founder of Belgian physiology. Born in Gand, Frédéricq (see Fig. No. 146) was already, as a child, fond of plants, animals, stones and physico-chemical phenomena. After studying natural sciences and medicine, he traveled extensively on the Continent, working with Hoppe-Seyler, Baumann, Bert, Marey and Du Bois-Reymond. As mentioned, he assumed the chair of physiology in 1880 vacated by Schwann in Liège. Fortunately, he was able to receive generous funds in order to build a new physiological institute in 1888.

Frédéricq had an extensive and successful career, discovering hemocyanide and designing methods for cross-circulation between two animals. In addition, he analyzed the gases in the blood, carried out an esophageal

Fig. 146. Léon Frédéricq (1851-1935).

registration of the cardiac function and the interruption of the bundle of His. Moreover, Frédéricq was the first to prove the existence of three varieties of protein in the blood: seroalbumin, seroglobulin and fibrinogen. In 1904, together with the physiologist *Paul Heger* (1846-1925) he founded the journal *Archives internationales de physiologie* in Brussels. Frédéricq was one of the last representatives of the classical period of physiology and received many awards and honors for his accomplishments.

Frédéricq's colleague, Paul Heger, also contributed significantly to the emergence of Belgian physiology. After his graduation in 1871, he went to study with the German histologist *Salomon Stricker* (1834-1898) in Vienna and Carl Ludwig in Leipzig, the latter becoming his acknowledged source of physiological knowledge and methodology. Heger took over the chair of physiology at Brussels in 1873, replacing the physiologist *Gottlieb Gluge* (1812-1898), and with the help of *Ernest Solvay* (1838-1922) he erected the Institut Solvay de Physiologie there. He was primarily interested in the various retinal elements and questions of circulation.

The pharmacologist *Jean-François Heymans* has already been cited in connection with the activities of Johannes Gad. Heymans founded in Gand the institute which carries his name. His son and successor, *Corneille Heymans* (1892-1968) perfected the technique of perfusing an isolated head through crossed circulation and also contributed to our knowledge of respiration and heat regulation. Heymans was awarded the Nobel Prize in physiology and medicine for 1938.

Another Belgian physiologist, *Jean Demoor*, believed in the existence of a cardiac hormone, as did Ludwig Haberlandt, and presented some data suggestive of a humoral stimulation of the heart. In another area of research, Demoor was concerned with the problem of synaptic transmission. The following figures belong to the present generation of prominent Belgian physiologists: P. Rijlant and F. Bremer (Brussels), Z.M. Bacq (Liège) and J.J. Bouckaert (Gent).

The development of physiology in the various European and non-European countries cannot be described here in detail because of the lack of space as well as source material in many instances. Moreover, a recent work[162] contains valuable information about the recent status of the discipline in many countries of the world. However, a short summary of the physiological development in *Japan* can be sketched here, based on some available sources.[163]

Until the nineteenth century, Japanese medicine maintained its traditional ties with ancient Chinese dogmas and procedures. A closer acquaintance with European medicine was only possible after the Dutch were allowed to trade with Japan and communicate European cultural elements and knowledge to their hosts. Thus, Dutch teachers provided medical instruction to Japanese students in Nagasaki in the early nineteenth century. For example, Blumenbach's physiology was translated into the Japanese language in 1837, and both Richerand's *Nouveaux Éléments de Physiologie* as well as Haller's *Physiologie* appeared in the same language in 1856.

After the year 1868 and beginning with the Meiji-period, German medical knowledge was also gradually introduced in Japan. A German physiologist, *Ernst Tiegel* founded in 1876 the first physiological institute in Tokyo. His most important disciple was *Kenji Osawa* (1852-1927), considered to be the "father of Japanese physiology." After carrying out some studies in Berlin and Strassburg, Osawa became Tiegel's successor in Tokyo in the year 1877. Another student of the German physiologist was *S. Amaya*, who was appointed to the chair of physiology at Kyoto in 1899 and was the first experimental physiologist in Japan.

Amaya was followed by *J. Ishikawa* (+1947), a student of Max Verworn at the University of Bonn, while *Makoto Ishihawa* (+1944) taught at Kiushu.

In turn, *Genichi Kato* (*1890) was the first physiologist at the Keio- University, an author who became famous for his investigations on nerve conduction (1920-1924) and support of the "all or none law." The great interest in neuro-muscular and electrophysiology still persists in Japan today.

In 1915, Osawa was followed by *Hisomu Nagai* (1876-1957), also a student of Max Verworn. One of Martin Gildemeister's disciples was *K. Hashida* (+1945), the founder of Japanese electrophysiology. In 1921, *Rinnosuke Shoji* (1886-1962), who studied with Barcroft and Bayliss in England, became professor of physiology at Kyoto, while *Jodishi Takamine* (1854-1922) isolated adrenalin from the adrenal glands in 1901. Takamine was one of the pioneers of modern biochemistry in Japan.

Since the first World War, the Anglo-Saxon influence has become stronger in Japan. Today, there are forty-five medical schools with physiology departments in that country, although their equipment and professional literature are still somewhat insufficient.

8.
PHYSIOLOGY IN THE TWENTIETH CENTURY

A. *General Lines of Development*

One of the indicators of the progress achieved in physiology is the increase of investigators studying the field. Until the second World War, good physiological laboratories were restricted to Europe, the United States and perhaps Japan. Since that time, however, more nations have gradually not only achieved political independence, but also reached scientific standards hitherto a monopoly of a few countries. New universities have arisen with chairs in physiology, and eventually the establishment of research laboratories.

The number of physiological investigators is reflected in their growing participation at the international meetings of the discipline.[164] This process can be illustrated as follows:

1st congress—Basel, Switzerland 1889: 13 nations, only European countries except for the United States represented with seven scientists. Total number of participants: 124

4th congress—Cambridge, England 1898: 18 nations, 129 participants

12th congress—Stockholm, Sweden 1926: 33 nations, 581 participants

17th congress—Oxford, England 1947: 37 nations, 1120 participants

19th congress—Montreal, Canada 1953: 41 nations, 2188 participants

23rd congress—Tokyo, Japan 1965: 37 nations, 2829 participants

24th congress—Washington, D.C., U.S. 1968: 51 nations, 4300 participants.

The distribution of those attending the Washington meeting was as follows:

United States: 2040	Switzerland: 25
Great Britain: 200	Mexico: 25
Canada: 155	Rumania: 24
Japan: 116	Denmark and Belgium: 22
France: 110	Brazil: 12
Germany: 98	Argentina: 17
Soviet Union: 59	Norway: 15
Italy: 55	India: 14
Sweden: 49	Venezuela: 13
Holland: 46	Poland: 11
Czechoslovakia: 46	Spain: 10
Australia: 33	

These figures[165] show that physiology, between 1889 and 1968, gradually became a worldwide discipline after having been purely European in its inception. Several countries which had hitherto merely imitated physiological research done elsewhere, have more recently exhibited some original investigators of their very own, such as Mexico's *Arturo S. Rosenblueth,* a student of Walter Cannon, and Argentina's Nobel Prize winner of 1947, *Bernardo A. Houssay* (1887-1971) who elucidated the role played by the hormone of the anterior pituitary lobe in carbohydrate metabolism.

In the United States, the growth of physiology is reflected in the following statistics[166]: At the time of its founding in 1887, the *American Physiological Society* had twenty-eight members. By 1937, the total membership had risen to 631 and in 1954 there were 1400. From the computations it was estimated that there were about 700 working professional physiologists in the United States during the year 1954.[167]

The number of those working in the field throughout the world has reached such proportions that any important individual investigator has been submerged in the large mass of researchers. Moreover, the newer results and conclusions tend to get lost in the anonymity of the general progress. Therefore, a future history of physiology will have to be primarily a history of physiological problems where the biographical elements will largely fade into the background. It is fair to say that the history of physiology will shift from a history of physiologists to an account of the development of physiological problems.

The increase in the number of investigators has been paralleled by an equally dramatic rise in the number of publications and journals, to the point that it constitutes a worrisome problem and a true flood of physiological literature. (See Fig. No. 147.) In 1950, two hundred and fifty such physiological journals were published with a total of 68,000 printed pages.[168] It was estimated that the hours required for reading this material exceeded three times the total available working time.

Thus, such an avalanche of scientific print has left many research results virtually unknown unless they were achieved in certain important areas of current research. Moreover, similar problems are often studied and duplicated in various laboratories with a great expenditure of resources because of such a lack of awareness of the foregoing research. In truth, human memory has actually surrendered to the mass of printed material and new techniques of storing and retrieving information must be devised if we are not to get completely lost in a jungle of knowledge.[169]

Another factor in modern physiology is the gradual specialization of physiological research with an impressive accumulation of details. Physiology is no longer a uniform and coherent field of investigation. Instead, the various special branches, such as biochemistry, neurophysiology, endocrinology, biophysics, cardiology and others, are being studied separately. Physiology has even ceased to be one whole and distinct teaching subject, a fact which virtually spells the end of the discipline as a certified field of scientific endeavor.[170] Previously, one chose and learned the pertinent methods in accordance with the research problems awaiting study or of interest to the re-

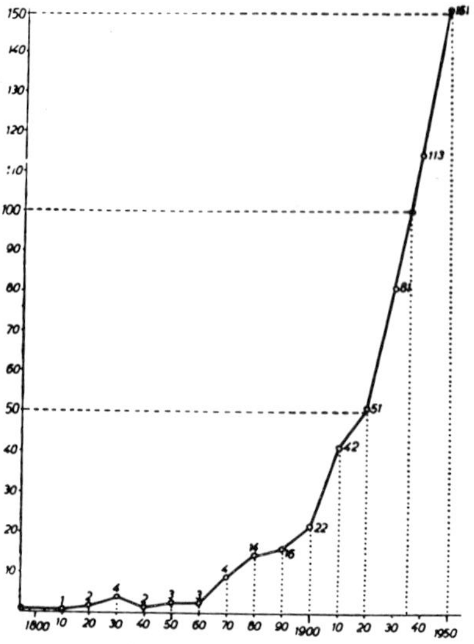

Fig. 147. The growth of periodicals dealing with physiological subjects (1795-1950).

searcher.[171] Today, to the contrary, one selects the questions in the light of the available methodology, reflecting the unusual importance and dimension which physiological methods have acquired in certain areas for the continuity of research.

From the above statements and comments, it is obvious that the entire field of physiology has been undergoing in recent times a gradual process of fragmentation. Moreover, new special disciplines have been added. While histology, embryology and physiological chemistry had already been separated from physiology proper during the nineteenth century, cardiology, electroencephalography and pathological physiology were largely absorbed by the clinical sciences. This fragmentation of the entire field and the subsequent separation of special areas can be viewed as debilitating factors for the physiological discipline as a whole. On the other hand, however, the physiological method and thought have gradually been connected more intimately with medicine and psychology.

Moreover, physicists and chemists have discovered in the biological sciences a fertile field for the application of their own methods. The significance of this development is that more non-medical scientists are now studying physiology than ever before, a fact which in some measure reflects and promotes the gradual estrangement occurring between medicine and physiology. This previously close association had been valuable since medicine was able to furnish a series of continuous questions which stimulated physiological research.

The lack of medical experience and empathy with the medical concerns exhibited by many modern physiologists lessens their understanding of the urgency involved in certain areas of research which are crucial to the advance of medicine. Whether the gradual divorce between medicine and physiology also decreases the attraction which the latter field has for prospective physiologists has been stressed by some older authors[172,173] but seems merely an impression rather than an established fact.

The methods employed in physiology since the second World War have drastically changed. Previously, most of the conventional armamentarium of a well-equipped laboratory consisted of surgical, mechanical and electromechanical devices and glassware. However, since 1940, electronics has furnished a myriad of better, faster and more sensitive instruments for measuring, stimulating and recording. Many of these contrivances are so extremely complicated that the investigator using them cannot understand their mode of action. Without the aid of electronic experts and the existence of special workshops, these implements become a set of mysterious "black boxes" for the researcher. Thus, teamwork gradually replaces more and more the individual scientist.

Another aspect of modern physiological research is the shift from a macroscopic organ-analysis to the elucidation of the functional implications of microscopical changes. Thus, the newer research is concerned with cells and macromolecules, and recent micro-techniques have been developed for such a purpose. Among them are micro-electrodes, histochemistry to study subcellular organelles, and methods to puncture microscopical structures. In addition, highly sensitive procedures have been evolved to demonstrate the presence of minute amounts of certain trace elements (photometer, spectrophotometer, etc.)

One author recently wrote that "the sagacity of the physiologist can be compared to a microtome, cutting increasingly thinner sections of the unknown and then submitting them to analysis."[174] Although the progress in physiology occurs on a wide front, the advancing steps become increasingly shorter when measured against the background of the major issues still awaiting elucidation. While the instruments necessary for the maintenance of life — biotechnology — are becoming better known, the fundamental mystery of the organic order — its *bionomy*[175] — has actually turned out to be more enigmatical than hitherto imagined.

B. *Special Discoveries in Twentieth Century Physiology.*

In view of the rather restricted space it is difficult to select those discoveries of our century which were especially important for the progress of physiology. In the first place, many of these achievements belong to biochemistry and biophysics while others were gathered in the field of electron microscopy. Therefore, those discoveries to be mentioned are insights which pertain to the more central questions studied in physiology.

The knowledge concerning some metabolic questions and the major groups of nutritional substances had already been completed around 1920. The same was true for the *energetic balances* through direct and indirect calorimetry and heat exchange. However, the field of thermoregulation became better understood thanks to the works of *Ludwig Aschoff* (1866-1947), *Rudolf Thauer* (*1906) and *Herbert Hensel* in Germany. In addition, many new facts became available in the field of vitamin research concerning the existence of various new vitamins, their origin, nature and mode of action. The Polish investigator *Casimir Funk* (*1884) designated these previously called "accessory food factors" with the term *vitamin* and his sharp distinction of these principles from other biological substances in 1912 created a new field of physiological investigation.[176]

Thus, the American biochemist *Elmer V. McCollum* (*1879) together with his co-worker *Marguerite Davis* (*1887) described between 1913 and 1919 the nature and action of vitamin A. The German biochemist *Wilhelm O. Stepp* (*1882) and the American chemist *Lafayette B. Mendel* were also involved in that research.

The American biochemist *Conrad A. Elvehjem* (1901-1962) and his coworkers carried out research which resulted in the recognition of vitamin B deficiency. Moreover, the German physician *Axel Holst* (1861-1931), the English biochemist *Sir Jack C. Drummond* (1891-1952) and the Hungarian investigator *Albert Szent-Győrgyi* (*1893) made significant discoveries regarding vitamin C between 1907 and 1927.

The relationship between vitamin D, rickets and ultraviolet radiation was clarified by the British pharmacologist Sir *Edward Mellanby* (1884-1955) in 1919, the American chemist *Thomas B. Osborne* (1859-1929) and the German chemist *Adolf O.R. Windaus* in 1927. Conversely, vitamin E was first discovered by the American researchers *Herbert McLean Evans* (*1882) and *Katherine Scott Bishop* (*1899) in 1922. Somewhat later, *Henrik Dam* (*1895) discovered the antihemorrhagic vitamin K for which he shared the Nobel Prize in physiology and medicine for the year 1943. Two other authors, *Ernest L. Smith* (*1904) and *Edward L. Rickes* (*1912) independently isolated the anti-pernicious anemia factor — named B12 — from the liver. The entire field was jointly developed by physiologists, biochemists and clinicians.

The area of intermediate metabolism and the knowledge of ferments was primarily developed by a series of biochemists. Next to the ferments existing in the digestive tract, these investigators discovered a series of new enzymes needed for anabolic and catabolic functions occurring at the tissue level, specifically in the metabolism of the cell. In this regard, the already mentioned German physiologists Otto Meyerhof and Otto Warburg (see Fig. No. 148) made notable contributions.

Likewise, the field of internal secretions experienced great progress in the last decades. Endocrinology has furnished physiology and clinical medicine with new and valuable insights. In addition to neurohumoral coordinations and regulatory functions, there is now an awareness of other hormonal control of a variety of organic functions. The whole chapter of the discovery of the various hormones is rich in detail.[177] The antecedents to this development have already been described: Berthold (1849), Bernard (1853), Brown-Séquard (1889) and others.

Fig. 148. Otto Warburg (1883-1970).

In 1905, William H. Bayliss and Ernest H. Starling defined the action at a distance produced by some internally secreted substances which Starling called *Hormones.* As in the case of the vitamin research, clinicians, physiologists and biochemists were at the forefront of the endocrinological investigations which culminated in 1921 with the isolation of insulin from islet tissue of the pancreas by the Canadian physiologist *Frederick Banting* (1891-1941) and the American *Charles H. Best* (*1899). In the same year, the American biologists Herbert M. Evans and *Joseph A. Long* (*1879), working on reproductive problems, discovered that extract of the anterior pituitary gland yielded a growth hormone.

The hormones associated with the male and female sexual organs and characteristics were isolated in the 1920's and 1930's. The work of *Edgar Allen* (1892-1943), *Edward A. Doisy* (*1893), *Hermann Zondek* (*1887) and *Selmar Aschheim* (*1878) was concerned with ovarian hormones. In turn, *Adolf F. Butenandt* discovered the testicular hormone in 1932. As already mentioned, the role of the pituitary gland in carbohydrate metabolism was elucidated by the recently deceased Argentine physiologist Bernardo A. Houssay (see Fig. No. 149). These discoveries led to the realization that the pituitary was a master gland influencing other endocrine structures in the body through the secretion of several distinct hormones.

The other successful hormonal research was devoted to the biological activity of various compounds secreted by the adrenal cortex, a task which the American physiological chemist *Edward C. Kendall* (1886 1972) and his co-workers at the Mayo Clinic carried out in the last four decades. The comprehensive investigations led to the discovery of the so-called "compound E" or corisone in 1949 which has become extremely useful in medical therapy.[178]

Fig. 149. Bernardo Alberto Houssay (1887-1971).

Finally, Walter Cannon's concept or principle of *homeostasis* should be cited. It was recognized that there existed in the organism a delicate equilibrium between the interaction of the many different endocrine secretions. Homeostasis addresses itself to such a balance and its disturbance can cause a series of endocrine diseases.

In the field of blood physiology, the German *Karl Landsteiner* (1868-1943) made two great discoveries. In 1900, he described the three basic blood groups: A, B, O, for which he received the Nobel Prize in physiology and medicine for the year 1930. Later, in 1940, Landsteiner also found the Rh factor in the blood. The knowledge concerning blood coagulation has become more detailed and vastly complex in the last decades, gaining greater clinical significance then ever.

Likewise, immunology has gradually evolved into a full-fledged field of clinical investigation. Conversely, the regulation of breathing and gas-exchange at the tissue level continue to be researched actively although the fundamental phenomena of respiration have already been known for quite some time. In this regard, Corneille Heymans discovered the role of the carotid sinus and the aortic mechanisms in the regulation of respiration.

The knowledge of cardiac excitability, automatism and the phenomenon of electric transmission throughout the heart has been greatly expanded thanks to the use of various electrophysiological methods and the cathode-ray oscillograph. The measurement of intracardiac pressure, first attempted by the German physician *Werner Forssmann* (*1904) through cardiac catheterization, has become a routine procedure thanks to the pioneering efforts of the French investigator *André F. Cournand* (*1895) and the American physiologist *Dickinson W. Richards* (*1895). All three authors shared the 1956 Nobel Prize in physiology and medicine for their accomplishments. The procedures have greatly facilitated the diagnosis of congenital and acquired cardiac defects and lesions.[179]

The older mechanical registration of the pulse has all but disappeared, but the measurement of blood pressure by the cuff method has been maintained since the first decade of this century as an essential diagnostic tool in clinical medicine.

The bases for an understanding of the blood circulation were already laid in the nineteenth century. However, new data had become available about the storage of blood in the liver, lungs and spleen (Barcroft 1923) as well as the microcirculation in the tissues. Moreover, our understanding of circulation and circulatory regulation mechanisms has greatly increased. Among the newly discovered functions are the Bainbridge reflex (1915), the carotid-sinus reflex (Hering 1924) and the Bezold-Jarisch reflex (1867 and 1937). In 1919, the Danish physiologist *August Krogh* was able to elucidate the relationships of capillary circulation and regulation, and accomplishment for which he received, as mentioned before, the Nobel Prize for 1920.

Much progress was made in the field of renal physiology. Ludwig's filtration hypothesis through a differential in pressure has been generally upheld. In fact, the theory was confirmed through a series of brilliant experiments using glomerular punctures which were executed by the physiologist *Joseph T. Wearn* (*1893) and the American pharmacologist *Alfred N. Richards* (1876-1966) in 1924. The role of the renal tubules has also been clarified and their function much better understood thanks to the new clearance methods (Møller, McIntosh and van Slyke 1928). Through the improvement of the microchemical methods, a new quantitative renal physiology has become available. Novel concepts such as the countercurrent principle and the active transport of substances across membranes have greatly aided in the understanding of renal function.

In the field of animal physiology, the advances of our century have been particularly impressive. The theory of muscular contraction has gone through a series of tentative hypotheses such as the swelling mechanism and the chemical theory of acid formation. The latter was experimentally disproven by the Danish biochemist *Christen Lundsgaard* (1883-1930) in 1929. Finally, *Andrew F. Huxley* (*1917) proposed in 1957 that the muscular contraction was due to a displacement of the active strings existing between those portions of myosin in the muscular fibrilles. For such a situation to occur, a complicated process takes place involving ionic potentials and the adenosine triphosphate as an immediate source of energy. Electromicroscopical examinations of the muscular fibrilles, as well as studies involving ionic transfers during stimulation and the reaction of the muscular proteins are all involved in the explanation of the entire phenomenon. The problem of muscular tone has not been completely solved for the striated nor the smooth muscle. The investigations of A.V. Hill regarding the thermic processes which occur during muscular contraction have already been noted.

The most important progress in physiology during the last fifty years has been achieved in the field of neurophysiology. One of the specific areas of research has been the excitation process occurring in the axon. The initial bases for the interpretation of the phenomena of excitation in muscles and nerves was laid by *Julius Bernstein* in 1902 and *Rudolf O.A. Höber* (1873-1953) in 1926 with their membrane theory. This hypothesis postulated a pre-existing membrane potential during rest whose balance was disrupted when depolarization ensued. For their extraordinary investigations concerning the excitation process with the help of new microelectrodes, *Alan L. Hodgkin* (*1914) (see Fig. No. 150) and *Andrew F. Huxley* received the Nobel Prize in physiology and medicine for the year 1963.

Fig. 150. Alan Lloyd Hodgkin (*1914).

Knowledge about the phenomena occurring at the level of the synapses and the independent control of the autonomic functions increased greatly. The participation of proprioreceptors located in the muscles and inner organs which regulate motion and posture was also recognized. Moreover, the concept of a functional control based on the model of feed-back mechanisms was verified.[180] Since the work of *Norbert Wiener* (1894-1964) in 1948, cybernetics or the study of human control functions has become a very popular field of endeavor. New information has become available about the distribution of central nervous functions in the cortex and the basal ganglia. The pathways and centers of the spinal cord are better known today, and the extrapyramidal tracts have been clearly separated from the pyramidal ones.

Fig. 151. Walter R. Hess (*1881).

In a related area, the role of the thalamus and the existence of a central activating and inhibitory system have become much more explicit and clear thanks to the work of Walter R.

Hess (see Fig. No. 151). New methods of localized stimulation proved the existence of vegetative centers for temperature control, the regulation of sleep and the innervation responsible for emotional responses in the mid-brain and areas of the hypothalamus.[181]

The mode of transmission of the excitement from the nerve to the muscle has been extensively discussed since the times of Du Bois-Reymond (1877). Today, the participation of acetylcholine in the process has been confirmed thanks to the work of Sir *Henry H. Dale* and *Wilhelm Feldberg* (*1900) in 1936. Similarly, the role of acetylcholine as an intermediary for the vagal effect on the heart was studied by Otto Loewi in 1921 while Walter Cannon became the first scientist to point out the release of adrenergic substances at the level of the sympathetic nerve endings in 1933.

9.
REVIEW—PROSPECTS—EPILOGUE

"History is the conscience of the present." (T. Litt)

When we look back now in 1971 to the many centuries of human endeavor to solve the riddle of vital organization and functions, we can do so with some justifiable pride. From rather modest beginnings, our understanding concerning the living actions has grown into a truly imposing body of knowledge. With the help of an anatomical comprehension of the underlying structures and a physico-chemical examination of the isolated functions, we have vastly increased our insights into the working elements of the body.[182]

The curve of such a growth follows a geometrical progression and has taken a steep ascent since the beginning of this century. The expansion of knowledge has been tremendous in all areas of physiology. Examples of this phenomenal development are the studies about the complicated physico-chemical structure of the blood, immuno-reaction, blood-groups and the phenomena of oxidation. Moreover, fields such as the circulatory and respiratory regulation processes and the chemical as well as thermodynamic occurrences in the muscle should also be included in the list of notable achievements. The afferent and efferent neurotransmissions and physico-chemical events in sensory organs are other topics which have been greatly expanded by modern research, together with the study of hormones, vitamins and ferments in intermediate metabolism. Finally, a multitude of regulations in the central nervous system and periphery have been discovered.

The triumphs of pure physiological and biochemical research are plainly noticeable and even awe inspiring. Moreover, the practical applications of these new discoveries have been so convincing, that medicine has shifted in most fields from a morphological to a functional kind of thinking. In truth, medical research has in most instances become a form of applied physiological investigation. This situation has prompted the acquisition of a series of useful practical insights into the causal factors and relationships within the diseased organism, which in turn facilitated the development of valuable diagnostic and therapeutic methods.

From such an enumeration of triumphs one could surmise that physiology will be even more successful in the future and thus the depiction of the historical evolution of the science could be concluded on a very hopeful and optimistic note. However, the present situation in physiology reveals a series of problems which could perhaps be more sharply delineated with the help of a broad historical perspective instead of merely existing in the narrow view of the researcher working in a specialized field. Thus, the following outlook should be understood as a critical examination of the present state of physiological research and knowledge.

Contemporary research in physiology will continue as long as physics and chemistry can furnish further new devices for the causal analysis of the physiological phenomena.[183] This method of thinking and experimenting will persist and still provide us with great new discoveries. However, it has only become readily apparent that the description of vital actions according to their "biotechnical" relationship offers only a one-sided view of them. These are abstractions, gained through the successful but eminently artificial isolation of processes from the general interaction of forces occurring at the level of the organism. Although these partial explanations are correct and important in their own right, they are necessarily incomplete and unsatisfactory when considered in the proper and total context of organic functions.

In truth, it becomes more and more obvious that those phenomena which are always reproducible in isolated experiments and strictly determined by certain causal factors, can be easily modified and controlled by the action of other organs or forces within the whole organism. Thus, the concept of regulation has clearly emerged and with it the meaning of the organic parts for the whole organism.

In the investigations on circulatory control, neuro- and hormonal regulations and metabolic adjustments, we have gradually realized the existence and importance of a multitude of correlations between the various functional systems of the organism. Such discernments have helped investigations directed towards the demonstration of regulative processes and balances within a great number of physiological functions. Such research has been feasible by combining the strict category of causality with another thought concerned about the relevance of the individual part to the whole biological unit. Hence, the teleological thinking — or more correctly "bionomic" thought — has returned to physiology, derived from the experience gathered precisely by those scientists working in the laboratory with the causal and analytical approach.

One can occasionally sense that many of these investigators are somewhat uncomfortable about such a teleological conclusion. Ernst Brücke's famous dictum about teleology being a lady without whom no biologist could live but is embarrassed to be seen with in public, is a rather pregnant statement of the modern state of affairs. The gradual realization that multiple correlations between individual physiological phenomena do exist, has provided a new image of the organism which represents a more sophisticated view than the hitherto predominant analytic level of comprehension.

The question which naturally arises is whether the contemporary viewpoint is the highest and last step in the development of physiological thought. This is highly improbable, and a few examples may illustrate the reasons for such a sceptical attitude. To be sure, the organism in terms of its working organs and their adjustment of the environment, is a finely adjusted system. Such an organization operates under a constant temperature control like a thermostat, regulating the amount of water and ionic content in fluids and organs much like a hydrostat and ionostat. Moreover, the organism keeps from the incoming foodstuffs all the necessary elements, and grows until a harmonic relationship between the various organs can be achieved. Only then, will the apposition of further protoplasmic substance come to an end.

Each change in the supply of substances will set into motion a number of regulatory mechanisms which control water, salt, protein and energy. At the same time each modification occurring in one part of the organism will trigger immediate homeostatic responses from all the other parts of the whole organization. For example, the absorption or formation of water in the body simultaneously changes the conditions of its heat regulation, osmotic pressure and ionic equilibrium. Thus, each regulating mechanism, such as the hydrostatic, osmotic, thermostatic and ionostatic one, cannot concern itself only with its own specific areas of control but must work in relative harmony with each other.

Finally, one may ask the question: who regulates whom? There has to be a system of afferent mechanisms, controls and efferent agencies of unimaginable complexity. It is conceivable that such multiple dependencies — veritable networks of causality — would have to be expressed scientifically some day in certain symbols rather than words. Nevertheless, the causal analysis will continue to elucidate some of these relationships under defined and known conditions capable of duplication.

However, the number of active coefficients has immensely grown for certain organic functions. Thus, highly complex relationships are established where only the initial and final links of the process are know. One of the experts in hormonal research has recently declared that in spite of the vast amount of isolated facts now available, the order inherent in the total system with its reciprocal stimulations and inhibitions has actually become less comprehensible than before. The same is true for the correlations and integrations occurring in the central nervous system where everything is linked with everything else. From the above sketch of the present situation in physiological research it is obvious that the multiple correlations of the various phenomena are beginning to hinder greatly the analysis as well as the synopsis of these physiological events.

At this point it may be permissible to take a look into the future development of physiology and attempt to sketch the next evolutionary level which appears plausible in the light of the present research efforts. A large portion of the present investigations are still concerned with macroscopic physiology — a level of visible and tangible dimensions. A reflex, the phenomenon of red-cell agglutination, the change of certain circulatory, respiratory or metabolic parameters still belong in this category. However, nobody would deny the existence of a microscopic world behind these events. In many areas we are already at the threshold of this micro-dimensional cosmos, such as in our physiological analysis of synapse-function, permeability and bio-rhythms. Unfortunately, we can only work in very narrow areas within certain fields where structure and function have no sharply defined boundaries and are inseparably linked together. The microscopical morphology is of little avail to the investigator, since he is dealing with colloidal and macromolecular dimensions where high electrostatic forces act in the smallest of spaces. In these intracellular functions, a level of awareness can be reached where the link and superposition of the various functions is extremely great. Such a situation precludes any real hope of an eventual insight into the broader interrelationships of all partial factors.

Thus, it appears that the causal-analytical approach has reached a rather soft and boggy bottom where a myriad of subtle balances dominate the situation and only one general correlation is really valid: everything is related to everything else. The field of macromolecular biology is an example of such a state, and physiological research is faced with peculiar and highly significant phenomena.

Those groups of substances which are essential participants in the frame and function of the body, such as the proteins, have a macromolecular structure. They are unstable from the thermodynamic standpoint and can only be formed and exist in a system where energy producing reactions are coupled with synthetic ones. These proteins are thus subjected to a continuous decomposition and transformation, existing only as intermediate steps of certain specific cyclic processes which have been called "elementary cycles" by H.V. Schultz.

Such a succession of periodically recurring events often overlap each other, and already at this lower stage of life every purely dynamic structure rests on specifically biological cooperation. It is estimated that to maintain such macromolecular structures in the human organism in relation to the body's proper protein balance, an energy of about 200 calories is needed daily.

Interestingly enough, the problem of a specific biological order is already present at this lowest level of organic function. The structure of the "correct" protein molecules in a given organism is, to a great extent, statistically improbable, and it appears, in the words of H.V. Schultz "that present physico-chemical laws cannot overcome such a statistical hurdle". Therefore, it understandable why some authors have expressed their reservations about the success of the causal-analytical thought and physico-chemical approach in a future physiology in spite of past successes.

Although such ideas and methods will undoubtedly uncover the quantitative and qualitative aspects of many subtle bodily functions, the control of the whole process may well remain an incomprehensible puzzle. One can, of course, declare that the object of physiology could be achieved by the elucidation of partial physico-chemical relationships, but such a limitation to the scope of physiological inquiry underestimates the thirst for knowledge characteristic of the human intellect.

The true object of physiology is to understand vitality. History shows that such a knowledge can be of two kinds: discernment of the need and the meaning of a given occurrence. The final goal can only be an understanding of the significance of those elements needed in the organism. Therefore, the development of physiological research is presently between its origin and its goal, comprising an infinite variety of pathways which all point towards that final destination.

We do not know whether the contemporary road and method will eventually stumble against some fundamental barrier. We only recognize single elements in this puzzle, but the body adds and orders them in a way which is difficult to understand. Occasionally, one would like to believe that the human intellect is incapable of comprehending the conditions of its own biological function. In any event, we are ignorant about the fortunes of our hitherto successful approach for the next hundred years and whether entirely new ways will have to be devised in order to continue our physiological studies.

We are at the end of our attempt to sketch the broad lines of physiological development and to describe the efforts of many great investigators and thinkers throughout the centuries. Physiology is a fascinating discipline devoted to the pursuit and explanation of the mystery of life. It is a field which continually offers the historian new insights into the factors which eventually determine success or failure, such as the individual genius and perseverance of the researcher, and the interest and support which he receives from his contemporaries.

REFERENCES

[1]H. Friedenwald, "Notes on Moritz Schiff (1823-1896)," *Bull. Inst. Hist. Med.* 5 (1937), pp. 589-602.

[2]Peter Riedo, *Der Physiologie Moritz Schiff (1823-1896) und die Innervation des Herzens,* Zürich, 1970.

[3]See biography by M.D. Grmek in *Dict. Scientif. Biogr.* 2 (1970), pp. 24-34.

[4]Claude Bernard, *An Introduction to the Study of Experimental Medicine,* transl. by H.C. Greene, New York, 1957.

[5]E. von Cyon, "Claude Bernard" in *Biographisches Lexikon der hervorragenden Ärzte,* Ed. Wernich u. Hirsch. Bd. I, Wien and Leipzig 1884, pp. 417-418.

[6]W. Riese, "Claude Bernard in the light of modern science," *Bull. Hist. Med.* 14 (1943), pp. 281-294.

[7]Ch. Robin, *Du microscope et des injections dans leur applications á l'anatomie et á la pathologie,* Paris, 1849.

[8]M. Klein, *Histoire des origines de la théorie cellulaire,* Paris, 1936.

[9]G. Canguilhem, "La philosophie biologique d'Auguste Comte et son influence en France au XIXe siècle," *Bull. Soc. Franc. de Philos.* (Paris, 1958).

[10]J.M.D. Olmstedt & E.H. Olmstedt, *Claude Bernard and the Experimental Method in Medicine,* New York, 1952.
M.D. Grmek, *Catalogue des manuscrits de Claude Bernard,* Paris, 1967.

[11]J.M.D. Olmstedt, "Claude Bernard as a dramatist," *Ann. Med. Hist.* 7 (1935), p. 253.

[12]R. Millet, *Claude Bernard ou l'aventure scientifique,* Paris, 1944.

[13]M.D. Grmek, ed., *Claude Bernard, Cahier de notes 1850-1860,* Paris, 1965.

[14]J. Schiller, "Claude Bernard and vivisection," *J. Hist. Med.* 22 (1967), pp. 246-260.

[15]M.D. Grmek, "First steps in Claude Bernard's discovery of the glycogenic function of the liver," *J. Hist. Biol.* 1 (1968), pp. 141-154.

[16]C.A. Culotta, "Tissue oxidation and theoretical physiology: Bernard, Ludwig and Pflüger," *Bull. Hist. Med.* 44 (1970), pp. 109-140.

[17]Reino Virtanen, *Claude Bernard and His Place in the History of Ideas,* Lincoln, Nebraska, 1960.

[18]J.M. Fataud, ed., *Claude Bernard: Introduction a l'étude de la Médecine expérimentale,* I. Partie, Bordas, 1966.

[19]G. Canguilhem, "Théorie et technique de l'expérimentation chez Claude Bernard," (=*Philos. et Methodol. de Cl. Bernard),* Paris, 1967, pp. 23-32.

[20]M.D. Grmek, "Évolution des conceptions de Cl. Bernard sur le milieu intérieur,"in *Ibid.,* pp. 117-150 and
F.L. Holmes, "The *milieu intérieur* and the cell theory," *Bull. Hist. Med.* 37 (1963), pp. 315-335.

[21]F.L. Holmes, "Claude Bernard and the milieu intérieur," *Arch. Int. Hist. Sci.,* 16 (1963), pp. 369-376 and
articles in *Claude Bernard and Experimental Medicine,* ed. by F. Grande & M.B. Visscher, Cambridge, Mass., 1967.

[22]H.E. Hoff & J.F. Fulton, "The centenary of the first American Physiological Society," *Bull. Inst. Hist. Med.* 5 (1937), pp. 687-734.

[23] See biography by N. Mani in *Dict. Scientif. Biogr.* 2 (1970), pp. 59-63.

[24] N. Mani, "Paul Bert als Politiker, Pädagoge und Begründer der Höhenphysiologie," *Gesnerus*, 23 (1966), pp. 109-116.
See also Charles A. Culotta, "A history of respiratory theory: Lavoisier to Paul Bert, 1777-1880," (unpbl. dissertation, University of Wisconsin, 1968).

[25] See biography by M.D. Grmek in *Dict. Scientif. Biog.* 2 (1970), pp. 524-526.

[26] J. Schiller, "Claude Bernard and Brown-Séquard. The chair of general physiology and the experimental method," *J. Hist. Med.* 21 (1966), pp. 260-270.

[27] J.M.D. Olmstead, *Charles-Edouard Brown-Séquard. A 19th Century Neurologist and Endocrinologist*, Baltimore, 1946.

[28] Ch. E. François Franck, *L'oeuvre de E.J. Marey*, Paris, 1905.

[29] Justus von Liebig, *Animal Chemistry or Organic Chemistry in its Application to Physiology and Pathology*, reprint, New York, 1964.

[30] W. Prandtl, *Humphry Davy, Jöns Jakob Berzelius*, Stuttgart, 1948.

[31] N. Mani, "Das Werk von Friederich Tiedemann und Leopold Gmelin," (Die Verdauung nach Versuchen) und seine Bedeutung für die Entwicklung der Ernährungslehre in der ersten Hälfte des 19. Jahrhunderts," *Gesnerus* 13 (1956), pp. 190-214.

[32] D. McKie, "Wöhler's synthetic urea and the rejection of vitalism: a chemical legend," *Nature* 153 (1944), p. 610.

[33] *J.J. Berzelius—J. von Liebig:* ihre Briefe von 1831-1845, ed. J. Carriere, München, 1898.
[34] Th. L.W. Bischoff, *Über den Einfluss Liebigs auf die Entwicklung der Physiologie*, München, 1874.

[35] A. Kohut, *Justus von Liebig*, Giessen, 1903 and
R. Blunck, *Justus von Liebig*, Berlin, 1938.

[36] T.O. Lipman, "The response to Liebig's vitalism," *Bull. Hist. Med.* 40 (1966), pp. 511-524.

[37] T.O. Lipman, "Vitalism and reductionism in Liebig's physiological thought," *Isis* 58 (1967), pp. 167-185.

[38] G.J. Goodfield, *The Growth of Scientific Physiology*, London, 1960.

[39] R. Vallery-Radot, *The Life of Pasteur*, transl. by R.L. Devonshire, Garden City, N.Y., 1924.

[40] Hilaire Cuny, *Louis Pasteur; the Man and His Theories*, transl. by P. Evans, New York, 1966.

[41] René Dubos, *Pasteur and Modern Science*, Garden City, N.Y., 1960.

[42] Fr. Lieben, *Geschichte der physiologischen Chemie*, Leipzig-Wien, 1935.

[43] H. Klickstein, "Charles Caldwell and the controversy over Liebig's 'animal chemistry'," *Chymia* 4 (1953), pp. 129-157.

[44] W. Steinhausen, *100 Jahre med. Forsch. in Greifswald*, Festschr. Greifswald, 1938.

[45] C.A. Culotta, "On the color of blood from Lavoisier to Hoppe-Seyler, 1777-1864: a theoretical dilemma," *Episteme* 4 (1970), pp. 219-233.

[46] E.F.J. Hoppe-Seyler, *Über die Entwicklung der physiologischen Chemie und ihre Bedeutung für die Medizin*, Rede, Straubing, 1884.
The speech was translated into English: "On the development of physiological chemistry and its significance for medicine," *N.Y. Med. J.* 40 (1884), pp. 169-171 and 197-199.

[47] E. Baumann and A. Kossel, "F. Hoppe-Seyler," *Z. physiol. Chem.* 21 (1895/96), pp. 1-61.

[48] A. Thierfelder, "F. Hoppe-Seyler," *Berl. Klin. Wschr.* (1895), p. 928.

[49] J.P. Greenstein, "Friedrich Miescher, 1844-1895," *Scientif. Mon.* 57 (1943), pp. 523-532.

[50]A. Kossel, "E. Baumann," *Z. physiol. Chem.* 23 (1897), p. 1.

[51]S. Edlbacher, "A. Kossel," *Z. physiol. Chem.* 177 (1928), p. 1.

[52]K. Spiro, "F. Hofmeister," *Erg. Physiol.* 22 (1923), p. 1.

[53]E. Abderhalden, "F. Hofmeister," *Med. Klinik* (1922), p. 1167 and C. Neuberg, "F. Hofmeister," *Biochem. Z.* 134 (1923), p. 1.

[54]Ph. Ellinger, "A. Ellinger," *Erg. Physiol.* 23 (1924), p. 139.

[55]D. Nachmanson, S. Ochoa & F.A. Lipmann, "Otto Meyerhof: 1884-1951," *Science* 115 (1952), pp. 365-368.

[56]See Nobel Foundation, Nobel lectures: *Physiology and Medicine,* including presentation speeches and laureates' biographies, vol. II 1922-1941, New York, 1965.

[57]In English appeared Meyerhof's *Chemical Dynamics of Life Phenomena,* Philadelphia, 1924.

[58]H.H. Weber, "Otto Meyerhof," *Die Naturwissenschaften* 39 (1952), p. 217.

[59]E. Klenk, "Hans Thierfelder," *Z. physiol. Chem.* 203 (1931), p. 1.

[60]See biographies by R. Abderhalden in *Z. Vitamin-, Hormon-, Fermentforschung* 4 (1951), p. 1 and K. Heyns in *Pflügers Archiv* 253 (1951), pp. 229-237.

[61]See biography by H. Schiefers in *Dict. Scientif. Biogr.* 2 (1970), pp. 560-563.

[62]See biographical sketches by H. Boruttau in *Dtsch. Med. Wschr.* (1908), p. 340 and O. Frank in *Z. Biol.* 93 (1933), p. 1.

[63]G. Lusk, "Carl von Voit, master and friend," *Ann. Med. Hist.* 3 (1931), pp. 583-594.

[64]O. Neustätter, *Max Pettenkofer,* Berlin, 1925.

[65]See biographical sketches by K. Thomas in *Klin. Wschr.* (1932), p. 926, and L. Englert in *Dtsch. Ärzteblatt* (1935), p. 1066.

[66]G. Lehmann, "Edgar Atzler," *Erg. Physiol.* 41 (1939), p. 5.

[67]A.E. Light, "Graham Lusk," *Yale J. Biol. and Med.* 6 (1934), p. 487.

[68]See biography by C.E. Rosenberg in *Dict. Scientif. Biogr.* 1 (1970), pp. 325-326.

[69]See biography by C.A. Culotta in *Dict. Scientif. Biogr.* 1 (1970), pp. 609-611.

[70]W.O. Atwater-F.G. Benedict, *A Respiration Calorimeter with Appliances for the Direct Determination of Oxygen,* Washington, 1905.

[71]See biographical sketch by W. Trendelenburg in *Erg. Physiol.* 37 (1935), p. 1.

[72]K. Wezler, "Otto Frank," *Z. Biol.* 103 (1950), p. 91.

[73]See biography by H. Kraft in *Münch. Med. Wschr.* (1902), p. 965.

[74]M.A.B. Brazier, "The historical development of neurophysiology," in *The Handbook of Physiology,* ed. by J. Field, Washington, D.C., 1959-68, I, pp. 1-58.

[75]See account by A. Bickel in *Dtsch. Med. Wschr.* (1902), p. 403.

[76]See biographical sketches by J.R. Ewald in *Pflügers Arch.* 94 (1903), p. 1.

[77]Accounts of A. Bethe can be found in *Klin. Wschr.* (1942), p. 511 & *Pflügers Arch.* 261 (1955), pp. 1-14.

[78]See S. Garten's account in *Pflügers Arch.* 170 (1918), p. 501 and A. v. Tschermak in *Münch. Med. Wschr.* (1934), p. 1230.

[79]C. Hess, "Ewald Hering," *Die Naturwissenschraften* (1918), p. 305.

[80]See account by F.N. Schulz in *Erg. Physiol.* 30 (1930), p. 11.

[81]See biographical sketches by W. Thörner in *Med. Klin.* (1922), p. 130, and F.W. Fröhlich in *Z. allg. Physiol.* 20 (1923), p. 185.

[82]M. Monnier, "Das wissenschaftliche Werk von W.R. Hess," *Schweiz. Med. Wschr.* 79 (1949), p. 1231.

[83]A. v. Tschermak, "F.B. Hofmann," *Erg. Physiol.* 26 (1928), p. 774.

[84]L.G. Stevenson, "In memoriam Bruno Kisch (1890-1966)," *J. Hist. Med.* 22 (1967), pp. 47-53.

[85]O. Frank, "Siegfried Garten," *Z. Biol.* 81 (1924), p. 1.

[86]See account by A. Jarisch in *Erg. Physiol.* 45 (1944), p. 1.

[87]Sir Edw. Sharpey-Schafer, *History of the Physiological Society during its First Fifty Years 1876-1926,* Cambridge, 1927.

[88]Gerald L. Geison, "Michael Foster and the rise of the Cambridge School of Physiology 1870-1900," (Diss., Yale Univ., 1970).

[89]D.W. Taylor, "The life and teaching of William Sharpey (1802-1880), 'father of modern physiology' in Britain," *Med. Hist.* 15 (1971), pp. 126-153 and 241-259.

[90]H. Dale, "Sir Michael Foster, K.C.B., F.R.S., a secretary of the Royal Society," *Notes Roy. Soc. London* 19 (1964), pp. 10-32.

[91]J.N. Langley, "M. Foster," *J. Physiol.* 35 (1906/07), p. 223.

[92]See biography by G.L. Geison in *Dict. Scientif. Biogr.* 2 (1970), pp. 598-599.

[93]Lady G. Burdon-Sanderson, *Sir John Burdon Sanderson: a Memoir,* ed. by J.S. and E.S. Haldane, Oxford, 1911.

[94]H. Hale Bellot, *University College London 1826-1926,* London, 1929.

[95]L. Hill, "Sir Edward Albert Sharpey-Schafer (1850-1935)," *Obituary Notices of Fellows of the Royal Society* 1 (1935), pp. 401-407.

[96]E. Riese, "Changing concepts of cerebral localization," *Clio Medica* 2 (1967), pp. 189-230 and
R.M. Young, "The functions of the brain: Gall to Ferrier (1808-1886), *Isis* 59 (1968), pp. 251-268.

[97]M. Brazier, *A History of the Electrical Activity of the Brain. The First Half Century,* London, 1961.

[98]*Hopkins and Biochemistry,* ed. by J. Needham & E. Baldwin, Cambridge, 1949.

[99]Archibald V. Hill, *First and Last Experiments in Muscle Mechanics,* Cambridge, 1970.

[100]Kenneth J. Franklin, *Joseph Barcroft (1872-1947),* Oxford, 1935.

[101]John S. Haldane, *The Philosophy of a Biologist,* Oxford, 1935.

[102]G.E. Allen, "J.S. Haldane: the development of the idea of control mechanisms in respiration," *J. Hist. Med.* 22 (1967), pp. 392-412.

[103]See biography by C.A.L. Evans in *Dict. Scientif. Biogr.* 1 (1970), pp. 535-538.

[104]Charles A.L. Evans, *Reminiscenses of Bayliss and Starling,* Cambridge, 1964.

[105]C.B. Chapman, "Ernest Henry Starling, the clinician's physiologist," *Ann. Int. Med. Suppl.* 2 (1962), pp. 1-43.

[106]C.B. Chapman & H.J. Mitchell, *Starling on the Heart,* London, 1965.

[107]A.V. Hill, "Bayliss and Starling and the happy fellowship of physiologists," *J. Physiol.* 204 (1969), pp. 1-13.

[108]R. Colp. Jr., "Ernest H. Starling — his contribution to medicine," *J. Hist. Med.* 7 (1952), pp. 280-294.

[109]R.D. French, "Some concepts of nerve structure and function in Britain, 1875-1885: background to Sir Charles Sherrington and the synapse concept," *Med. Hist.* 14 (1970), pp. 154-165.

[110]J.F. Fulton, "Sir Charles Scott Sherrington (1857-1952)," *J. Neurophysiol.* 15 (1952), pp. 167-190.

[111]J.C. Eccles, "Some aspects of Sherrington's contribution to neurophysiology," *Notes Roy. Soc. London,* 12 (1957), pp. 216-225.

[112]Ragnar Granit, *Charles Scott Sherrington. An Appraisal,* London, 1966.

[113]Henry Cohen, *Sherrington: Physiologist, Philosopher and Poet,* Liverpool, 1958.

[114]Donald Fleming, *William H. Welch and the Rise of Modern Medicine,* Boston, 1954.

[115]Simon Flexner-James T. Flexner, *William Henry Welch and the Heroic Age of American medicine,* New York, 1941.

[116]H. Sewall, "Henry Newell-Martin," *Johns Hopkins Hosp. Bull.* 22 (1911), p. 327.
See also G.H. Daniels, "The process of professionalization in American science: the emergent period, 1820-1860," *Isis* 58 (1967), pp. 151-166 and by the same author
"Finalism and positivism in 19th century American physiological thought," *Bull. Hist. Med.* 38 (1964), pp. 343-363.

[117]W.B. Cannon, "Henry Pickering Bowditch," *Mem. Nat. Acad. Sci.* 17 (1924), pp. 183-196, and biography by E. Mendelsohn in *Dict. Scientif. Biogr.* 2 (1970), pp. 365-368.

[118]G. Rosen, "Carl Ludwig and his American students," *Bull. Hist. Med.* 4 (1936), pp. 609-650.

[119]H. Sewall, "The beginnings of physiological research in America," *Science* 58 (1923), p. 187.

[120]C.S. Breathnach, "Henry Newell-Martin (1848-1893), a pioneer physiologist," *Med. Hist.* 13 (1969), pp. 271-279.

[121]W.J. Meek, *History of the American Physiological Society,* Semicentennial 1887-1937, Baltimore. 1938.

[122]A.M. Chesny, "An anniversary tribute to the memory of the late W.H. Howell," *Bull. Johns Hopkins Hospital* 109 (1961), pp. 1-19.

[123]Warren P. Lombard, *Directions for Laboratory Work in Physiology, for the Use of Medical Students,* Ann Arbor, Mich., 1906.

[124]See biography by E.H. Thomson in *Dict. Scientif. Biogr.* 3 (1971), pp. 516-519.

[125]Elizabeth H. Thomson, *Harvey Cushing, Surgeon, Author, Artist,* New York, 1950.

[126]H.H. Dale, "Walter Bradford Cannon (1871-1945)," *Obituary Notices Fell. Roy. Soc.* 5 (1947), pp. 407-423.

[127]Walter B. Cannon, *The Way of an Investigator. A Scientist's Experiences in Medical Research,* New York, 1945.

[128]Gerald B. Webb-Desmond Powell, *Henry Sewall, Physiologist and Physician,* Baltimore, 1946.

[129]O. Zmija, "Kurze Geschichte der russischen Physiologie mit besonderer Berücksichtigung ihrer Beziehungen zum europäischen Raum," (Med. Diss., Münster, 1962).

[130]Khachatur S. Koshtoiants, *Essays on the History of Physiology in Russia,* transl. by D. Boder et al., Washington, D.C., 1964.

[131]P.M. Amacher, "Thomas Laycock, I.M. Sechenov and the reflex arc concept," *Bull. Hist. Med.* 38 (1964), pp. 168-183.

[132]Ivan Sechenov, *Selected Works,* ed. by A.A. Subkov, Moscow, 1935.

[133]S.A. Česnakova and M. Lindemann, "Wissenschaftliche Kontakte zwischen Physiologen aus Russland und aus Deutschland in der zweiten Hälfte des vorigen Jahrhunderts," *NTM* 7

(1970), pp. 85-98, and by the same authors "Russische Physiologen in den Laboratorien von Carl Ludwig," *Z. d. Karl-Marx-Univ. Leipzig. Math. Naturwiss. R.* 19 (1970), pp. 315-322.

[134]Hilaire Cuny, *Ivan Pavlov: The Man and His Theories,* transl. by P. Evans, New York, 1965.

[135]Boris P. Babkin, *Pavlov: a Biography,* Chicago, Ill., 1949.

[136]Ivan P. Pavlov, *Conditoned Reflexes: An Investigation of the Physiological Activity of the Cerebral Cortex,* New York, 1960.

[137]Ivan P. Pavlov, *Selected Works,* ed. by J. Gibbons and transl. by S. Belsky, Moscow, 1955.

[138]Y.P. Frolov, *Pavlov and His School,* London, 1937.

[139]Gregory H.S. Razran, *A Bibliography of Soviet Experiments in I) Interoceptive Conditioning and II) Unconditioned Interoception in the Last Ten Years,* Bethesda, Md., 1959.

[140]W.M. Gafafer, "Bibliographical biography of Peter Ludwig Panum (1820-1885), epidemiologist and physiologist," *Bull. Inst. Hist. Med.* 2 (1934), pp. 259-280.

[141]See biographical sketch of Bohr by R. Tigerstedt in *Skand. Arch. Physiol.* 25 (1911), p. 5.

[142]G. Liljestrand, "August Krogh," *Acta Physiol. Scand.* 20 (1950), p. 109.

[143]J. Bernstein, "F. Holmgren," *Naturwiss. Rundschau* 12 (1897), p. 579.

[144]W. Kensmann, "Die Geschichte der Schwedischen Physiologie unter besonderer Berücksichtigung der deutschschwedischen Beziehungen," (Med. diss., Münster, 1961).

[145]Th. Thunberg, "O. Hammarsten," *Erg. Physiol.* 35 (1933), p. 13.

[146]See biography by A.J. Ihde in *Dict. Scientif. Biogr.* 4 (1971), pp. 485-486.

[147]Theodore L. Sourkes-Lloyd G. Stevenson, *Nobel Prize Winners in Medicine and Physiology 1901-1965,* New York, 1966.

[148]See biography by R.J. Ch. V. ter Laage in *Dict. Scientif. Biogr.* 4 (1971), pp. 162-164.

[149]Frans C. Donders, *On the Anomalies of Accommodation and Refraction of the Eye,* Amsterdam, 1963.

[150]N.S.R. Maluf, "Carson, Ludwig and Donders and the negative interpleural pressure," *Bull. Hist Med.* 16 (1944), pp. 417-418.

[151]J. Moleschott, *F. C. Donders,* Giessen, 1888.

[152]H. Kingreen, "Theodor Wilhelm Engelmann (1843-1909). Ein bedeutender deutscher Physiologe an der Schwelle zum 20. Jahrhundert," (Med. diss., Münster, 1968).

[153]J. Röntgen, *Johannes Brahms im Briefwechsel mit Th. Engelmann,* Leipzig, 1918.

[154]M. Verworn, "Th. W. Engelmann," *Z. allg. Physiol.* 10 (1910), p. 1.

[155]See biography by S. Hoogerwerf in *Dict. Scientif. Biogr.* 4 (1971), pp. 333-335.

[156]George E. Burch-Nicholas P. Depasquale, *A History of Electrocardiography,* Chicago, Ill., 1964.

[157]A.K.M. Noyons, "Hendrik Zwaardemaker," *Erg. Physiol.* 33 (1931), p. 5.

[158]G. Liljestrand, "Rudolf Magnus," *Erg. Physiol.* 29 (1929), p. 646, and W. Heubner, "Rudolf Magnus," *Klin. Wschr.* (1927), p. 2022.

[159]M. Florkin, "Léon Frédéricq et les débuts de la physiologie en Belgique," *Coll. Nat. Ser.* 3 1943, Nr. 36.

[160]Léon Frédéricq, *Théodore Schwann, sa vie et ses travaux,* Liège, 1884

[161]M. Florkin, "Une autobiographie inédite de Théod. Schwann," *Bull. Acad. Roy. Méd. Belg.* 16 (1951), p. 445.

[162]*Perspectives in Physiology, an International Symposium, 1953,* ed. by I. Veith, Washington, D.C., 1954.

[163]Chandler McC. Brooks-Kiyomi Koizumi, *Japanese Physiology, Past and Present,* Tokyo, 1965.

[164]*History of the International Congresses of Physiological Sciences 1889-1968,* ed. by W.O. Fenn, Baltimore, 1968.

[165]W. Blasius, "Das Denken in Zahlen und seine Folgen für die Menschheit," in *Gefährdete Schöpfung,* Bad Homburg, 1970, pp. 43-62.

[166]R.W. Gerard, "United States," in *Perspectives in Physiology,* Washington, D.C., 1954, pp. 160-168.

[167]See also Wallace O. Fenn, *History of the American Physiological Society: the Third Quarter Century 1937-1962,* Washington, D.C., 1963.

[168]K.E. Rothschuh-A. Schäfer, "Quantitative Untersuchungen über die Entwicklung des physiologischen Fachschrifttums (Periodical) in den letzten 150 Jahren," *Centaurus* 4 (1955), pp. 63-66.

[169]A. Muralt, "Development and limitations of physiological research" in *Perspectives in Physiology,* pp. 9-14.

[170]C. McC. Brooks, "The development of physiology in the last fifty years," *Bull. Hist. Med.* 33 (1959), pp. 249-262.

[171]O. Loewi, "Reflections on the study of physiology," *Annual Review of Physiology* 16 (1954), p. 1.

[172]E.D. Adrian, "Summing up," in *Perspectives in Physiology,* 1954, pp. 3-8.

[173]H. Schaefer, "Die Physiologie des Herzens in ihrer Entwicklung in den letzten 100 Jahren," *Münch. Med. Wschr.* 100 (1958), pp. 9-12.

[174]R. Wagner, "Aktuelle Betrachtungen zur Lage der Physiology," *Münch. Med. Wschr.* 100 (1958), pp. 1637-1640.

[175]K.E. Rothschuh, *Theorie des Organismus. Bios, Psyche, Pathos.,* 2. ed. München-Berlin, 1963.

[176]B. Harrow, *Casimir Funk, Pioneer in Vitamins and Hormones,* New York, 1955.

[177]C. McC. Brooks & H.A. Levey, "Humorally-transported integrators of body functions and the development of endocrinology," in *The Historical Development of Physiological Thought,* New York, 1959, pp. 184-238.

[178]Amos Kendall, *Autobiography,* Magnolia, Mass., 1971.

[179]Nobel Foundation, *Nobel Lectures: Physiology or Medicine,* vol. III, 1942-62, New York, 1964.

[180]R. Wagner, "Über die Zusammenarbeit der Antagonisten bei der Willkürbewegung," *Z. Biol.* 83 (1925), pp. 59-93.

[181]W.R. Hess, *Die Methodik der lokalisierten Reizung und Ausschaltung subkortikaler Hirnabschnitte, I.* Teil., Leipzig, 1932.

[182]Archibald V. Hill, *Trails and Trials in Physiology, a Bibliography 1909-1964,* Baltimore, 1965.

[183]Harold Himsworth, *The Development and Organization of Scientific Knowledge,* London, 1970.

Index

Abdenhalden, Emil 290, 291
Abel, John J. 319
Abernethy, John 180
Achillini, Allesandro 49
Aetios 27
Africanus, Constantinus 32
Albinus, Bernard S. 115
Alcmaeon 3, 5, 7
Alderotti, Thaddeo 37
Alexander 11, 12, 15
Alexander of Tralles 27
Alfanus 32
Alhazen (Ibn al Haitham) 30, 31
Alighieri, Dante 33, 39
Alison, William Pulteney 186
Allen, Edgar 353
Almén, August T. 338
Altdorfer, Albrecht 47
Amaya, S. 347
Anaxagoras 4
Anaximander 2
Anaximenes 2, 4
Andreae, Johann V. 57
an Nafis, Ibn 64
Anrep, G.V. 49, 315
Anthony, Mark 15
Aquapendente 51
Aquinas, Thomas 33, 34, 35
Archimedes 12
Aristarchus 12
Aristotle 2, 9, 10, 11, 16, 39
Arnaldo of Villanova 32, 55, 56
Arnold, Phillip F. 245
Arrhenius, Svante 283, 338
Aschheim, Selmar 353
Aschoff, Ludwig 352
Asclepiades 15, 20
Aselli, Gasparo 72
Asher, Leon 253, 254
Atwater, Wilber O. 295, 324, 325
Atzler, Edgar 295
Augustin, Fredrich L. 156
Aubert, Hermann 230
Aurelius, Marcus 16
Avenzoar (Ibn Zuhr) 29
Averroes 29
Avicenna (Ibn Sina) 29
Avogadro, Amedeo 277

Babkin, Boris P. 305, 315, 334
Babuchin, Alexander I. 328
Bacon, Francis 64, 74, 99, 183
Bacon, Roger 34, 36, 37, 39, 55
Baglivi, Giorgio 72, 76, 82, 83
Bang, Christian Ivar 338
Banting, Frederick 353
Barcroft, Joseph 312, 313
Bartels, Ernst D. 171
Barthez, Paul J. 122
Bartholin, Thomas 71, 72, 73, 93
Bartholomew 35
Basso, Sebastian 82
Batelli, Frédéric 258
Bauhin, Gaspard 51
Baumann, Eugen 285, 286, 287
Baxt, N.N. 327
Bayliss, William M. 154, 307, 313, 314, 315, 353
Beaumont, William 187
Becher, Johann J. 122
Bechterev, Vladimir H. 328, 334
Beck, Adolf 309
Belezky, N.F. 327
Bell, Charles, Sir 140, 167, 181, 184
Bellini, Lorenzo 76, 83
Benedict, Francis G. 294, 295
Bernard, Claude 152, 264, 265, 266, 267, 268, 269, 270, 271, 272
Bernoulli, Daniel 81, 96
Bernstein, Julius 217, 218, 356
Bert, Paul 272
Berthelot, Marellin 282
Berthold, Arnold A. 179, 180
Berzelius, Jöns J. 176, 277, 334
Best, Charles H. 353

Bethe, Albrecht 240, 298, 299
Beyer, Henry G. 323
Bichat, François X. 122, 166, 167
Bidder, Friedrich H. 176
Bidder, Heinrich F. 201, 244, 293
Bidloo, Gottfried 115
Biedermann, Wilhelm 301, 302
Biot, Jean B. 204
Bishoff, Theodor L.W. 175, 176
Bishop, Katherine Scott 352
Bizzozero, Guilio 238
Black, James 185
Black, Joseph 135
Blasius, Gerhard 93
Bleibtreu, Max 227, 228
Blix, Magnus G. 250, 340
Blumenbach, Johann F. 122
Boccaccio, Giovanni 39
Boerhaave, Hermann 114, 115, 116 117, 118, 119, 120
Boethius 33
Bőhm, Rudolf 250
Bőhme, Jakob 57
Bohn, Johannes 62
Bohr, Christian 210, 256, 335
Boldyreff, W.N. 236, 238
Boll, Franz C. 224, 240
Borel, Pierre 99
Borelli, Giovanni 75, 76, 79, 80, 81, 82
Borgia, Cesare 38
Boruttau, Heinrich 244
Botazzi, Filippo 40, 258
Botkin, Sergei P. 331
Boussingault, Jean Baptiste 281
Bowditch, Henry Pickering 187, 209, 210, 256, 319, 320, 321, 322
Bowman, William, Sir 186, 247
Boyle, Robert 56, 76, 82, 89, 90, 91, 92, 93, 99, 182
Bramante, Donato 39
Brandis, August 196
Breuer, Joseph 236
Brodie, Benjamin C. 167, 184
Brőmser, Phillip 297
Brown, John 131, 157
Brown-Sếquard, Charles E. 175, 272, 273
Brunings, Wilhelm 298
Bruno, Giordano 71
Buchner, Eduard 291
Budge, Julius L. 175
Buess 22
Bunsen, Robert W. 204, 254
Burdach, Karl F. 156, 167, 170
Burdon-Sanderson, John 210, 239, 307, 308
Burian, Richard 305
Bűrker, Karl 233, 254
Butenandt, Adolf F. 291, 353
Bykov, K.M. 334

Caldani, Leopoldo·M.A. 140
Calvin, John 65, 66, 67
Canano, Giambattista 67
Cannon, Walter B. 322, 323, 354
Carnot, Nicholas L.S. 204
Carpenter, Willian B. 186
Carrel, Alexis 275
Carus, Carl G. 156, 162, 166, 201
Castelli, Benedetto 79
Caton, Richard 309
Cavendish, Henry 122, 123
Cesalpino, Andrea 67
Cesi, Federigo 99
Chaternikov, Mikhail 330
Charleton, Walter 82
Chauveau, J.B. Auguste 274
Cheselden, William 123
Chevreul, Michael E. 276
Chittenden, Russell H. 241, 324, 325
Chladni, Ernst F.F. 204
Cicero 15
Cohnheim-Kestner, Otto 240, 241
Colombo, Matteo Realdo 50, 66,

67
Commodus 16
Cooper, Sir Astley 186
Copernicus, Nicholas 12, 48
Corti, Alfonso 247
Cournand, André F. 354
Crassus 15
Cremer, Max 295
Croone, William 87, 89
Cruikshank, William C. 138
Cullen, William 131
Cushing, Harvey 322
Cushny, Arthur R. 315, 316
Cuvier, Georges L.C.F. 176, 265
Czermak, Johannes N. 173, 174, 175

da Carpi, Berengario 49
Dale, Sir Henry 309, 310, 357
D'Alembert, Jean le Rond 112
Dalton, John 277
Dalton, John C., Jr. 187, 189, 271
Dam, Henrik 352
Danilevsky, Wassilij J. 328
D'Arsonval, Jacques A. 272
Darwin, Charles 150
Dastré, Albert J. 272
da Varignana, Bartholomeo 37
da Vinci, Leonardo 39, 40, 41, 42, 43, 49, 64, 66
Davis, Marguerite 352
Davy, John 186
de Bordeu, Théophile 122
de Condillac, Etienne 122
de Cyon, Elie 209, 271, 327
de Fourcroy, Antoine 198, 276
de Graaf, Regnier 61
Dekkers, Frederich 115
de la Mettrie, Julien Offroy 117
de Laplace, Pierre S. 136, 204
della Porta, Giambattista 84
della Torre, Marc Antonio 40
de Luzzi, Mondino 37
Democritus 4, 5, 13, 15
Demoor, Jean 346
Denis, Jean B. 89
de Réaumur, René A.F. 132
de Rouelle, Hilary M. 276
des Sauvages, François Boissier de la Croix 122
de Tourquemada, Tomás 47
Descartes, René 74, 77, 78, 79
de Waal, Jan (Walaeus) 71, 72
Diderot, Denis 112
Diocles 9
Diogenes 4, 5, 7
Dogiel, Ivan M. 331
Doisy, Edward A. 353
Döllinger, Ignaz 156, 159, 171
Dömling, Johann J. 156
Donaldson, Henry H. 320, 323
Donders, Franciscus 340, 341
d'Oresme, Nicholas 39
Douglas, James 123
Draper, John W. 187, 189
Drechsel, Edmund 254
Driesch, Hans 11
Drummond, Sir Jack C. 352
Du Bois-Reymond, Emil H. 152, 153, 179, 199, 205, 206, 220, 221, 222, 223, 224, 225, 232, 233, 235
Du Fay, Charles F.C. 140
Dumas, Jean-Baptiste A. 276
Dunglison, Robley 187, 188
du Petit, François Pourfour 138
Dürer, Albrecht 47
Durig, Arnold 236, 239
Detrochet, René J.H. 200
Duverney, Joseph G. 119

Eccles, Sir John 317
Eckhard, Conrad 176, 251
Ehrenberg, Christian G. 213
Ehrlich, Paul 309
Eijkman, Christian 312, 343
Einbrodt, P.P. 331

Einthoven, Willem 308, 341, 342, 343
Eller, Johann T. 117, 119
Ellinger, Alexander 289
Ellis, Frederick W. 321
Elvehjem, Conrad A. 352
Embden, Gustav 289
Empedocles 3, 5, 10
Engelmann, Theodor W. 341, 342
Erasistratus 12, 13, 14, 20
Erasmus, Desiderius 48
Estienne, Charles 51
Euclid 12
Euripides 4
Eustachio, Bartholommeo 51
Evans, Herbert McLean 352
Ewald, Richard 298, 299
Ewell, Thomas 187
Exner, Sigmund 217, 236, 242

Fabricius, Hieronymus 67
Fallopio, Gabriello 50, 51
Faraday, Michael 204, 277
Feldberg, Wilhelm 357
Fernel, Jean 52
Ferrier, Sir David 316
Fichte, Johann G. 156
Fick, Adolf 177, 248, 249, 250
Fick Franz Ludwig 206
Filomafitsky, A.M. 326
Fischer, Emil 283, 290, 291
Fischer, Hans 291
Flateau, E. 233
Fletcher, Walter M. 309, 312
Flint, Austin 187, 189, 190, 271
Flourens, Marie-Jean P. 167
Fludd, Robert 71
Folin, Otto 338
Fontana, Felice 140, 143
Forssmann, Werner 354
Foster, Michael 154, 306, 307
Fracastoro, Girolamo 40
Francis of Assissi, Saint 33
François-Franck, Charles E. 274
Frank, Johann P. 171
Frank, Otto 254, 292, 295, 296, 297
Frédéricq, Leon 345, 346
Freud, Sigmund 236
Froriep, Robert 246
Fulton, John F. 317
Funk, Casimir 352

Gad, Johannes 224, 233, 250
Galen 11, 15, 16, 17, 18, 19, 20, 21, 22, 27, 30, 63, 64
Galilei, Galileo 74, 76, 99
Galvani, Aloisius 141, 142
Garcia, Manuel 175
Garrey, Walter E. 325
Garten, Siegfried 304
Gaskell, Walter H. 210, 256, 306, 342
Gassendi, Pierre 82
Gaub, Jerome D. 115, 117
Gaule, Georg J. 259
Gaule, Justus 298
Gauss, Carl F. 204
Gay-Lussac, Joseph L. 276
"Geber" (Ibn Haiyan) 56
Gerard 32
Gildemeister, Martin 230, 231
Girtanner, Christoph 144
Glauber, Johann R. 58
Glax, Julius 236
Glebov, I.T. 327
Glisson, Francis 86, 87
Gluge, Gottlieb 346
Gmelin, Leopold 277
Goldscheider, Alfred 233
Goltz, Friedrich L. 292, 297, 298, 299
Görres, Joseph 162, 163, 164, 165
Gotch, Francis 307, 308
Grainger, Richard D. 186
Granit, Ragnar 317, 339
Gréhant, Nestor 272

Grew, Nehemiah 100
Grützner, Paul 232, 233
Gscheidlen, Richard 232
Guidi, Guido 51
Gullstrand, Allvar 340
Gürber, August 250
Gutenberg, Johannes 39

Haberlandt, Ludwig 236, 238, 305
Haeckel, Ernst H. 203, 247, 248
Haldane, John S. 307, 313
Hales, Stephen 95, 96, 185, 186
Hall, Granville S. 321
Hall, Marshall 167, 185, 186
Hallmann, Eduard 203, 220
Halsted, William S. 322
Hamburger, Hartog J. 341, 342, 343
Hammersten, Olof 210, 256, 337, 338
Hare, Hobert A. 324
Harless, Emil 293
Harvey, William 62, 63, 67, 68, 69, 70, 71, 73, 85
Hashida, K. 347
Hasselbach, Karl A. 335
Hassenfratz, Jean H. 137
Hayem, Georges 238
Head, Henry 301
Heger, Paul 346
Heidenhain, Rudolph P.H. 172, 200, 229, 231, 232, 233
Heister, Lorenz 117
Henle, Jakob 201, 244, 245, 246
Henriques, Valdemar 335, 336
Henry of Mondeville 32
Hensel, Herbert 352
Heraclitus 2
Herder, Johann G. 157
Hering, Ewald 255, 292, 299, 300, 301, 304
Hering, Heinrich E. 305
Hermann, Ludimar 222, 224, 228, 229
Herophilus 12
Hess, Walter R. 303, 304, 357
Hewson, William 134
Heymans, Corneille 346, 354
Heymans, Jean-François 233, 346
Hildegard 32, 33
Hill, Archibald V. 290, 312
Hill Leonard E. 308, 312
His, Wilhelm Jr. 259
His, Wilhelm Sr. 258
Hippocrates 4, 5, 7, 20
Hobbes, Thomas 85, 112, 183
Höber, Rudolf O.A. 298, 356
Hodgkin, Alan L. 356
Hoffman, Friedrich 82, 89, 114, 115, 119
Hofmann, Franz B. 304
Hofmann, Paul 219
Hofmeister, Franz 283, 287
Holmes, Oliver Wendell 319
Holmgren, Alarik F. 210, 240, 256, 336
Holst, Axel 352
Hooke, Robert 90
Hoppe-Seyler, Felix 254, 283, 284, 285
Hopkins, Sir Frederick Gowland 309, 312
Horn, Franz 160
Horsley, Victor A. 256, 307, 308
Howell, William H. 319, 320, 321
Houssay, Bernardo A. 349, 353
Hufeland, Christoph W. 131
Hüfner, Gustav 254
Hunain, Ibn Ishak 29
Hunter, John 122, 133, 134, 180, 181
Huppert, Karl Hugo 288
Hutchinson, John 172, 187
Hürthle, Karl 232, 233
Huxley, Andrew F. 355, 356
Huygens, Christiaan 75
Hyrtl, Joseph 234

Ishikawa, J. 347
Ishikawa, Makoto 347

Jaffé, Max 289
Jakubovitch, Nicolai M. 327
Johansson, Jöns E. 339
Jones, Thomas Wharton 187
Jordan, Hermann J. 302, 344
Junge, E.A. 331
Jurin, James 83, 95
Justinian 27

Kahn, Richard H. 233
Kalenitchenko, I.O. 327
Kant, Immanuel 113
Kastner, Karl W.G. 196
Kato, Genichi 347
Keill, James 76, 83, 95
Kendall, Edward C. 353
Kepler, Johannes 74, 83, 84, 85
Kielmeyer, Karl F. 131, 157
Kisch, Bruno 304
Klemensiewicz, Rudolf 236, 238
Kohlrausch, Arndt 219, 295
Kohnstamm, Otto 233
König, Arthur 217, 220
Konrad of Megenberg 35
Kossel, Albrecht 287
Kovalesky, Nicolai O. 328
Kreidl, Alois 236, 241
Krogh, August 335, 336, 355
Kronecker, Hugo 207, 240, 251, 252, 253
Kühne, Willy 239, 240, 251, 325
Külz, Eduard 287

Laghi, Tommasseo 140
Lagrange, Joseph L. 137, 204
Laker, Karl 238
Lamarck, Jean Baptiste 170
Landsteiner, Karl 354
Landois, Leonard 228
Langendorff, Oscar 230
Langerhans, Paul 247
Langley, John N. 154, 306, 309, 312
Laplace, Pierre S. 136, 204
Lauder-Brunton, Sir Thomas 210, 256
Laurent, Auguste 276
Lavoisier, Antoine L. 113, 135, 136
Lawrence, William 180, 181
Lea, Sheridan 306
Lee, Frederick S. 320, 324
Legallois, Julien J.C. 167
Leibniz, Gottfried W. 112
Lewis, Sir Thomas 316
Liebig, Justus 176, 183, 277, 280, 281, 282, 291, 293
Lieberkühn, Johann N. 103, 117
Liljestrand, Goran 339
Lindhard, Johannes 336
Locke, John 112
Loeb, Jacques 250, 325
Loewi, Otto 288
Lohmann, Karl 290
Löhner, Leopold 236, 238, 303
Lombard, Warren P. 208, 210, 320, 321, 322
Long, Joseph A. 353
Longet, François A. 175
Lorry, Anne C. 138
Lotze, Rudolph H. 177
Lovén, Christian 256, 336, 337
Lower, Richard 89, 90
Luchsinger, Johann B. 230
Luciani, Luigi 210, 256, 257
Ludwig, Carl Friedrich Wilhelm 152, 174, 179, 204, 205, 206, 207, 208, 209, 210, 211, 212, 235, 248, 249, 254, 256, 257, 258, 322
Lundsgaard, Christen 355
Lusk, Graham 295, 325
Luther, Martin 47
Lykus 20

McCollum, Elmer 352
Macullum, Archibald B. 320, 324
Magendie, François 140, 151, 154, 167, 168, 169, 265, 266, 268
Magnus, Albertus 33, 34, 35, 36, 55
Magnus, Heinrich G. 282
Magnus, Rudolf 317, 344
Malassez, Louis C. 272
Mall, Franklin P. 319, 324
Malpighi, Marcello 62, 75, 99, 100, 101
Maly, Richard 238
Mangold, Ernst 302
Marcus, Adalbert F. 159
Marey, Etienne J. 252, 274
Marinesco, George 233
Marinus 14, 16, 20
Matteucci, Carlo 220
Maurolyco, Francesco 83
Maxwell, James C. 204
Mayer, Julius Robert 213
Mayer, Sigmund 242
Mayow, John 90, 91, 92
Meckel, Johann F. 154, 156
Medici, Cosimo 38
Meissner, Georg 176, 243, 244
Mellanby, Sir Edward 352
Mendel, Lafayette B. 325, 352
Metchnikoff, Ilya I. 331
Meltzer, Samuel J. 252, 324
Metzner, Rudolf 219
Meyer, Lothar 283
Meyerhof, Otto 289, 290, 298, 312, 352
Michaelangelo 39
Miescher, Friedrich Jr. 203, 285, 286
Miescher, Johannes F. 259
Mill, John Stuart 183
Milne-Edwards, Henri 200
Milton, John 85
Minot, Charles S. 320, 323
Minot, George R. 210
Mislavsky, Nicolai A. 328
Mitchell, Silas Weir 187, 188, 271
Mitscherlich, Karl Gustav 277
Moleschott, Jacob 176, 177
Morat, Jean-Pierre 272
Morgan, Charles E. 189
Moser, Adolf 170
Mosso, Angelo 210, 256, 257, 258
Mulder, Gerardus J. 283, 340
Müller, Heinrich 177
Müller, Johannes P. 131, 134, 140, 146, 153, 154, 170, 195, 196, 197, 198, 199, 200, 201, 202, 203, 211, 212, 242
Munk, Hermann 224, 233
Munk, Immanuel 234
Munzer, Egmont 305
Mynsicht, Hadrian 59

Nafis, Ibn an 30
Nagai, Hisomu 347
Nagel, Willibald A. 218, 219
Nasse, Hermann 176, 277
Nasse, Otto 277
Navalychin, Ivan G. 327
Newell-Martin, Henry 230, 306, 319, 320, 321, 323
Nicholas of Cusa 39, 76

Oersted, Hans C. 204
Oesterlen, Friedrich 183
Oken, Lorenz 144, 153, 156
Orbeli, Leon A. 334
Oribasios 27
Osterreicher, Johann H. 160
Osawa, Kenji 347
Osborne, Thomas B. 352
Ott, Isaac 321
Overton, Charles Ernest 250, 251
Ovsianikov, Filip W. 327

Pachutin, Victor W. 251, 256, 329

Panum, Peter L. 271, 334, 335
Paracelsus (Von Hohenbeim, Bombastus) 56, 57, 58
Pascal, Blaise 74
Passavant, Daniel 97
Pasteur, Louis 267, 271, 276, 283, 291
Paul of Aegina 27
Pavlov, Ivan P. 232, 256, 326, 332, 333, 334
Pecquet, Jean 72
Pelouze, Théophile J. 276
Pericles 8
Petoule, Théophile J. 276
Petrova, Marija K. 334
Petrarch 39
Pfaff, Christoph H. 144
Pflüger, Eduard F.W. 224, 225, 226, 227
Philip, Alexander P.W. 183
Philistion 9
Piper, Hans 219
Pirogov, Nicholai I. 326
Plato 8, 9, 17, 63
Platter, Felix 51
Politzer, Adam 343
Polybus 5
Porter, William Townsend 154, 233, 322
Potter, Francis 89
Praxagoras 12, 14
Prevost, Jean L. 276
Preyer, Wilhelm 301
Priestley, Joseph 122, 135
Primerose, James 71
Pringle, John 117
Prochaska, Georg 139, 140, 156
Proust, Joseph L. 276
Prout, William 181, 186
Ptolemy 12
Purkinje, Johann E. (Purkyně) 146, 152, 153, 171, 172, 173
Pythagoras of Samos 3, 5
Ranke, Otto F. 297
Ranvier, Louis-Antoine 265, 271
Rasi, Ibn Zakariah (Rhazes) 29, 30
Rathke, Martin H. 170
Redi, Francesco 75
Reenpää, Yrjö R. 339
Regius, Henricus 82
Reichert, Karl B. 203
Reil, Johann C. 154
Reimarus, Johann A.H. 182
Remak, Robert 246
Retzius, Anders 334
Reuchlin, Johann 48
Richards, Alfred N. 355
Richards, Dickinson W. 354
Richet, Charles 274, 275
Richter, Georg G. 117
Richter, Jeremias B. 123, 277
Rickes, Edward L. 352
Riesser, Otto 289
Ringer, Sidney 308
Riolan, Jean Jr. 71
Ritter, Johann W. 144
Robin, Charles 188
Roget, Peter M. 182
Rollet, Alexander 236, 237
Röntgen, Wilhelm C. 177
Röschlaub, Andreas 159
Rosemann, Rudolf 228
Rosenblueth, Arturo S. 323, 349
Rosenthal, Isidor 224, 233
Rousseau, Jean-Jacques 113
Rubner, Max 254, 294
Rudbeck, Olaf 72
Rudolphi, Carl A. 156, 167, 169, 196
Ruysch, Frederick 103

Sachs, Karl S. 224
Sagorsky, Alexander P. 326
Salah-ad-Din 31
St. Clair, Andrew 117
Salkowski, Ernest 285, 287

Samojlov, Alexander F. 330
Santorio, Santorio 76, 77
Saucerotte, Nicholas 138
Scarpa, Antonio 171
Scheele, Carl W. 122, 334
Scheiner, Christoph 85
Schelling, Carl E. 156
Scheminsky, Ferdinand 236, 239
Schenk, Friedrich 250
Schiff, Moritz 264
Schleiden, Matthias 247
Schmidt, Alexander 284, 287
Schmidt, Carl 244, 287, 293
Schultze, Carl A.S. 152, 172
Schultze, Max J.S. 246
Schwann, Theodor 201, 245, 345
Schweiger-Seidel, Franz W. 232
Sechenov, Ivan M. 236, 238, 251, 256, 328, 329
Sedgwick, William T. 320, 323
Seeman, John 295
Séguin, Armand 136
Sennert, Daniel 58, 82
Servetus, Michael 48, 65, 66
Severino, Marc A. 99
Sewall, Henry 320, 323
Sforza, Francesco 38
Shakespeare, William 85
Sharpey-Schaefer, Edward 210, 256, 306, 307, 308
Sharpey, William 306
Sherrington, Charles S. 154, 219, 316, 317, 318
Shoji, Rinnosuke 347
Simon, J. Franz 277, 278
Singer, Jacob 305
Skoda, Josef 234
Smith, Ernest L. 352
Snell, Willebord (Snellius) 77
Snellen, Herman 341
Socrates 4, 8
Sokolov, Alexej A. 327
Solvay, Ernst 346
Sondén, Klas 339
Sophocles 4
Soranus 15
Spallanzani, Lazzaro 132, 133, 137
Speransky, A.D. 334
Spiro, Karl A. 238
Stahl, George E. 114, 120, 121, 123
Stannius, Hermann F. 173
Starling, Ernest H. 154, 232, 241, 313, 314, 315, 316, 353
Steinach, Eugen 305
Stensen, Niels 79, 93, 94, 95
Stephan, Jan 50
Stepp, Wilhelm 352
Stigler, Robert 238
Stirling, William 210, 256
Strato 12, 13
Straub, Hermann 297
Stricker, Salomon 346
Stuart, Alexander 138, 183
Sumner, James B. 292
Suslova, Nadeshda P. 330
Swammerdam, Jan 102
Sylvius, Franciscus (Frans de le Boë) 60, 61
Sylvius, Jacques Dubois 49, 67
Szent-Györgyi, Albert 292, 352

Takamine, Jodishi 347
Tarchanoff, Ivan Romanovitch —(or nov) 271, 329
Tchagovez, Vassily J. 328
Tertullian 12
Thales 2
Thauer, Rudolf 299, 352
Themison 15
Theophrastus 9, 12
Thessalos 15
Thierfelder, Hans 290
Thiry, Ludwig 244
Thomas, B.W. 331
Thomas of Cantimpré 35

Thomson, Thomas 204
Thucydides 4
Thudicum, Ludwig J.W. 309, 312
Tiedemann, Friedrich 167, 201, 277
Tiegel, Ernst 347
Tigerstedt, Carl 339
Tigerstedt, Robert A. 339
Tigerstedt, Robert A.A. 210, 256, 338, 339
Timirjasen, Kliment A. 331
Todd, Robert B. 186
Toricelli, Evangelista 75
Trembley, Abraham 122
Trendelenburg, Wilhelm 219, 220
Treviranus, Gottfried 156, 169
Troxler, Ignaz P.V. 156
Tűrck, Ludwig 175

Uchtomsky, Alexsei A. 330

Valentin, Gabriel G. 170, 173, 174
Valentine, Basil 56
Valli, Eusebio 143
van Deen, Isaac Abraham 341
van den Spieghel Spigelius, Andriaan 51
van der Kolk, Jacques S. Schroeder 340
van Hoff, Jacobus Henricus 283
van Helmont, Johannes Baptista 57, 59, 60
van Leeuwenhoek, Anton 98, 102
van Swieten, Gerhard 117
Vaugham, Victor C. 325
Vedensky, Nicolai E. 330
Vella, Luigi 271
Vellensky, Danilo Mikhailovitch 326
Verigo, Borislav F. 330
Verworn, Max 247, 302, 303
Vesalius, Andreas 48, 49, 50, 51, 52, 65
Vincent of Beauvais 35
Virchow, Rudolf 246, 247, 284
Viviani, Vincenzo 75
Vogt, Karl 177
Voit, Carl 280, 281, 292, 295
Voit, Erwin 295
Volkmann, Alfred W. 176, 179, 231, 244
Volta, Allessandro 142, 144
von Andernach, Gűnther 49
von Autenrieth, Johann H.F. 154, 156, 169
von Baer, Karl E. 170
von Bezold, Albert 153, 177, 224, 225
von Brűcke, Ernst Wilhelm Ritter 152, 174, 206, 207, 234, 235, 241
von Brűcke, Theodor 304, 305
von Bunge, Gustav 287
von Ebner, Viktor 236, 238
von Euler-Chelpin, Hans K.A. 340
von Euler-Chelpin, Ulf S. 340
von Fraunhofer, Joseph 204
von Frey, Max 254, 255
von Gőrres, Joseph 155
von Goethe, Johann W. 144, 145, 146
von Graefe, Albrecht 341
von Guericke, Otto 75, 93
von Haller, Albrecht 117, 123, 124, 125, 126, 127, 128, 129, 130, 131, 137
von Hardenburg, Friedrich (Navalis) 155
von Hebra, Ferdinand 234
von Helmholtz, Hermann L.F. 152, 179, 205, 206, 212, 213, 214, 215, 216, 217
von Hess, Carl 301
von Hősslin, Heinrich 295
von Humboldt, Alexander 143, 153, 165, 166, 200, 279
von Hutten, Ulrich 48

von Kőlliker, Rudolf A. 153, 177, 245
von Krehl, Rudolf 290
von Kries, Johannes 217, 218, 219
von Liebig, Justus 204, 275, 278, 279
von Linné, Carl (Linnaeus) 334
von Marxow, Ernst Fleischl 236, 309
von Müller, Friedrich 295
von Pettenkofer, Max 292
von Pfeüfer, Karl 246
von Rokitanski, Carl 234
von Schelling, Friedrick W.J. 155, 156, 157, 158, 159
von Schlegel, August W. 156
von Schlegel, Friedrich 156
von Siebold, Carl T.E. 173
von Skramlik, Emil 219, 297
von Stradonitz, Friedrich A. Ké-kulé 283
von Tschermak-Seysenegg, Armin 304
von Uexküll, Jacob J. 240, 241
von Vierordt, Herman 243
von Vierordt, Karl 242
von Vintschgau, Maxmillian R. 236, 238, 239
von Voit, Karl 154, 176, 280, 281, 292, 293
von Walther, Philip F. 156, 159, 196
von Wolff, Christian 182
Vulpian, E. Felix A. 273

Wagner, Johann J. 155, 160
Wagner, Richard 239, 297
Wagner, Rudolph 170, 176, 177
Waldeyer, Wilhelm 232
Walker, Alexander 183
Waller, Augustus Désiré 308
Waller, Augustus Volney 175
Walther, Anton A. 334
Warburg, Otto 289, 290, 292, 352
Warren, Joseph W. 321
Wearn, Joseph T. 355
Weber, Arthur E. 297
Weber, Edouard F. 151, 178
Weber, Ernst H. 151, 176, 178
Weber, Wilhelm E. 179, 204
Weigel, Valentin 57
Weiss, Otto 230, 244
Welch, William H. 210, 318, 319
Wenzel, Carl F. 123
Wezler, Karl 297
Wharton, Thomas 101
Whytt, Robert 117, 138, 143, 183
Wiener, Norbert 356
Wiggers, Carl J. 297
Willbrand, Johann B. 156, 161, 162
William of Occam 39
Willis, Thomas 62, 86, 87, 88, 89
Wilson, Alexander Philip 167
Windaus, Adolf O.R. 291, 352
Winkelmann, August 161
Winkelmann, Stephan A. 155
Winslow, Jacques B. 123
Witelo 30, 35, 36, 37
Wlassak, Rudolf 298
Wőhler, Friedrich 204, 277, 280, 283
Wooldridge, Leonard C. 256
Wren, Christopher 90
Wundt, Wilhelm 251

Young, John Richardson 187
Young, Thomas 183, 204

Zeno 28
Zimmerman, Johann G. 138
Zinn, Johann G. 138
Zondek, Hermann 353
Zoth, Oscar 236, 237, 238
Zuntz, Nathan 227, 234
Zwaardemaker, Hendrik 240, 343